DESIGN OF
THERMAL SYSTEMS

Also available from McGraw-Hill

Schaum's Outline Series in Mechanical and Industrial Engineering

Each outline includes basic theory, definitions, and hundreds of solved problems and supplementary problems with answers.

Current List Includes:

Acoustics
Basic Equations of Engineering
Continuum Mechanics
Engineering Economics
Engineering Mechanics, 4th edition
Fluid Dynamics
Fluid Mechanics & Hydraulics
Heat Transfer
Introduction to Engineering Calculations
Lagrangian Dynamics
Machine Design
Mechanical Vibrations
Operations Research
Strength of Materials, 2d edition
Theoretical Mechanics
Thermodynamics

Available at Your College Bookstore

DESIGN OF THERMAL SYSTEMS

Third Edition

W. F. Stoecker

Professor Emeritus of Mechanical Engineering
University of Illinois at Urbana-Champaign

McGraw-Hill Book Company

New York   St. Louis   San Francisco   Auckland   Bogotá   Caracas
Colorado Springs   Hamburg   Lisbon   London   Madrid   Mexico   Milan
Montreal   New Delhi   Oklahoma City   Panama   Paris   San Juan
São Paulo   Singapore   Sydney   Tokyo   Toronto

This book was set in Times Roman by Publication Services.
The editors were Anne T. Brown, Lyn Beamesderfer, and John M. Morriss;
the production supervisor was Janelle S. Travers.
The cover was designed by Fern Logan.
Project supervision was done by Publication Services.
R. R. Donnelley & Sons Company was printer and binder.

DESIGN OF THERMAL SYSTEMS

1 2 3 4 5 6 7 8 9 0   DOCDOC   8 9 4 3 2 1 0 9

ISBN   0-07-061620-5

Library of Congress Cataloging-in-Publication Data

Stoecker, W. F. (Wilbert F.), (date).
    Design of thermal systems.

    Includes bibliographies and index.
    1. Heat engineering.   2. Systems engineering.
3. Engineering design.   I. Title.
TJ260.S775 1989   621.402   88-13281
ISBN 0-07-061620-5

ABOUT THE AUTHOR

Wilbert F. Stoecker is Professor Emeritus of Mechanical Engineering at the University of Illinois at Urbana-Champaign where he continues to teach part time. He received his undergraduate education at the University of Missouri at Rolla and graduate degrees from the University of Illinois and Purdue University. Dr. Stoecker is the author of *Industrial Refrigeration*, *Refrigeration and Air Conditioning*, with J. D. Jones, and *Microcomputer Controls of Thermal and Mechanical Systems*, with P. A. Stoecker. He is the author of over 50 technical papers, has lectured internationally, and serves as an industrial consultant. He is a member of ASME, ASEE, and ASHRAE, and several international refrigeration and air conditioning organizations and is also the recipient of numerous teaching awards.

CONTENTS

PREFACE
TO THE THIRD EDITION

The field of thermal system design and analysis continues to develop. The number of workers is growing, technical papers appear in greater numbers, and new textbooks are being written.

The major objective of this third edition is to organize some of the new approaches that are now available and to provide more flexibility to instructors who use Design of Thermal Systems as a text. The changes to the twelve chapters of the second edition are modest and mainly constitute the inclusion of some additional end-of-chapter problems. Chapters 13 through 19, however, are all new. One possible use of the text is to cover the first twelve chapters in an advanced-level undergraduate course and the remaining seven chapters as a graduate course. In some engineering schools students already have some kind of optimization course prior to taking the thermal design course. For those classes certain chapters of the first twelve (usually the ones on search methods, dynamic programming, and linear programming) can be omitted and material can be supplemented from the new seven chapters.

Several of the new chapters are extensions of the introductions offered in the first twelve chapters, especially mathematical modeling, steady-state system simulation, and search methods. Chapter 14 addresses some of the challenges that arise when simulating large thermal systems. New material appears in Chapter 15 on dynamic behavior, in Chapter 18 which introduces calculus of variations as a companion to dynamic programming, and in Chapter 19 on probabilistic approaches to design, which is exploratory.

The author thanks colleagues both at the University of Illinois at Urbana-Champaign and at other engineering schools for continued input and suggestions during the past several years on how to keep the book fresh.

McGraw-Hill would also like to thank the following reviewers for their many useful comments: John R. Biddle, California State Polytechnic

University, Pomona; Theodore F. Smith, The University of Iowa; Edward O. Stoffel, California State Polytechnic University, San Luis Obisbo; John A. Tichy, Rensselaer Polytechnic Institute; Daniel T. Valentine, Clarkson College; and William J. Wepfer, Georgia Institute of Technology.

W. F. Stoecker

PREFACE
TO THE SECOND EDITION

The field of thermal system design has begun to mature. The origin of the discipline was probably the University of Michigan project sponsored in the mid-1960s by the National Science and Ford Foundations. The motivation at that time might have been considered artificial, because the participants in that program were seeking ways of using digital computers in engineering education. The topics and techniques identified in the project, modeling, simulation, and optimization, proved to be significant.

The first edition of *Design of Thermal Systems* appeared in the early 1970s and concentrated on the applications to thermal systems of modeling, simulation, and optimization. At that time the industrial applications were somewhat rare, essentially limited to large chemical and petroleum facilities.

The emergence of the energy crisis about 1973 provided the impetus for the industrial application of system simulation. System simulation has become an accepted tool for energy analysis of power generating, air conditioning, refrigeration, and other thermal processing plants. Simulation is often used in the design or development stage to evaluate energy requirements of the proposed system or to explore potential savings in first cost. The acceptance of optimization techniques as an industrial tool is moving less rapidly, but many engineers feel that it is only a matter of time and increased familiarity with the power of sophisticated optimization techniques before acceptance becomes widespread.

The reason for preparing a second edition of *Design of Thermal Systems* is that the field has advanced during the years since the appearance of the first edition. Some of the techniques of simulation and optimization have become more stabilized. The author also believes that some of the topics can now be explained more clearly and that additional examples and problems can vitalize the use of these topics. Since the comprehensive designs at the end of the text have been attractive to many instructors, their number has been increased.

W. F. Stoecker

PREFACE
TO THE FIRST EDITION

The title, *Design of Thermal Systems,* reflects the three concepts embodied in this book: *design, thermal,* and *systems.*

DESIGN

A frequent product of the engineer's efforts is a drawing, a set of calculations, or a report that is an abstraction and description of hardware. Within engineering education, the cookbook approach to design, often practiced during the 1940s, discredited the design effort so that many engineering schools dropped design courses from their curricula in the 1950s. But now design has returned. This reemergence is not a relapse to the earlier procedures; design is reappearing as a creative and highly technical activity.

THERMAL

Within many mechanical engineering curricula the term *design* is limited to *machine design.* In order to compensate for this frequent lack of recognition of thermal design, some special emphasis on this subject for the next few years is warranted. The designation *thermal* implies calculations and activities based on principles of thermodynamics, heat transfer, and fluid mechanics.

The hardware associated with thermal systems includes fans, pumps, compressors, engines, expanders, turbines, heat and mass exchangers, and reactors, all interconnected with some form of conduits. Generally, the working substances are fluids. These types of systems appear in such industries as power generation, electric and gas utilities, refrigeration, air conditioning and heating, and in the food, chemical, and process industries.

SYSTEMS

Engineering education is predominantly *process oriented,* while engineering practice is predominantly *system oriented.* Most courses of study in engineering provide the student with an effective exposure to such processes as the flow of a compressible fluid through a nozzle and the behavior of hydrodynamic and thermal boundary layers at solid surfaces. The practicing engineer, however, is likely to be confronted with a task such as designing an economic system that receives natural gas from a pipeline and stores it underground for later usage. There is a big gap between knowledge of individual processes and the integration of these processes in an engineering enterprise.

Closing the gap should not be accomplished by diminishing the emphasis on processes. A faulty knowledge of fundamentals may result in subsequent failure of the system. But within a university environment, it is beneficial for future engineers to begin thinking in terms of systems. Another reason for more emphasis on systems in the university environment, in addition to influencing the thought patterns of students, is that there are some techniques—such as simulation and optimization—which only recently have been applied to thermal systems. These are useful tools and the graduate should have some facility with them.

While the availability of procedures of simulation and optimization is not a new situation, the practical application of these procedures has only recently become widespread because of the availability of the digital computer. Heretofore, the limitation of time did not permit hand calculations, for example, of an optimization of a function that was dependent upon dozens or hundreds of independent variables. This meant that, in designing systems consisting of dozens or hundreds of components, the goal of achieving a *workable* system was a significant accomplishment and the objective of designing an *optimum* system was usually abandoned. The possibility of optimization represents one of the few facets of design.

OUTLINE OF THIS BOOK

The goal of this book is the design of optimum thermal systems. Chapters 6 through 11 cover topics and specific procedures in optimization. After Chap. 6 explains the typical statement of the optimization problem and illustrates how this statement derives from the physical situation, the chapters that follow explore optimization procedures such as calculus methods, search methods, geometric programming, dynamic programming, and linear programming. All these methods have applicability to many other types of problems besides thermal ones and, in this sense, are general. On the other hand, the applications are chosen from the thermal field to emphasize the opportunity for optimization in this class of problems.

If the engineer immediately sets out to try to optimize a moderately complex thermal system, he is soon struck by the need for predicting the performance of that system, given certain input conditions and performance characteristics of components. This is the process of *system simulation*. System simulation not only may be a step in the optimization process but may have a usefulness in its own right. A system may be designed on the basis of some maximum load condition but may operate 95 percent of the time at less-than-maximum load. System simulation permits an examination of the operating conditions that may pinpoint possible operating and control problems at non-design conditions.

Since system simulation and optimization on any but the simplest problems are complex operations, the execution of the problem must be performed on a computer. When using a computer, the equation form of representation of the performance of components and expression of properties of substances is much more convenient than tabular or graphical representations. Chapter 4 on mathematical modeling presents some techniques for equation development for the case where there is and also where there is not some insight into the relationships based in thermal laws.

Chapter 3, on economics, is appropriate because engineering design and economics are inseparable, and because a frequent criterion for optimization is the economic one. Chapter 2, on workable systems, attempts to convey one simple but important distinction—the difference between the design process that results in a workable system in contrast to an optimum system. The first chapter on engineering design emphasizes the importance of design in an engineering undertaking.

The appendix includes some problem statements of several comprehensive projects which may run as part-time assignments during an entire term. These term projects are industrially oriented but require application of some of the topics explained in the text.

The audience for which this book was written includes senior or first-year graduate students in mechanical or chemical engineering, or practicing engineers in the thermal field. The background assumed is a knowledge of thermodynamics, heat transfer, fluid mechanics, and an awareness of the performance characteristics of such thermal equipment as heat exchangers, pumps, and compressors. The now generally accepted facility of engineers to do basic digital computer programming is also a requirement.

ACKNOWLEDGMENTS

Thermal system design is gradually emerging as an identifiable discipline. Special recognition should be given to the program coordinated by the University of Michigan on Computers in Engineering Design Education, which in 1966 clearly delineated topics and defined directions that have since proved to be productive. Acknowledgment should be given to activities within the chemical engineering field for developments that are closely related, and in some cases identical, to those in the thermal stem of mechanical engineering.

Many faculty members during the past five years have arrived, often independently, at the same conclusion as the author: the time is opportune for developments in thermal design. Many of these faculty members have shared some of their experiences in the thermal design section of *Mechanical Engineering News* and have, thus, directly and indirectly contributed to ideas expressed in this book.

This manuscript is the third edition of text material used in the Design of Thermal Systems course at the University of Illinois at Urbana-Champaign. I thank the students who have worked with me in this course for their suggestions for improvement of the manuscript. The second edition was an attractively printed booklet prepared by my Department Publication Office, George Morris, Director; June Kempka and Dianne Merridith, typists; and Don Anderson, Bruce Breckenfeld, and Paul Stoecker, draftsmen. Special thanks are due to the Engineering Department of Amoco Chemicals Corporation, Chicago, for their interest in engineering education and for their concrete evidence of this interest shown by printing the second edition.

Competent colleagues are invaluable as sounding boards for ideas and as contributors of ideas of their own. Professor L. E. Doyle offered suggestions on the economics chapter and Prof. C. O. Pedersen, a coworker in the development of the thermal systems program at the University of Illinois at Urbana-Champaign, provided advice at many stages. Mr. Donald R. Witt and a class of architectural engineering students at Pennsylvania State University class-tested the manuscript and provided valuable suggestions from the point of view of a user of the book. Beneficial comments and criticisms also came from the Newark College of Engineering, where Prof. Eugene Stamper and a group of students tested the manuscript in one of their classes. Professor Jack P. Holman of Southern Methodist University, consulting editor of McGraw-Hill Book Company, supplied perceptive comments both in terms of pedagogy as well as in the technical features of thermal systems.

The illustrations in this book were prepared by George Morris of Champaign, Illinois.

By being the people that they are, my wife Pat and children Paul, Janet, and Anita have made the work on this book, as well as anything else I do, seem worthwhile.

W. F. Stoecker

DESIGN OF
THERMAL SYSTEMS

CHAPTER
1

ENGINEERING DESIGN

1.1 INTRODUCTION

Typical professional activities of engineers include sales, construction, research, development, and design. Design will be our special concern in this book. The immediate product of the design process is a report, a set of calculations, and/or a drawing that are abstractions of hardware. The subject of the design may be a process, an element or component of a larger assembly, or an entire system.

Our emphasis will be on *system* design, where a system is defined as a collection of components with interrelated performance. Even this definition often needs interpretation, because a large system sometimes includes subsystems. Furthermore, we shall progressively focus on *thermal* systems, where fluids and energy in the form of heat and work are conveyed and converted. Before adjusting this focus, however, this chapter will examine the larger picture into which the technical engineering activity blends. We shall call this larger operation an engineering *undertaking*, implying that engineering plays a decisive role but also dovetails with other considerations. Engineering undertakings include a wide variety of commercial and industrial enterprises as well as municipally, state-, and federally sponsored projects.

1.2   Decisions in an Engineering Undertaking

In recent years an appreciable amount of attention has been devoted to the *methodology* or *morphology* of engineering undertakings. Studies on these topics have analyzed the steps and procedures used in reaching decisions. One contribution of these studies has been to stimulate engineers to reflect on the thinking processes of themselves and others on the project team. Certainly the process and sequence of steps followed in each undertaking is different, and no one sequence, including the one described in this chapter, is universally applicable. Since the starting point, the goal, and the side conditions differ from one undertaking to the next, the procedures must vary.

The advantage of analyzing the decision process, especially in complex undertakings, is that it leads to a more logical coordination of the many individual efforts constituting the entire venture. The flow diagram in Fig. 1-1 shows typical steps followed in the conception, evaluation, and execution of the plan. The rectangular boxes, which indicate actions, may represent considerable effort and expenditures on large projects. The diamond boxes represent decisions, e.g., whether to continue the project or to drop it.

The technical engineering occurs mostly in activities 5 and 7, product or system design and research and development. Little will be said in this chapter about product or system design because it will be studied in the chapters to follow. The flow diagram shows only how this design procedure fits into the larger pattern of the undertaking. The individual nondesign activities will be discussed next.

1.3   NEED OR OPPORTUNITY (STEP 1)

Step 1 in the flow diagram of Fig. 1-1 is to define the need or opportunity. It may seem easy to state the need or opportunity, but it is not always a simple task. For example, the officials of a city may suppose that their need is to enlarge the reservoir so that it can store a larger quantity of water for municipal purposes. The officials may not have specified the actual need but instead may have leaped to one possible solution. Perhaps the need would better have been stated as a low water reserve during certain times of the year. Enlargement of the reservoir might be one possible solution, but other solutions might be to restrict the consumption of water and to seek other sources such as wells. Sometimes possible solutions are precluded by not stating the need properly at the beginning.

The word "opportunity" has positive connotations, whereas "need" suggests a defensive action. Sometimes the two cannot be distinguished. For example, an industrial firm may recognize a new product as an opportunity, but if the company does not then expand its line of products, business is likely to decline. Thus the introduction of a new product is also a need.

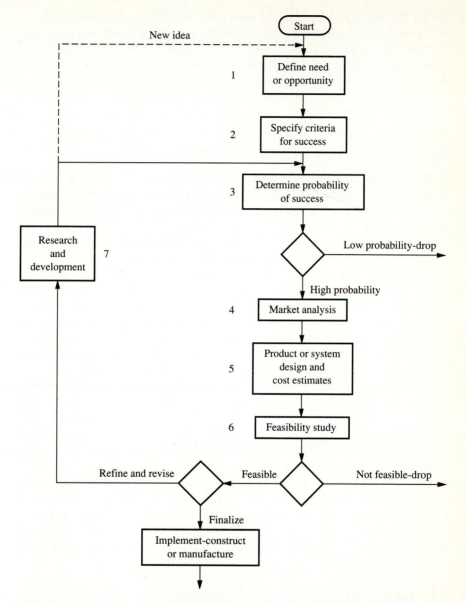

FIGURE 1-1
Possible flow diagram in evaluating and planning an engineering undertaking.

In commercial enterprises, typical needs or opportunities lie in the renovation or expansion of facilities to manufacture or distribute a current product. Opportunity also arises when the sale of a product not manufactured by the firm is rising and the market potential seems favorable. Still a third form in which an opportunity arises is through research and development within the organization. A new product may be developed intentionally or accidentally. Sometimes a new use of an existing product can be found by making a slight modification of it. An organization may know how to manufacture a gummy, sticky substance and assign to the research and development department the task of finding some use for it.

Of interest to us at the moment is the need or opportunity that requires engineering design at a subsequent stage.

1.4   CRITERIA OF SUCCESS (STEP 2)

In commercial enterprises the usual criterion of success is showing a profit, i.e., providing a certain rate of return on the investment. In public projects the criterion of success is the degree to which the need is satisfied in relation to the cost, monetary or otherwise.

In a profit-and-loss economy, the expected earning power of a proposed commercial project is a dominating influence on the decision to proceed with the project. Strict monetary concerns are always tempered, however, by human, social, and political considerations to a greater or lesser degree. In other words, a price tag is placed on the nonmonetary factors. A factory may be located at a more remote site at a penalty in the form of transportation costs so that its atmospheric pollution or noise affects fewer people. As an alternative, the plant may spend a lot on superior pollution control in order to be a good neighbor to the surrounding community.

Sometimes a firm will design and manufacture a product that offers little opportunity for profit simply to round out a line of products. The availability of this product, product A, permits the sales force to say to a prospective customer, "Yes, we can sell you product A, but we recommend product B," which is a more profitable item in the company's line and may actually be superior to product A.

Often a decision, particularly in an emergency, appears outside the realm of economics. If a boiler providing steam for heating a rental office building fails, the decision whether to repair or replace the boiler may seem to be outside the realm of economics. The question can still be considered an economic one, however, the penalty for not executing the project being an overpowering loss.

1.5   PROBABILITY OF SUCCESS (STEP 3)

Plans and designs are always directed toward the future, for which only probability, not certainty, is applicable. There is no absolute assurance that the plant will meet the success criteria discussed in Sec. 1.4, only a likelihood or probability that it will do so.

The mention of *probability* suggests the normal distribution curve (Fig. 1-2), an excellent starting point for expressing uncertainty in the decision-making process. The significance of the distribution curve lies particularly

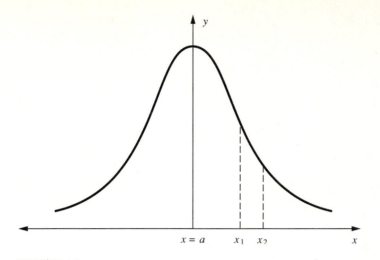

FIGURE 1-2
Probability distribution curve.

in the evaluation of the area beneath the curve. The area under the curve between x_1 and x_2, for example, represents the probability P of the event's occurring between values x_1 and x_2. Thus,

$$P = \int_{x_1}^{x_2} y\,dx$$

Since the probability of the event's occurring somewhere in the range of x is unity, the integration over the entire range of x is equal to 1.0:

$$\int_{-\infty}^{\infty} y\,dx = 1$$

The equation for the probability distribution curve is

$$y = \frac{h}{\sqrt{\pi}}e^{-h^2(x-a)^2} \qquad (1.1)$$

The maximum value of the ordinate is $h/\sqrt{\pi}$, which occurs when $x = a$. This fact suggests that increasing the value of h alters the shape of the distribution curve, as shown in Fig. 1-3. If h_1 is greater than h_2, the peak of the h_1 curve rises higher than that of the h_2 curve.

To extend the probability idea to decision making in an engineering undertaking, suppose that a new product or facility is proposed and that the criterion for success is a 10 percent rate of return on the investment for a 5-year life of the plant. After a preliminary design, the probability distribution curve is shown as indicated in Fig. 1-4. Since rough figures were used throughout the evaluation, the distribution curve is flat, indicating no great confidence in an expected percent of return of investment of, say,

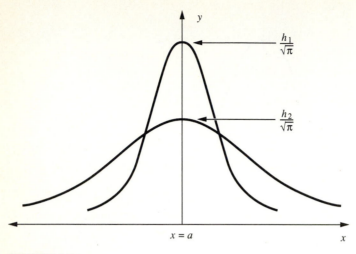

FIGURE 1-3
Several different shapes of the probability distribution curve.

18 percent. The expected rate of return is attractive enough, however, to proceed to a complete design, including cost estimates. If the most probable return on investment after this complete design were 16 percent, for example, the confidence in this figure would be greater than the confidence in the 18 percent figure after the preliminary design because costs have now

FIGURE 1-4
Distribution curves at various stages of decision making.

been analyzed more carefully and marketing studies have been conducted more thoroughly.

The probability distribution curves at two other stages, after construction and after 1 year of operation, show progressively greater degrees of confidence in the rate of return after a 5-year life. After 5 years, the rate of return is known exactly, and the probability distribution curve degenerates into a curve that is infinitesimally thin and infinitely high.

The recognition that prediction of future behavior is not deterministic, so that only one set of events or conditions will prevail, has spawned a new probabilistic approach to design (see the additional readings at the end of the chapter and Chapter 19). One of the activities of this new study is that of quantifying the curves shown in Fig. 1-4. It is valuable for the decision maker to know not only the most likely value of the return on investment but also whether there is a high or low probability of achieving this most likely value.

1.6   MARKET ANALYSIS (STEP 4)

If the undertaking is one in which a product or service must eventually be sold or leased to customers, there must be some indication of favorable reaction by the potential consumer. An ideal form of the information provided by a market analysis would be a set of curves like those in Fig. 1-5. With an increase in price, the potential volume of sales decreases until such a high price is reached that no sales can be made. The sales-volume to price relationship affects the size of the plant or process because the unit price is often lower in a large plant. For this reason, the marketing and plant capabilities must be evaluated in conjunction with each other.

FIGURE 1-5
End result of a market analysis.

Because the sales and advertising effort influences the volume of sales for a given price, a family of curves is expected. Since a cost is associated with the sales and advertising effort, and since a continuous increase of this effort results in diminishing improvement in sales, there exists an optimum level of sales and advertising effort. A marketing plan should emerge simultaneously with the technical plans for the undertaking.

1.7 FEASIBILITY (STEP 6)

The feasibility study, step 6, and the subsequent feasibility decision refer to whether the project is even possible. A project may be feasible, or possible, but not economical. Infeasibility may result from unavailability of investment capital, land, labor, or favorable zoning regulations. Safety codes or other regulatory laws may prohibit the enterprise. If an undertaking is shown to be infeasible, either alternatives must be found or the project must be dropped.

1.8 RESEARCH AND DEVELOPMENT (STEP 7)

If the product or process is one new to the organization, the results from research and development (R&D) may be an important input to the decision process. Research efforts may provide the origin or improvement of the basic idea, and development work may supply working models or a pilot plant, depending upon the nature of the undertaking.

Placing R&D in a late stage of decision making, as was done in Fig. 1-1, suggests that an idea originates somewhere else in the organization or in the field and eventually is placed at the doorstep of R&D for transformation into a workable idea. The possibility of the idea's originating in the research group should also be exploited and is indicated by the dashed line in Fig. 1-1. Research people often learn of new ideas in other fields which might be applied to their own activity.

1.9 ITERATIONS

The loop in Fig. 1-1 emphasizes that the decision-making process involves many iterations. Each pass through the loop improves the amount and the quality of information and data. Eventually a point is reached where final decisions are made regarding the design, production, and marketing of the product. The substance that circulates through this flow diagram is information, which may be in the form of reports and conversations and may be both verbal and pictorial. The iterations are accomplished by communication between people, and this communication is interspersed by go-or-no-go decisions.

1.10 OPTIMIZATION OF OPERATION

The flow diagram of Fig. 1-1 terminates with the construction or beginning of manufacture of a product or service. Actually another stage takes over at this point, which seeks to optimize the operation of a given facility. The facility was designed on the basis of certain design parameters which almost inevitably change by the time the facility is in operation. The next challenge, then, is to operate the facility in the best possible manner in the light of such factors as actual costs and prices. A painful activity occurs when the project is not profitable and the objective becomes that of minimizing the loss.

1.11 TECHNICAL DESIGN (STEP 5)

Step 5 in Fig. 1-1, the product or system design, has not been discussed. The reason for this omission is that the system design is the subject of this book from this point on. This step is where the largest portion of engineering time is spent. System design as an activity lies somewhere between the study and analysis of individual processes or components and the larger decisions, which are heavily economic. Usually one person coordinates the planning of the undertaking. This manager normally emerges with a background gained from experience in one of the subactivities. The manager's experience might be in finance, engineering, or marketing, for example. Whatever the original discipline, the manager must become conversant with all the fields that play a role in the decision-making process.

The word "design" encompasses a wide range of activities. Design may be applied to the act of selecting a single member or part, e.g., the size of a tube in a heat exchanger; to a larger component, e.g., the entire shell-and-tube heat exchanger; or to the design of the system in which the heat exchanger is only one component. Design activities can be directed toward mechanical devices which incorporate linkages, gears, and other moving solid members, electrical or electronic systems, thermal systems, and a multitude of others. Our concentration will be on thermal systems such as those in power generation, heating and refrigeration plants, the food-processing industry, and in the chemical and process industries.

1.12 SUMMARY

The flow diagram and description of the decision processes discussed in this chapter are highly simplified and are not sacred. Since almost every undertaking is different, there are almost infinite variations in starting points, goals, and intervening circumstances. The purpose of the study is to emphasize the advantage of systematic planning. Certain functions are common in the evaluation and planning of undertakings, particularly the iterations and the decisions that occur at various stages.

ADDITIONAL READINGS

Introductory books on engineering design

Alger, J. R. M., and C. V. Hays: *Creative Synthesis in Design*, Prentice-Hall, Englewood Cliffs, N. J., 1964.

Asimow, M.: *Introduction to Design*, Prentice-Hall, Englewood Cliffs, N. J., 1962.

Beakley, G. C., and H. W. Leach: *Engineering, An Introduction to a Creative Profession*, Macmillan, New York, 1967.

Buhl, H. R.: *Creative Engineering Design*, Iowa State University Press, Ames, 1960.

Dixon, J. R.: *Design Engineering: Inventiveness, Analysis, and Decision Making*, McGraw-Hill, New York, 1966.

Harrisberger, L.: *Engineersmanship, A Philosophy of Design*, Brooks/Cole, Belmont, Calif., 1966.

Krick, E. V.: *An Introduction to Engineering and Engineering Design*, Wiley, New York, 1965.

Middendorf, W. H.: *Engineering Design*, Allyn and Bacon, Boston, 1968.

Mischke, C. R.: *An Introduction to Computer-Aided Design*, Prentice-Hall, Englewood Cliffs, N. J., 1968.

Morris, G. E.: *Engineering, A Decision-Making Process*, Houghton Mifflin Company, Boston, 1977.

Woodson, T. T.: *Introduction to Engineering Design*, McGraw-Hill, New York, 1966.

Probabilistic approaches to design

Ang, A. H-S., and W. H. Tang: *Probability Concepts in Engineering Planning and Design*, Wiley, New York, 1975.

Haugen, E. B.: *Probabilistic Approaches to Design*, Wiley, New York, 1968.

Rudd, D. F., and C. C. Watson: *Strategy of Process Engineering*, Wiley, New York, 1968.

CHAPTER
2

DESIGNING
A WORKABLE
SYSTEM

2.1 WORKABLE AND OPTIMUM SYSTEMS

The simple but important point of this chapter is the distinction between designing a workable system and an optimum system. This chapter also continues the progression from the broad concerns of an undertaking, as described in Chapter 1, to a concentration on engineering systems and, even more specifically, on thermal systems.

It is so often said that "there are many possible answers to a design problem" that the idea is sometimes conveyed that all solutions are equally desirable. Actually only one solution is the optimum, where the optimum is based on some defined criterion, e.g., cost, size, or weight. The distinction then will be made between a workable and an optimum system. It should not be suggested that a workable system is being scorned. Obviously, a workable system is infinitely preferable to a nonworkable system. Furthermore, extensive effort in progressing from a workable toward an optimum system may not be justified because of limitations in calendar time, cost of engineering time, or even the reliability of the fundamental data on which the design is based. One point to be explored in this chapter is how superior solutions may be ruled out in the design process by prematurely eliminating some system concepts. Superior solutions may also be precluded by fixing interconnecting parameters between components and selecting the components based on these parameters instead of letting the parameters float until the optimum total system emerges.

2.2   A WORKABLE SYSTEM

A workable system is one that

1. Meets the requirements of the purposes of the system, e.g., providing the required amount of power, heating, cooling, or fluid flow, or surrounding a space with a specified environment so that people will be comfortable or a chemical process will proceed or not proceed
2. Will have satisfactory life and maintenance costs
3. Abides by all constraints, such as size, weight, temperatures, pressure, material properties, noise, pollution, etc.

In summary, a workable system performs the assigned task within the imposed constraints.

2.3   STEPS IN ARRIVING AT A WORKABLE SYSTEM

The two major steps in achieving a workable system are (1) to select the concept to be used and (2) to fix whatever parameters are necessary to select the components of the system. These parameters must be chosen so that the design requirements and constraints are satisfied.

2.4   CREATIVITY IN CONCEPT SELECTION

Engineering, especially engineering design, is a potentially creative activity. In practice creativity may not be exercised because of lack of time for adequate exploration, discouragement by supervision or environment, or the laziness and timidity of the engineer. It is particularly in selecting the concept that creativity can be exercised. Too often only one concept is ever considered, the concept that was used on the last similar job. As a standard practice, engineers should discipline themselves to review all the alternative concepts in some manner appropriate to the scope of the project. Old ideas that were once discarded as impractical or uneconomical should be constantly reviewed. Costs change; new devices or materials on the market may make an approach successful today that was not attractive 10 years ago.

2.5   WORKABLE VS. OPTIMUM SYSTEM

The distinction between the approaches used in arriving at a workable system and an optimum system can be illustrated by a simple example. Suppose that the pump and piping are to be selected to convey 3 kg/s from one location to another 250 m away from the original position and 8 m higher. If the design is approached with the limited objective of achieving a workable system, the following procedure might be followed:

1. The elevation of 8 m imposes a pressure difference of

$$(8 \text{ m})(1000 \text{ kg/m}^3)(9.807 \text{ m/s}^2) = 78.5 \text{ kPa}$$

Arbitrarily choose an additional 100 kPa to compensate for friction in the 250 m of pipe.

2. According to the foregoing decision, select a pump which delivers 3 kg/s against a pressure difference of 178.5 kPa. Finally, select a pipe size from a handbook such that the pressure drop in 250 m of length is 100 kPa or less. A pipe size of 50 mm (2 in) satisfies the requirement.

Appoaching the same problem with the objective of achieving an optimum system presupposes agreement on a criterion to optimize. A frequently chosen criterion is cost (sometimes first cost only in speculative projects, and sometimes the lifetime cost, consisting of first plus lifetime pumping and maintenance costs).

In designing the optimum pump and piping system for minimum lifetime cost, the pressure rise to be developed by the pump is not fixed immediately but left free to float. If the three major contributors to cost are (1) the first cost of the pump, (2) the first cost of the pipe, and (3) the lifetime pumping costs, these costs will vary as a function of pump pressure, as shown in Fig. 2-1. As the pump-pressure rise increases, the cost of the pump probably increases for the required flow rate of 3 kg/s because of the need for higher speed and/or larger impeller diameter. With the increase in pressure rise, the power required by the pump increases and is reflected in

FIGURE 2-1
Contributions to costs by pump and piping systems.

a higher lifetime pumping cost. The first cost of the pipe, the third contributor to the total cost, becomes enormously high as the pressure available to overcome friction in the pipe reduces to zero. The available pressure for the pipe is the pump-pressure rise minus 78.5 kPa needed for the difference in elevation. An appropriate optimization technique can be used to determine the optimal pump-pressure rise, which in Fig. 2-1 is approximately 150 kPa. Finally the pump can be selected to develop 150 kPa pressure rise, and a pipe size can be chosen such that the pressure drop due to friction is 71.5 kPa or less.

The tone of the preceding discussion indicates a strong preference toward designing optimum systems. To temper this bias, several additional considerations should be mentioned. If the job is a small one, the cost of the increased engineering time required for optimization may devour the savings, if any. Not only the engineer's time but pressure of calendar time may not permit the design to proceed beyond a workable design.

2.6 DESIGN OF A FOOD-FREEZING PLANT

Large-scale engineering projects are extremely complex, and decisions are often intricately interrelated; not only do they influence each other in the purely technical area but also cross over into the technoeconomic, social, and human fields. To illustrate a few of the decisions involved in a realistic commercial undertaking and to provide a further example of the contrast between a workable system and an optimum system, consider the following project.

A food company can buy sweet corn and peas from farmers during the season and sell the vegetables as frozen food throughout the year in a city 300 km away. What are the decisions and procedures involved in designing the plant to process and freeze the crops?

The statement of the task actually starts at an advanced stage in the decision process, because it is already assumed that a plant will be constructed. This decision cannot realistically be made until some cost data are available to evaluate the attractiveness of the project. Let us assume, therefore, that an arbitrarily selected solution has been priced out and found to be potentially profitable. We are likely, then, to arrive at a solution that is an improvement over the arbitrary selection.

Some major decisions that must be made are (1) the location, (2) size, and (3) type of freezing plant. The plant could be located near the producing area, in the market city, or somewhere between. The size will be strongly influenced by the market expectation. The third decision, the type of freezing plant, embraces the engineering design. These three major decisions are interrelated. For example, the location and size of plant might reasonably influence the type of system selected. The selection of the type of freezing plant includes choosing the concept on which the freezing-plant design will be based. After the concept has been decided, the internal design of the plant can proceed.

An outline of the sequence of tasks and decisions by which a workable design could be arrived at is as follows:

1. Decide to locate the plant in the market city adjacent to a refrigerated warehouse operated by the company.
2. Select the freezing capacity of the plant on the basis of the current availability of the crop, the potential sale in the city, and available financing.
3. Decide upon the concept to be used in the freezing plant, e.g., the one shown in Fig. 2-2. In this system the food particles are frozen in a fluidized bed, in which low-temperature air blows up through a conveyor chain, suspending the product being frozen. This air returns from the fluidized-bed conveyor to a heat exchanger that is the evaporator of a refrigerating unit. The refrigerating unit uses a reciprocating compressor and water-cooled condenser. A cooling tower, in turn, cools the condenser water, rejecting heat to the atmosphere.
4. The design can be quantified by establishing certain values. Since the throughput of the plant has already been determined, the freezing capacity in kilograms per second can be computed by deciding upon the number of shifts to be operated. Assume that one shift is selected, so that now the refrigeration load can be calculated at, say, 220 kW. To proceed with the design, the parameters shown in Table 2.1 can be pinned down.

FIGURE 2-2
Schematic flow diagram of freezing plant.

TABLE 2.1

	Temperature, °C
Air, chilled supply	−30
Return	−23
Refrigerant, evaporation	−38
Condensation	45
Condenser, cooling water, inlet	30
Outlet	35

5. After these values have been fixed, the individual components can be selected. The flow of chilled air can be calculated to remove 220 kW with a temperature rise of 7°C. The conveyor length and speed must now be chosen to achieve the required rate of heat transfer. The air-cooling evaporator can be selected from a catalog because the airflow rate, air temperatures, and refrigerant evaporating temperature fix the choice. The compressor must provide 220 kW of refrigeration with an evaporating temperature of −38°C and a condensing temperature of 45°C, which is adequate information for selecting the compressor or perhaps a two-stage compression system. The heat-rejection rate at the condenser exceeds the 220-kW refrigeration capacity by the amount of work added in the compressor and may be in the neighborhood of 300 kW. The condenser and cooling tower can be sized on the basis of the rate of heat flow and the water temperature of 30 and 35°C. Thus, a workable system can be designed.

Unlike the above procedure, an attempt to achieve on optimum system returns to the point where the first decisions are made. Such decisions as the location, size, and freezing concept should be considered in connection with each other instead of independently. The choice of fluidized-bed freezing with a conventional refrigeration plant is only one of the commercially available concepts, to say nothing of the possibility (admittedly remote) of devising an entirely new concept. Other concepts are a freezing tunnel, where the air blows over the top of the product; packaging the product first and immersing the package in cold brine until frozen; or freezing the product with liquid nitrogen purchased in liquid form in bulk. An example of the interconnection of decisions is that the location of the plant that is best for one concept may not be best for another concept. A compression refrigeration plant may be best located in the city as an extension of existing freezing facilities, and it may be unwise to locate it close to the producing area because of lack of trained operators. The liquid-nitrogen freezing plant, on the other hand, is simple in operation and could be located close to the field; furthermore, it could be shut down for the idle off-season more conveniently than the compression plant. If the possibility of two or even three shifts were considered, the processing rate of the plant could be reduced by a factor of 2 or 3, respectively, for the same daily throughput.

Within the internal design of the compression refrigeration plant, the procedure was to select reasonable temperatures and then design each component around those temperatures and resulting flow rates. When one approaches the design with the objective of optimization, all those interconnecting parameters are left free to float and one finds the combination of values of these parameters which results in the optimum (probably the economic optimum).

2.7 PRELIMINARIES TO THE STUDY OF OPTIMIZATION

Any attempt to apply optimization theory to thermal systems at this stage is destined for frustration. There are a variety of optimization techniques available, some of which are introduced in Chapters 8 to 12. The first attempts to optimize a thermal system with a dozen components, however, will be detoured by the need to predict the performance of the system with given input conditions. This assignment, called *system simulation*, must be studied first and will be considered in Chapters 6 and 14. For complex systems, system simulation must be performed with a computer. For this purpose the performance characteristics making up the system could possibly be stored as tables, but a far more efficient and useful form is equation-type formulation. Translating catalog tables into equations, called *component simulation* or *mathematical modeling*, is a routine preliminary step to system simulation and will be treated in Chapters 4, 5, and 13.

Finally, since optimization presupposes a criterion, which in engineering practice is often an economic one, a review of investment economics in Chapter 3 will be appropriate.

The sequence to be followed in the ensuing studies, then, will be (1) economics, (2) mathematical modeling, (3) system simulation, and (4) optimization.

PROBLEMS

2.1. Location S in Fig. 2-3 is an adequate source of water, and location A, B, and C are points at which water must be provided at the following rates of flow:

Location	A	B	C
L/s	2.5	3.5	1.5

Points S, A, B, and C are all at the same elevation. The demands for water at A and C occur intermittently and only during the working day, and they may coincide. The demand for water at B occurs only during nonworking hours and is also intermittent. Ground-level access exists in a 3-m border surrounding the building. Access is not permitted over, through, or under the building.

FIGURE 2-3
Supply and consumption points in water-distribution system.

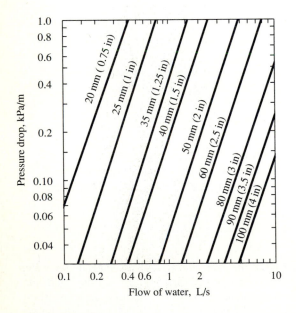

FIGURE 2-4
Pressure drop in pipe.

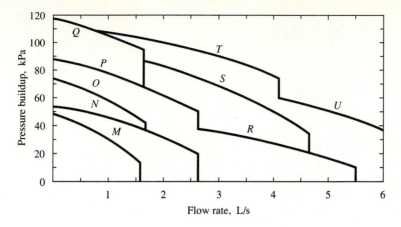

FIGURE 2-5
Pump performance curves.

(*a*) Describe all the concepts of workable methods you can devise to fulfill the assignment.

(*b*) The influence of such factors as the expected life of the system has resulted in the decision to use a system in which a pump delivers water into an elevated storage tank, which supplies the piping system. A water-level switch starts and stops the pump. Design the system; this includes sketching the pipe network chosen, listing all the pipe sizes, selecting the pump, and specifying the elevation of the storage tank. Use pressure-drop data from Fig. 2-4 and pump performance from Fig. 2-5. (Neglect the pressure drop in the pipe fittings and pressure conversions due to kinetic energy.) Fill out Table 2.2.

TABLE 2.2
Design data for Prob. 2.1

Pipe section	Pipe size, mm	Design flow, L/s		Δp, kPa	
		Day	Night	Day	Night
(*S* to *A*, for example)					

Pump selected _____   Elevation of water in storage tank _____

(c) Review the design and list the decisions that preclude possible optimization later in the design.

2.2 A heating and ventilating system for a public indoor swimming pool is to be designed.

Specifications
Pool water temperature, 26°C
Indoor air dry-bulb temperature, 27°C
Outdoor design temperature, −12°C
Outdoor design humidity ratio, 0.00105 kg/kg
For odor control minimum rate of outdoor air for ventilation, 5.8 kg/s
For comfort, temperature of air introduced, 38 to 65°C

Construction features
Pool dimensions, 15 × 45 m
Essentially the structure has masonry walls and a glass roof:
   Glass area, 1850 m²
   Wall area, 670 m²
   U value of wall, 1.2 W/(m² · K)

Supplementary information
   Rate of evaporation of water from pool, g/s = (0.04) (area, m²)$(p_w - p_a)$

where p_w = water-vapor pressure at pool temperature, kPa
   p_a = partial pressure of water vapor in surrounding air, kPa

   A choice of the type of window must be made (single, double, or triple pane). They have the following heat-transfer coefficients:

	Heat-transfer coefficent, W/(m² · K)		
	Single	Double	Triple
Outside air film	34	34	34
Glass (between external and internal surfaces)	118	8	3.4
Inside film	11.4	11.4	11.4

The basic purpose of the system is as follows:

1. To maintain the indoor temperature when design conditions prevail outdoors
2. To abide by the constraints
3. To prevent condensation of water vapor on the inside of the glass
(a) Describe at least two different concepts for accomplishing the objectives of the design. Use schematic diagrams if useful.
(b) Assume that the concept chosen is one where ourdoor air is drawn in, heated, and introduced to the space and an equal quantity of moist room air is exhausted. Should additional heating be required, it is provided

either by heating recirculated air or by using convectors around the perimeter of the building. Perform the design calculations in order to specify the following:

Flow rate of ventilation air___kg/s
Temperature of air entering space___°C
Type of glass selected___
Temperature of inside surface of glass___°C
Condition of indoor air: dew point___°C
Rate of water evaporated from pool___ kg/s
Heat loss by conduction through walls and glass___ kW
Heat supplied to raise temperature of ventilation air from room temperature to supply temperature___ kW
Heat to recirculated air or perimeter convectors___ kW

(c) Review the design and list the decisions that precluded possible optimization later in the design.

2.3. You have just purchased a remote uninhabited island where you plan to give parties lasting till late at night. You need to install an electric power system that will provide at least 8 kW of lighting.

(a) List and describe, in several sentences, three methods of generating power in this remote location, assuming that equipment and supplies can be transported to the island.

(b) Assume that the decision has been made to produce the electric power by means of an engine-driven generator. Specifically, an engine will be direct-connected to a generator that delivers power to bulbs. Available choices are as follows:

Bulbs. 100-W bulbs at 115 V. The bulbs may be connected in parallel, in series, or in combination, and the current flowing through a bulb must be above 70 percent of the rated current to obtain satisfactory lighting efficiency but below 110 percent of the rated current to achieve long life.

Engines. A choice can be made between two engines whose power deliveries at wide-open throttle are as follows:

Engine	Power delivery, kW, at given speed, r/s				
	10	20	30	40	50
1	2.2	5.5	9.0	11.2	12.7
2	3.4	7.1	11.1	13.8	15.2

Generators. Single phase, alternating current. No adjustment of output voltage is possible except by varying the speed. The frequency must be greater than 35 Hz in order to prevent flickering of the lights. Neglect the voltage drop and power loss in the transmission lines. A choice can be made between the following two generators:

Generator	Poles	Efficiency	Maximum allowable current, A	Output voltage, V, at given speed, r/s				
				10	20	30	40	50
1	2	0.72	100	38	71	106	140	174
2	4	0.83	50	59	117	170	224	270

Make the following design selections and specifications:

> Engine number___
> Generator number___
> Engine-generator speed___ r/s
> Frequency___ Hz
> Power delivered by engine___ kW
> Power delivered by generator___ kW
> Voltage___ V
> Current___ A
> Number of bulbs and circuiting___

(*c*) Review the design and list the decisions that preclude possible optimization later in the design.

2.4. Crude oil is to be transported overland in Alaska in such a way that the environment is not adversely affected.[1]

(*a*) Describe two workable methods of transporting this oil.

(*b*) Of the workable methods, a pipeline is the method chosen for further examination. The inside diameter of the pipe is 600 mm, and the pipe will carry a crude oil flow rate of 44 kg/s. The distance between pumping stations is 32 km. To facilitate pumping, a heater will be installed at each pumping station, as shown in Fig. 2-6. The pipeline is to be buried in permafrost whose temperature at design conditions is $-4°C$. The permafrost is not to be melted. Insulation may be used on the pipe, and the external surface temperature of the pipe or insulation in contact with the permafrost must be maintained at $0°C$ or below. Heat-transfer data applicable to the pipe and insulation are shown in Fig. 2-7. The

---

[†]Numbered references appear at the end of the chapter.

FIGURE 2-6
Oil pipeline.

Coefficient of heat transfer between the surface
and permafrost = 7.2 W/(m²·K)

Insulation, k = 0.036 W/(m·K)

Insulation thickness, x m

Pipe, neglect thermal resistance

Convection coefficient of oil
to pipe = 113 W/(m²·K)

FIGURE 2-7
Heat-transfer data.

available thicknesses of insulation are 25, 50, 75, and 100 mm. The overall
heat-transfer coefficient U between the oil and permafrost in watts per square
meter per kelvin based on the inside pipe area A_i is

$$\frac{1}{UA_i} = \frac{1}{113A_i} + \frac{x}{k[(A_i + A_o)/2]} + \frac{1}{7.2\, A_o}$$

where A_o is the outside area of the insulation. The temperature change dt in
a differential length of pipe (see Fig. 2-8) is expressed by

$$wc_p(-dt) = U\pi D[t - (-4°C)]dL$$

where w = mass rate of flow = 44 kg/s
c_p = specific heat of oil = 1930 J/(kg · K)
t = oil temperature, °C
dt = change in oil temperature in length dL
U = overall heat-transfer coefficient between oil and permafrost,
W/(m² · K)
D = pipe diameter = 0.6 m
L = length of pipe, m

FIGURE 2-8
Differential length of pipe.

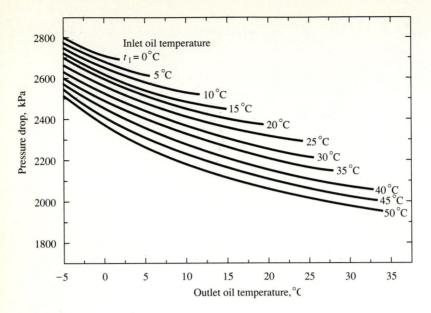

FIGURE 2-9
Oil pressure drop in 32-km length of pipe.

The pressure drop in the 32-km section of pipe is a function of the inlet and outlet temperatures of the oil because of the influence of these temperatures on viscosity. Figure 2-9 shows pressure drops for 32 km. The maximum pressure the pipe can withstand is 2350 kPa gauge. Specify the following:

> Insulation___ mm
> Inlet oil temperature t_1___°C
> Outlet oil temperature t_2___°C
> Pressure drop___ kPa
> Temperature of surface in contact with permafrost (highest in 32-km run)___°C

(c) Review the design and list the decisions that preclude possible optimization later in the design.

2.5. The tube spacing and fin height are to be selected for the steam-generating section of a furnace, shown in Fig. 2-10. The furnace section is 1.8 m wide; the tubes are 2.5 m long and are arranged in a square array, six rows high. The tubes have an OD of 75 mm, and they can be either bare or equipped with fins. The fins are 2 mm thick and are spaced 6 mm apart along the length of the tube.

The steam temperature in the boiler is 175°C, and the entering temperature of the stack gases is 560°C. The minimum spacing between the centerlines of the tubes, which is dictated by the smallest U bend available, is

---

†Problem based on a suggestion of O.B. Taliaferro, Exxon Research and Engineering Company.

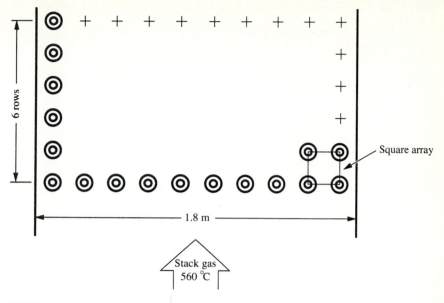

FIGURE 2-10
Steam-generating section in a furnace.

125 mm. There are two restrictions on the design: (1) there must be 3 mm clearance between the fins of adjacent tubes and between the fins and the walls and (2) the maximum metal temperature, which occurs at the tip of the fins, must not exceed 415°C in order to limit oxidation.

The essential requirement of the furnace is that the product of UA for the boiler section be 18,500 W/K or higher. The U value, based on steam-side area, is

$$\frac{1}{U} = R_{\text{total}} = R_{\text{steam}} + R_{\text{tube}} + R_{\text{gas}}$$

where the resistances are

$$R_{\text{steam}} = \frac{1}{5700} = 0.0001754 (\text{m}^2 \cdot \text{K})/W$$

$$R_{\text{tube}} = \frac{0.004\text{m thick}}{k = 44} = 0.0000909 (\text{m}^2 \cdot \text{K})/W$$

$$R_{\text{gas}} = \frac{1}{h_g(A_{\text{prime}}/A_{\text{steam}}) + h_g(A_{\text{fin}}\eta/A_{\text{steam}})}$$

where h_g = gas-side heat-transfer coefficient = 70 W/(m² · K)
A_{prime} = 0.157 m² per meter of tube length
A_{steam} = 0.210 m² per meter of tube length
A_{fin} = fin area per meter of tube length
η = fin effectiveness

TABLE 2.3
Fin data

Fin height, mm, measured from outside of tube	A_{fin}, m²/m	η	$\dfrac{560 - t_{tip}}{560 - t_{root}}$
0	0.079	1.00	1.00
12	1.199	0.94	0.88
18	1.872	0.86	0.77
25	2.754	0.75	0.67
31	3.591	0.64	0.58
37	4.503	0.55	0.49
44	5.665	0.46	0.41

Table 2.3 gives fin areas per meter of tube length, the fin effectiveness, and values of the expression for the fin-tip temperature as a function of the root temperature. In turn the root temperature can be computed from

$$\frac{t_{root} - 175}{560 - 175} = \frac{R_{steam} + R_{tube}}{R_{total}}$$

Specify the number of tubes in each horizontal row and the fin height so that the assembly fits into the space available (abiding by the necessary clearances), provides a UA of 18,500 W/K or more, and maintains a fin-tip temperature of 415°C or less.

REFERENCES

1. D. F. Othmer and J. W. E. Griemsmann, "Moving the Arctic Oil: Pipelines and the Pour Point," *Mech. Eng.*, vol. 93, no. 11, pp. 27-32, November 1971, describes further approaches to moving oil under Arctic conditions.

3

ECONOMICS

3.1 INTRODUCTION

The basis of most engineering decisions is economic. Designing and building a device or system that functions properly is only part of the engineer's task. The device or system must, in addition, be economic, which means that the investment must show an adequate return. In the study of thermal systems, one of the key ingredients is optimization, and the function that is most frequently optimized is the potential profit. Sometimes the designer seeks the solution having minimum first cost or, more frequently, the minimum total lifetime cost of the facility.

Hardly ever are decisions made solely on the basis of monetary considerations. Many noneconomic factors affect the decisions of industrial organizations. Decisions are often influenced by legal concerns, such as zoning regulations, or by social concerns, such as the displacement of workers, or by air or stream pollution. Aesthetics also have their influence, e.g., when extra money is spent to make a new factory building attractive. Since these social or aesthetic concerns almost always require the outlay of extra money, they revert to such economic questions as how much a firm is willing or able to spend for locating a plant where the employees will live in a district with good schools.

This chapter first explains the practice of charging interest and then proceeds to the application of interest in evaluating the worth of lump sums, of series of uniform payments, and of payments that vary linearly with time. Numerous applications of these factors will be explored, including such standard and important ones as computing the value of bonds. Methods of making economic comparisons of alternatives, the influence of taxes, several methods of computing depreciation, and continuous compounding will be explained.

3.2 INTEREST

Interest is the rental charge for the use of money. When renting a house, a tenant pays rent but also returns possession of the house to the owner after the stipulated period. In a simple loan, the borrower of money pays the interest at stated periods throughout the duration of the loan, e.g., every 6 months or every year, and then returns the original sum to the lender.

The existence of interest gives money a time value. Because of interest it is not adequate simply to total all the expected lifetime receipts and in another column total all the expected lifetime expenditures of a facility and subtract the latter from the former to determine the profit. A dollar at year 4 does not have the same value as a dollar at year 8 (even neglecting possible inflation) due to the existence of interest. A thought process that must become ingrained in anyone making economic calculations is that the worth of money has two dimensions, the dollar amount and the time. Because of this extra dimension of time, equations structured for solution of economic problems must equate amounts that are all referred to a common time base.

The most fundamental type of interest is *simple interest*, which will be quickly dismissed because it is hardly ever applied.

> **Example 3.1.** Simple interest of 8 percent per year is charged on a 5-year loan of $500. How much does the borrower pay to the lender?
>
> *Solution.* The annual interest is ($500)(0.08) = $40, so at the end of 5 years the borrower pays back to the lender $500 + 5($40) = $700.

3.3 LUMP SUM, COMPOUNDED ANNUALLY

Annual compounding means that the interest on the principal becomes available at the end of each year, this interest is added to the principal, and the combination draws interest during the next year. In Example 3.1 if the interest were compounded annually, the value at the end of the first year would be

$$\$500 + (\$500)(0.08) = \$540$$

TABLE 3.1

Year	Interest during year	Amount S at end of year
1	Pi	$P + Pi = P(1 + i)$
2	$P(1 + i)i$	$P(1 + i) + P(1 + i)i = P(1 + i)^2$
. . . .
n	$P(1 + i)^{n-1}i$	$P(1 + i)^{n-1} + P(1 + i)^{n-1}i = P(1 + i)^n$

During the second year the interest is computed on $540, so that at the end of the second year the value is

$$\$540 + (\$540)(0.08) = \$583.20$$

At the end of the third year the value is

$$\$583.20 + (\$583.20)(0.08) = \$629.86$$

The pattern for computing the value of an original amount P subjected to interest compounded annually at a rate i is shown in Table 3.1.

Example 3.2. What amount must be repaid on the $500 loan in Example 3.1 if the interest of 8 percent is compounded annually?

Solution. Amount to be repaid after 5 years $= (\$500)(1 + 0.08)^5 = \$734.66$.

3.4 LUMP SUM COMPOUNDED MORE OFTEN THAN ANNUALLY

In most business situations compounding is often semiannually, quarterly, or even daily. At compounding periods other than annual a special interpretation is placed on the designation of the interest rate in that i refers to *nominal annual interest rate*. If interest is compounded semiannually, an interest of $i/2$ is assessed every half year.

Example 3.3. What amount must be repaid on a 5-year $500 loan at 8 percent annual interest compounded quarterly?

Solution

$$\text{Amount to be repaid} = (\$500)\left(1 + \frac{0.08}{4}\right)^{20} = \$742.97$$

The interpretation of the 8 percent interest in Example 3.3 is that one-fourth of that rate is assessed each quarter and there are 20 quarters, or compounding periods, in the 5-year span of the loan.

3.5   COMPOUND-AMOUNT FACTOR (f/p) AND PRESENT-WORTH FACTOR (p/f)

The factors that translate the value of lump sums between present and future worths are

Compound-amount factor (CAF or f/p)
Present-worth factor (PWF or p/f)

The application of these factors is

$$\text{Future worth} = (\text{present worth})(f/p)$$
$$\text{Present worth} = (\text{future worth})(p/f)$$

The two factors are the reciprocal of each other, and the expressions for the factors are

$$f/p = \left(1 + \frac{i}{m}\right)^{mn} \quad \text{and} \quad p/f = \frac{1}{(1 + i/m)^{mn}}$$

where i = nominal annual interest rate
n = number of years
m = number of compounding periods per year

Example 3.4.   You invest $5000 in a credit union which compounds 5 percent interest quarterly. What is the value of the investment after 5 years?

Solution

$$\text{Future amount} = (\text{present amount})\left(f/p, \frac{0.05}{4}, 20 \text{ periods}\right)$$

where the meaning of the convention is (factor, rate, period).

$$\text{Future amount} = (\$5000)\left(1 + \frac{0.05}{4}\right)^{20} = (\$5000)(1.2820) = \$6410$$

Example 3.5.   A family wishes to invest a sum of money when a child begins elementary school so that the accumulated amount will be $10,000 when the child begins college 12 years later. The money can be invested at 8 percent, compounded semiannually. What amount must be invested?

Solution

$$\text{Present amount} = (\text{future amount})\left(p/f, \frac{0.08}{2}, 24 \text{ periods}\right)$$

$$= \frac{\$10,000}{(1 + 0.08/2)^{24}} = \frac{10,000}{2.5633} = \$3901.20$$

3.6 FUTURE WORTH (f/a) OF A UNIFORM SERIES OF AMOUNTS

The next factor to be developed translates the value of a uniform series of amounts into the value some time in the future. The equivalence is shown symbolically in Fig. 3-1, where R is the uniform amount at each time period. The arrow indicating future worth is the equivalent value in the future of that series of regular amounts with the appropriate interest applied.

Suppose that in Fig. 3-1 the magnitude of R is $100 and that this amount appears on an annual basis. Further assume that the interest rate is 6 percent, compounded annually. At the end of the first year the accumulated sum S is the $100 that has just become available. At the end of 2 years the accumulated sum is the $100 from the first year with its interest plus the new $100 amount:

$$S = (\$100)(1 + 0.06) + \$100 = (\$100)[(1 + 0.06) + 1]$$

At the end of 7 years

$$S = (\$100)[(1 + 0.06)^6 + (1 + 0.06)^5 + \cdots + (1 + 0.06) + 1]$$
$$= (\$100)(8.3938) = \$839.38$$

In general, the future worth S of a uniform series of amounts, each of which is R, with an interest rate i that is compounded at the same frequency as the R amounts is

$$S = R[(1 + i)^{n-1} + (1 + i)^{n-2} + \cdots + (1 + i) + 1]$$

The term in the brackets is called the *series-compound-amount factor* (SCAF or f/a)

$$f/a = (1 + i)^{n-1} + (1 + i)^{n-2} + \cdots + (1 + i) + 1 \qquad (3.1)$$

A closed form for the series can be developed by first multiplying both sides of Eq. (3.1) by $1 + i$

$$(f/a)(1 + i) = (1 + i)^n + (1 + i)^{n-1} + \cdots + (1 + i) \qquad (3.2)$$

Subtracting Eq. (3.1) from Eq. (3.2) yields

$$(f/a)[(1 + i) - 1] = (1 + i)^n - 1$$

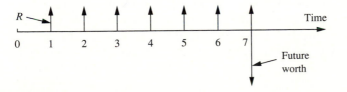

FIGURE 3-1
Future worth of a series of amounts.

and so

$$f/a = \frac{(1 + i)^n - 1}{i} \tag{3.3}$$

The inverse of f/a, called the sinking-fund factor (SFF or a/f), is the multiplying factor that provides the regular amount R when the future worth is known:

Regular amount $R =$ (future worth)(SFF) $=$ (future worth)(a/f)

The term a/f is the reciprocal of f/a:

$$a/f = \frac{i}{(1 + i)^n - 1} \tag{3.4}$$

A point to emphasize is the convention for when the first regular amount becomes available. Figure 3-1 shows that the first amount R is available at the end of the first period and not at time 0. Sometimes when a series is to be evaluated the first amount may be available at time 0; such conversions can be made and will be explained later in the chapter. The important point is that whenever Eq. (3.3) or (3.4) is used, it applies to the convention by which the first of the regular amounts occurs at the end of the first time period.

Example 3.6. A new machine has just been installed in a factory, and the management wants to set aside and invest equal amounts each year starting 1 year from now so that $16,000 will be available in 10 years for the replacement of the machine. The interest received on the money invested is 8 percent, compounded annually. How much must be provided each year?

Solution. The annual amount R is

$$R = (\$16,000)(a/f, 8\%, 10) = (\$16,000) \frac{0.08}{(1 + 0.08)^{10} - 1}$$

$$= (\$16,000)(0.06903) = \$1104.50$$

3.7 PRESENT WORTH (p/a) OF A UNIFORM SERIES OF AMOUNTS

The *series-present-worth factor* (SPWF or p/a) translates the value of a series of uniform amounts R into the present worth, as shown symbolically in Fig. 3-2. As in the previous section, the convention is that the first payment occurs at the end of the first time period and not at time 0. The present worth of the series can be found by applying the p/f factor to each of the R amounts:

$$\text{Present worth} = \frac{R}{(1 + i)^1} + \frac{R}{(1 + i)^2} + \cdots + \frac{R}{(1 + i)^n}$$

FIGURE 3-2
Present worth of a series of amounts.

or

$$\text{Present worth} = R\left[\frac{1}{(1+i)^1} + \frac{1}{(1+i)^2} + \cdots + \frac{1}{(1+i)^n}\right] \quad (3.5)$$

If both sides of Eq. (3.5) are multiplied by $(1 + i)^n$, the expression in the brackets after multiplying by $(1 + i)^n$ is the same as f/a in Eq. (3.1). Thus

$$(\text{Present worth})(1+i)^n = R(\text{f/a}) = R\frac{(1+i)^n - 1}{i}$$

The series-present-worth factor p/a is the one which when multiplied by R yields the present worth. Thus

$$\text{p/a} = \frac{(1+i)^n - 1}{i(1+i)^n} \quad (3.6)$$

It is not surprising to find from a comparison of Eqs. (3.3) and (3.6) that p/a is (f/a) $[1/(1 + i)^n]$. Thus

$$\text{p/a} = (\text{f/a})(\text{p/f})$$

The reciprocal of the series-present-worth factor a/p is called the *capital-recovery factor*.

$$\text{a/p} = \frac{i(1+i)^n}{(1+i)^n - 1} \quad (3.7)$$

Example 3.7. You borrow $1000 from a loan company that charges 15 percent nominal annual interest compounded monthly. You can afford to pay off $38 per month on the loan. How many months will it take to repay the loan?

Solution

$$\$1000 = (\$38)(\text{p/a},\ 1.25\%,\ n\ \text{months})$$

$$1000 = (38)\frac{(1.0125)^n - 1}{(0.0125)(1.0125)^n}$$

$$(1.0125)^n = 1.490$$

$$n = 32.1\ \text{months}$$

3.8   GRADIENT-PRESENT-WORTH FACTOR

The factors p/a, f/a, and their reciprocals apply where the amounts in the series are *uniform*. Instead the present worth of a series of *increasing* amounts may be sought. Typical cases that the series of increasing amounts approximates are maintenance costs and energy costs. The cost of maintenance of equipment is expected to increase progressively as the equipment ages, and energy costs may be projected to increase as fossil fuels become more expensive.

The gradient-present-worth factor (GPWF) applies when there is no cost during the first year, a cost G at the end of the second year, $2G$ at the end of the third year, and so on, as shown graphically in Fig. 3-3. The straight line in Fig. 3-3 starts at the end of year 1. The present worth of this series of increasing amounts is the sum of the individual present worths

$$\text{Present worth} = \frac{G}{(1 + i)^2} + \frac{2G}{(1 + i)^3} + \cdots + \frac{(n - 1)G}{(1 + i)^n}$$

This series can be expressed in closed form as

$$\text{Present worth} = G(\text{GPWF}) = G\left\{\frac{1}{i}\left[\frac{(1 + i)^n - 1}{i(1 + i)^n} - \frac{n}{(1 + i)^n}\right]\right\}$$

Example 3.8. The annual cost of energy for a facility is $3000 for the first year (assume payable at the end of the year) and increases by 10 percent, or $300, each year thereafter. What is the present worth of this 12-year series of energy costs, as shown in Fig. 3-4, if the interest rate is 9 percent compounded annually?

FIGURE 3-3
Gradient series.

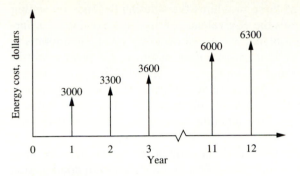

FIGURE 3-4
Progressively increasing energy costs in Example 3.8.

Solution. The series of amounts shown in Fig. 3-4 could be reproduced by a combination of two series, a uniform series of $3000 and a gradient series in which G = 300. Thus

Present worth = (3000)(p/a, 9%, 12) + (300)(GPWF, 9%, 12)

$$\text{PW} = (\$3000)\frac{(1.09)^{12} - 1}{(0.09)(1.09)^{12}} + (\$300)\left\{ \frac{1}{0.09}\left[\frac{(1.09)^{12} - 1}{(0.09)(1.09)^{12}} - \frac{12}{(1.09)^{12}} \right] \right\}$$

PW = $21,482 + $9648 = $31,130

3.9 SUMMARY OF INTEREST FACTORS

The factors p/f, f/a, p/a, their reciprocals, and the GPWF are tools that can be judiciously applied and combined to solve a variety of economic problems. These factors are summarized in Table 3.2.

Sections 3.10 to 3.13 and a number of problems at the end of the chapter illustrate how these factors can be combined to solve more compli-

TABLE 3.2
Summary of interest factors

Factor	Formula
f/p	$(1 + i)^n$
f/a	$\dfrac{(1 + i)^n - 1}{i}$
p/a	$\dfrac{(1 + i)^n - 1}{i(1 + i)^n}$
GPWF	$\dfrac{1}{i}\left[\dfrac{(1 + i)^n - 1}{i(1 + i)^n} - \dfrac{n}{(1 + i)^n} \right]$

cated situations. In most of these problems the solution depends on setting up an equation that expresses the equivalence of amounts existing at different times. Many of these problems are thus solved by translating amounts to the same basis, such as the future worth of all amounts, the present worth, or the annual worth. One point to emphasize is that the influence of interest means that a given sum has differing values at different times. As a consequence the equations just mentioned should never add or subtract amounts applicable to different times. The amounts must always be translated to a common base first.

One of the important financial transactions is the issuance, selling, and buying of bonds. Bonds will be the first application of the combination of factors shown in Table 3.2.

3.10 BONDS

Commercial firms have various methods for raising capital to conduct or expand their business. They may borrow money from banks, sell common stock, sell preferred stock, or issue bonds. A bond is usually issued with a face value of $1000, which means that at maturity the bond's owner surrenders it to the firm that issued it and receives $1000 in return. In addition, interest is paid by the firm to the owner at a rate specified on the bond, usually semiannually. The owner of a bond therefore receives $(i/2)(\$1000)$ every 6 months plus $1000 at maturity. The original purchaser does not have to keep the bond for the life of the bond but may sell it at a price agreed on with a purchaser. The price may be higher or lower than $1000.

Interest rates fluctuate, depending upon the availability of investment money. A firm may have to pay 10 percent interest in order to sell the bond at the time it is issued, but 2 years later, for example, the going rate of interest may be above or below 10 percent. The owner of a 10 percent bond can demand more than $1000 for the bond if the going interest rate is, say, 8 percent, but would have to sell the bond for less than $1000 if the going rate of interest were 12 percent. In other words, the current price of the bond is such that the total investment earns the current rate of interest.

The principle of setting up an equation that reflects all values to a common basis will now be applied by equating what should be paid for the bond to what is acquired from the bond. Arbitrarily choose the future worth as the basis of all amounts

$$P_b\left(\text{f/p}, \frac{i_c}{2}, 2n\right) = \left(\text{f/a}, \frac{i_c}{2}, 2n\right)\frac{i_b}{2}(1000) + 1000 \tag{3.8}$$

where P_b = price to be paid for bond now
$\quad i_c$ = current rate of interest
$\quad i_b$ = rate of interest on bond
$\quad n$ = years to maturity

The terms in Eq. (3.8) are as follows. On the left side of the equation is the future worth of the investment, which is the price to be paid for the bond translated to the future. The interest applicable is the current rate of interest, which is assumed to continue from now until maturity of the bond.

The first term on the right of Eq. (3.8) is the future worth of a uniform series of the semiannual interest payments on the bond. It is assumed that the investor immediately reinvests the interest payments at the going rate of interest i_c. The other term on the right of the equation is $1000, which is the amount the firm will pay back to the owner of the bond at maturity. This $1000 is already at the future time, so no correction need be made on it.

Example 3.9. A $1000 bond that has 10 years to maturity pays interest semiannually at a nominal annual rate of 8 percent. An investor wishes to earn 9 percent on her investment. What price could she pay for the bond in order to achieve this 9 percent interest rate?

Solution. Application of Eq. (3.8) gives

$$P_b(\text{f/p}, 0.045, 20) = (\text{f/a}, 0.045, 20)(0.04)(1000) + 1000$$

$$P_b(2.4117) = (31.371)(0.04)(1000) + 1000$$

$$P_b = \$934.96$$

3.11 SHIFT IN TIME OF A SERIES

In the series shown in Fig. 3-1 for future worth and Fig. 3-2 for present worth, the convention is that the first regular amount appears at the end and not the beginning of the first period. In an actual situation the first amount may appear at time 0, and no amount appears at the end, as illustrated in Fig. 3-5. If the future worth of this series is sought, for example, Eq. (3.1) would be modified by multiplying each of the terms by $1 + i$. The closed form in Eq. (3.3) could be multiplied by $1 + i$, with the result that the f/a for the series in Fig. 3-5 is

$$(\text{f/a})_{\text{shift}} = (1 + i)\frac{(1 + i)^n - 1}{i} \tag{3.9}$$

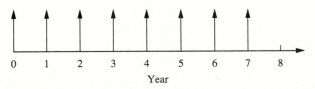

FIGURE 3-5
First amount in a series appearing at time 0.

3.12 DIFFERENT FREQUENCY OF SERIES AMOUNTS AND COMPOUNDING

The formulas for f/a and p/a were developed on the basis of identical compounding and payment frequencies, e.g., annual or semiannual. A situation often arises where the periods are different, e.g., quarterly compounding with annual series amounts. More frequent compounding than payments can be accommodated by determining an equivalent rate of interest applicable between the payment periods.

> **Example 3.10** Annual investments of $1200 are to be made at a savings and loan institution for 10 years beginning at the end of the first year. The institution compounds interest quarterly at a nominal annual rate of 5 percent. What is the expected value of the investment at the end of 10 years?
>
> *Solution.* If an amount X starts drawing interest at the beginning of a year and is compounded quarterly at an annual rate of 5 percent, the value Y at the end of the year is
>
> $$Y = X\left(1 + \frac{0.05}{4}\right)^4 = X(1.0509) = X(1 + 0.0509)$$
>
> The equivalent annual rate of interest due to the quarterly compounding is therefore 5.09 percent. This rate can be used in the f/a factor for the annual amounts
>
> Future worth $= (f/a, 5.09\%, 10 \text{ years})(\$1200)$
>
> $$= \frac{(1.0509)^{10} - 1}{0.0509}(1200) = (12.631)(1200) = \$15,157$$

3.13 CHANGES IN MIDSTREAM

A plan involving a series of payments (or withdrawals) is set up on the basis of certain regular amounts and expected interest rates. During the life of the plan the accepted interest rate may change and/or the ability to make the regular payments may change, and an alteration in the plan may be desired. This class of problems can generally be solved by establishing the worth at the time of change and making new calculations for the remaining term.

> **Example 3.11.** A sinking fund is established such that $12,000 will be available to replace a facility at the end of 10 years. At the end of 4 years, following the fourth uniform payment, management decides to retire the facility at the end of 9 years of life. At an interest rate of 6 percent, compounded annually, what are the payments during the first 4 and last 5 years?
>
> *Solution.* According to the original plan, the payments to be made for 10 years were

$$(\$12,000)(a/f, 6\%, 10) = (12,000)(0.075868) = \$910.42$$

The worth of the series of payments at the end of 4 years is

$$(\$910.42)(f/a, 6\%, 4) = (910.42)(4.3746) = \$3982.72$$

This accumulated value will draw interest for the next 5 years, but the remainder of the $12,000 must be provided by the five additional payments plus interest

$$R(f/a, 6\%, 5) = \$12,000 - (3982.72)(f/p, 6\%, 5)$$

and so the payments during the last 5 years must be

$$R = \frac{12,000 - (3,982.72)(1.3382)}{5.6371} = \$1183.30$$

3.14 EVALUATING POTENTIAL INVESTMENTS

An important function of economic analyses in engineering enterprises is to evaluate proposed investments. A commercial firm must develop a rate of return on its investment that is sufficient to pay corporation taxes and still leave enough to pay interest on the bonds or dividends on the stock that provide the investment capital. The evaluations can become very intricate, and only the basic investment situations will be explained. This fundamental approach is, however, the starting point from which modifications and refinements can be made in more complicated situations.

Four elements will be considered in investment analyses: (1) first cost, (2) income, (3) operating expense, and (4) salvage value. The *rate of return* is treated as though it were interest.

Example 3.12. An owner-manager firm of rental office buildings has a choice of buying building A or building B with the intent of operating the building for 5 years and then selling it. Building A is in an improving location, so that the expected value is to be 20 percent higher in 5 years, while building B is in a declining neighborhood, with an expected drop in value of 10 percent in 5 years. Other data applicable to these buildings are shown in Table 3.3. What will be the rate of return on each building?

Solution. The various quantities must be translated to a common basis, future worth, annual worth, or present worth. If all amounts are placed on a present-worth basis, the following equations apply:

Building A:

$$800,000 = (160,000 - 73,000)(p/a, i\%, 5) + (960,000)(p/f, i\%, 5)$$

Building B:

$$600,000 = (155,000 - 50,300)(p/a, i\%, 5) + (540,000)(p/f, i\% , 5)$$

TABLE 3.3
Economic data for Example 3.12

	Building *A*	Building *B*
First cost	$800,000	$600,000
Annual income from rent	160,000	155,000
Annual operating and maintenance cost	73,000	50,300
Anticipated selling price	960,000	540,000

The unknown in each of the two present-worth equations is the rate of return *i*. Since the expressions are not linear in *i*, an iterative technique must be used to find the value of *i*:

$$i = \begin{cases} 13.9\% & \text{building } A \\ 16.0\% & \text{building } B \end{cases}$$

Building *B* provides a greater rate of return, and the firm must now decide whether even the 16 percent is adequate to justify the investment.

3.15 TAXES

The money for operating the government and for financing services provided by the government comes primarily from taxes. The inclusion of taxes in an economic analysis is often important because in some cases taxes may be the factor deciding whether to undertake the project or not. In certain other cases the introduction of tax considerations may influence which of two alternatives will be more attractive economically.

In most sections of the United States, property taxes are levied by a substate taxing district in order to pay for schools, city government and services, and perhaps park and sewage systems. Theoretically, the real estate tax should decrease as the facility depreciates, resulting in lower real estate taxes as the facility ages. Often on investments such as buildings, the tax, as a dollar figure, never decreases. It is therefore a common practice to plan for a constant real estate tax when making the investment analysis. The effect of the tax is to penalize a facility which has a high taxable value.

Federal corporation income taxes on any but the smallest enterprises amount to approximately 50 percent of the profit. In Example 3.12 the rate of return on the investment in building B was 16 percent, which may seem very favorable compared with the usual ranges of interest rates of 5 to 10 percent. The rate of return of 16 percent is before taxes, however; after the corporation income tax has been extracted, the rate of return available for stock or bond holders in the company is of the order of 8 percent. Since income tax is usually a much more significant factor in the economic analysis than property tax, income tax will be discussed further. An ingredient of income tax calculations is depreciation, explored in the next section.

3.16 DEPRECIATION

Depreciation is an amount that is listed as an annual expense in the tax calculation to allow for replacement of the facility at the end of its life. Numerous methods of computing depreciation are permitted by the Internal Revenue Service, e.g., straight line, sum-of-the-year's digits, and double-rate declining balance. The first two will be explained.

Straight-line depreciation consists simply of dividing the difference between the first cost and salvage value of the facility by the number of years of tax life. The result is the annual depreciation. The tax life to be used is prescribed by the Internal Revenue Service and may or may not be the same as the economic life used in the economic analysis.

In the sum-of-the-year's-digits (SYD) method, the depreciation for a given year is represented by the formula

$$\text{Depreciation, dollars} = 2\,\frac{N - t + 1}{N\,(N + 1)}(P - S) \qquad (3.10)$$

where N = tax life, years
t = year in question
P = first cost, dollars
S = salvage value, dollars

If the tax life is 10 years, for example, the depreciation is

$$\text{First year} = 2(10/110)(P - S)$$
$$\text{Second year} = 2(9/110)(P - S)$$
$$\cdots\cdots\cdots\cdots\cdots\cdots\cdots\cdots\cdots\cdots\cdots$$
$$\text{Tenth year} = 2(1/110)(P - S)$$

A comparison of the depreciation rates calculated by the straight-line method with those calculated by the SYD method shows that the SYD method permits greater depreciation in the early portion of the life. With the SYD method the income tax that must be paid early in the life of the facility is less than with the straight-line method, although near the end of the life the SYD tax is greater. The total tax paid over the tax life of the facility is the same by either method, but the advantage of using the SYD method is that more of the tax is paid in later years, which is advantageous in view of the time value of money.

The straight-line method has an advantage, however, if it is likely the tax rate will increase. If the rate jumps, it is better to have paid the low tax on a larger fraction of the investment.

3.17 INFLUENCE OF INCOME TAX ON ECONOMIC ANALYSIS

To see the effect of depreciation and federal income tax, consider the following simple example of choosing between alternative investments A and B, for which the data in Table 3.4 apply.

A calculation of the annual cost of both alternatives without inclusion of the income tax is as follows:

Alternative A:
First cost on annual basis (200,000)(a/p, 9%, 20)	$21,910
Annual operating expense	14,000
Real estate tax and insurance	10,000
Total	$45,910

Alternative B:
First cost on annual basis (270,000)(a/p, 9%, 30)	$26,280
Annual operating expense	6,200
Real estate tax and insurance	13,500
Total	$45,980

The economic analysis of alternatives A and B shows approximately the same annual costs and incomes (in fact, the example was rigged to accomplish this).

In the computation of profit on which to pay income tax, the actual interest paid is listed as an expense, and if straight-line depreciation is applied, the expenses for the first year for the two alternatives are as shown in Table 3.5.

A higher income tax must be paid on A than on B during the early years. In the later years of the project, a higher tax will be paid on B. The example shows that even though the investment analysis indicated equal profit on the two alternatives, the inclusion of income tax shifts the preference to B. The advantage of B is that the present worth of the tax payments is less than for A.

TABLE 3.4
Alternative investments

	Alternative A	Alternative B
First cost	$200,000	$270,000
Life	20 years	30 years
Salvage value	0	0
Annual income	$60,000	$60,000
Annual operating expense	$14,000	$6,200
Real estate tax and insurance		
(5% of first cost)	$10,000	$13,500
Interest	9%	9%

TABLE 3.5
Income tax on two alternatives

First-year expenses	Alternative *A*	Alternative *B*
Depreciation	$10,000	$ 9,000
Interest, 9% of unpaid balance	18,000	24,300
Operating expense, tax and insurance	24,000	19,700
Total expenses	$52,000	$53,000
Profit = income − expenses	8,000	7,000
Income tax (50% of profits)	4,000	3,500

3.18 CONTINUOUS COMPOUNDING

Frequencies of compounding such as annual, semiannual, and quarterly have been discussed. Even shorter compounding periods are common, e.g., the daily compounding offered investors by many savings and loan institutions. High frequency of compounding is quite realistic in business operation, because the notion of accumulating money for a quarterly or semiannual payment is not a typical practice. Businesses control their money more on a flow basis than on a batch basis. The limit of compounding frequency is *continuous compounding* with an infinite number of compounding periods per year. This section discusses three factors applicable to continuous compounding: (1) the continuous compounding factor corresponding to f/p, (2) uniform lump sums continuously compounded, and (3) continuous flow continuously compounded.

The f/p term with a nominal annual interest rate of i compounded m times per year for a period of n years is

$$f/p = \left(1 + \frac{i}{m}\right)^{mn}$$

The f/p factor for continuous compounding $(f/p)_{cont}$ is found by letting m approach infinity

$$(f/p)_{cont} = \left(1 + \frac{i}{m}\right)^{mn} \Big|_{m \to \infty} \tag{3.11}$$

To evaluate the limit, take the logarithm of both sides

$$\ln (f/p)_{cont} = mn \left[\ln\left(1 + \frac{i}{m}\right)\right]_{m \to \infty}$$

Express $\ln (1 + i/m)$ as an infinite series of i/m

$$\ln (f/p)_{cont} = mn \left[0 + \frac{i}{m} + (\text{const})\frac{i^2}{m^2} + \cdots\right]_{m \to \infty}$$

Cancel m and let $m \to \infty$

$$\ln \, (f/p)_{\text{cont}} = in$$

Therefore

$$(f/p)_{\text{cont}} = e^{in} \tag{3.12}$$

Example 3.13. Compare the values of (f/p, 8%, 10) and [(f/p)$_{\text{cont}}$, 8%, 10].

Solution

$$(f/p, \, 8\%, \, 10) = (1 + 0.08)^{10} = 2.1589$$

$$[(f/p)_{\text{cont}}, \, 8\%, \, 10] = e^{0.8} = 2.2255$$

The next factor presented is that of the future worth of a uniform series of lumped amounts compounded continuously. The continuous f/a factor for annual amounts R can be derived by modifying Eq. (3.1) to continuous compounding by replacing $1 + i$ with e^i

$$(f/a)_{\text{cont,lump}} = (e^i)^{n-1} + (e^i)^{n-2} + \cdots + e^i + 1$$

In the closed-form expression of Eq. (3.3) $1 + i$ can be replaced by e^i and i by $e^i - 1$, yielding

$$(f/a)_{\text{cont, lump}} = \frac{e^{in} - 1}{e^i - 1} \tag{3.13}$$

The final continuous-compounding factor to be presented is the *continuous-flow future-worth factor* (f/a)$_{\text{flow}}$, which expresses the future worth of a continuous flow compounded continuously. If \$1 per year is divided into m equal amounts and spread uniformly over the entire year, and if each of these m amounts begins drawing interest immediately, the future worth of this series after n years is

$$\frac{\$1}{m}\left(1 + \frac{i}{m}\right)^{mn-1} + \frac{\$1}{m}\left(1 + \frac{i}{m}\right)^{mn-2} + \cdots + \frac{\$1}{m}\left(1 + \frac{i}{m}\right)^{1} + \frac{\$1}{m} \tag{3.14}$$

If we use the analogy between Eqs. (3.1) and (3.3), Eq. (3.14) can be translated into the closed form

$$(f/a)_{\text{flow}} \text{ for \$1 per year} = \frac{1}{m} \frac{(1 + i/m)^{mn} - 1}{i/m} = \frac{(1 + i/m)^{mn} - 1}{i}$$

As m approaches infinity, the term $(1 + i/m)^{mn}$ approaches e^{in}, so that

$$(f/a)_{\text{flow}} \text{ for \$1 per year} = \frac{e^{in} - 1}{i} \tag{3.15}$$

Example 3.14. Location *A* for a factory is expected to produce an annual profit that is $10,000 per year greater than if location *B* is chosen. This additional profit is spread evenly over the entire year and constitutes a continuous flow of $10,000 per year. The profit is continuously reinvested at the rate of return of 14 percent per year. At the end of 8 years, the expected economic life of the factories, what is the difference between the worths of the two investments?

Solution. Since the $10,000 per year difference is a continuous flow continuously compounded, the difference between future worths is

$$(\$10,000)\frac{e^{(0.14)(8)} - 1}{0.14} = \$147,490$$

3.19 SUMMARY

There are several levels of economic analysis higher than that approached by this chapter. The complications in accounting, financing, and tax computations involve sophistications beyond those presented here. The stage achieved by this chapter might be described as the second level of economic analysis. The first level would be a trivial one of simply totaling costs with no consideration of the time value of money. The second level introduces the influences of interest, which imposes the dimension of time as well as amount in assessing the value of money.

The methods of investment analyses explained in this chapter are used repeatedly in engineering practice, and in most cases engineers are not required to go beyond these principles. These methods also are the base for extensions into more complex economic analyses.

PROBLEMS

3.1. Using a computer, calculate your personal set of tables for the factors f/p, f/a, p/a, and GPWF. Devote a separate page to each of the factors, label adequately, and calculate at the interest rates of 4, 5, 6, 7, 8, 9, 10, 11, 12, 14, 16, 18, 20, 25 percent. Calculate for the following interest periods: 1 to 20 by ones, 22 to 30 by twos, and 30 to 60 by fives. Print out the factors to four places after the decimal point.

3.2. Annual investments are being made so that $20,000 will be accumulated at the end of 10 years. The interest rate on these investments is initially expected to be 4 percent compounded annually. After 4 years, the rate of interest is unexpectedly increased to 5 percent, so that payments for the remaining 6 years can be reduced. What amounts should be invested annually for the first 4 years and what sums for the last 6?

    Ans.: Final payment, $1547.

3.3. A firm wishes to set aside equal amounts at the end of each of 10 years, beginning at the end of the first year, in order to have $8000 in maintenance funds available at the end of the seventh, eighth, ninth, and tenth years. What is the required annually payment if the money is invested and draws 6 percent compounded annually?

Ans.: $2655.

3.4. A home mortgage extends for 20 years at 8 percent interest compounded monthly. The payments are also made monthly. After how many months is half of the principal paid off?

Ans.: 164 months.

3.5. A lender offers a 1-year loan at what he calls 8 percent interest but requires the interest to be paid at the beginning rather than at the end of the year, as the usual practice is. To what interest rate computed in the conventional manner does this interest charge correspond?

Ans.: 8.7%.

3.6. A loan of $50,000 at 8 percent compounded annually is to be paid off in 25 years by uniform annual payments beginning at the end of the first year. These annual payments proceed on schedule until the end of the eighth year, when the borrower is unable to pay and misses the payment. He negotiates with the lender to increase the remaining 17 payments in such a way that the lender continues to receive 8 percent. What is the amount of the original and the final payments in the series?

Ans.: Final payments, $5197.44.

3.7. An $18,000 mortgage on which 8 percent interest is paid, compounded monthly, is to be paid off in 15 years in equal monthly installments. What is the total amount of interest paid during the life of this mortgage?

Ans.: $12,964.

3.8. What will be the future worth of a series of 15 annual $1000 payments if the nominal annual interest rate is 8 percent and the interest is compounded quarterly?

Ans.: $27,671.

3.9. A sum of sufficient magnitude is to be invested now so that starting 10 years from now an amount of $2000 per year can be paid in each of 8 succeeding years. The unexpended money remains invested at 8 percent compounded annually. How much must be allocated now?

Ans.: $5749.50.

3.10. A mortgage that was originally $20,000 is being paid off in regular quarterly payments of $500. The interest is 8 percent compounded quarterly. How much of the principal remains after 9 years, or 36 payments?

Ans.: $14,800.60.

3.11. A 20-year mortgage set up for uniform monthly payments with 6 percent interest compounded monthly is taken over by a new owner after 8 years. At that time $12,000 is still owed on the principal. What was the amount of the original loan?

Ans.: $16,345.

3.12. An investor buys common stock in a firm for $1000. At the end of the first year and every year thereafter, she receives a dividend of $100, which she immediately invests in a savings and loan institution that pays 5 percent interest compounded annually. At the end of the tenth year, just after receiving her dividend, she sells the stock for $1200. What is the rate of interest (on an annual compounding basis) yielded by this investment program?

 Ans.: 9.41%.

3.13. A sum of $20,000 is borrowed at an interest rate of 8 percent on the unpaid balance compounded semiannually. The loan is to be paid back with 10 equal payments in 20 years. The payments are to be made every 2 years, starting at the end of the second year. What is the amount of each biennial payment?

 Ans.: $4291.

3.14. The packing in a cooling tower that cools condensing water for a power plant progressively deteriorates and results in gradually rising costs due to reduced plant efficiency. These costs are treated as lump sums at the end of the year, as shown in Fig. 3-6. The cost is zero for the first year and then increases $1000 per year until the packing is 16 years old, when replacement is mandatory. At a point 8 years into the life of the packing (just after the $7000 annual cost has been assessed), a decision is to be made on the plan for the next 8 years, i.e., whether to replace the packing or to continue with the existing packing. Money can be borrowed at 9 percent interest, compounded annually. What is the maximum amount that could be paid for the packing in order to justify its replacement?

 Ans.: $44,279.

3.15. The anticipated taxes on a facility for its 10-year tax life decline in a straight-line fashion as follows:

At end of year 1	$10,000
At end of year 2	9,000
. . . .	. . .
At end of year 10	1,000

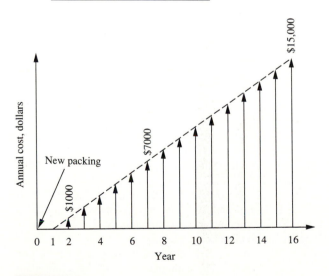

FIGURE 3-6
Increasing costs due to cooling-tower deterioration.

In an economic analysis of the facility the present worth of this series must be computed on the basis of 6 percent interest compounded annually.

(a) Using a combination of available factors, determine a formula for the present worth of a declining series like this one.

(b) Using the formula from part (a), compute the present worth of the above series.

   Ans.: (b) $43,999.

3.16. Calculate the uniform annual profit on a processing plant for which the following data apply:

Life	12 years
First cost	$280,000
Annual real estate tax and insurance	4% of first cost
Salvage value at end of 12 years	$50,000
Annual cost of raw materials, labor, and other supplies	$60,000
Annual income	$140,000
Maintenance costs, during first year	0
At end of second year	$1000
At end of third year	$2000
. . . .	
At end of twelfth year	$11,000

The interest rate applicable is 6 percent compounded annually.

   Ans.: $33,560.

3.17. A $1000 bond was issued 5 years ago and will mature 5 years from now. The bond yields an interest rate of 5 percent, or $50 per year. The owner of the bond wishes to sell the bond, but since interest rates have increased, a prospective buyer wishes to earn a rate of 6 percent on his investment. What should the selling price be? Remember that the purchaser receives $50 per year, which is reinvested, and receives the $1000 face value at maturity. Interest is compounded annually.

   Ans.: $957.88.

3.18. Equation (3.8) relates the value of a bond P_b to the bond interest and current rate of interest by reflecting all values to a future worth. Develop an equation that reflects all values to a uniform semiannual worth and solve Example 3.9 with this equation.

3.19. Using a computer program, calculate tables of the price of a $1000 bond that will yield 5.0, 5.5, 6.0, 6.5, 7.0, 7.5, 8.0, 8.5, 9.0, 9.5, 10.0, 10.5, and 11.0 percent interest when the interest rates on the bond are 5.0, 5.5, 6.0, 6.5, 7.0, 7.5, 8.0, 8.5, 9.0, 9.5, and 10.0 percent. Compute the foregoing table for each of the following number of years to maturity: 2, 3, 4, 5, 6, 7, 8, 9, and 10. Interest is compounded semiannually.

3.20. A municipality must build a new electric generating plant and can choose between a steam or a hydro facility. The anticipated cost of the steam plant is $10 million. Comparative data for the two plants are

Plant	Generating cost, including maintenance, per kWh	Equipment life, years	Salvage value, percentage of first cost
Steam	$0.004	20	10
Hydro	0.002	30	10

The expected annual consumption of power is 300,000 MWh. If money is borrowed at 5 percent interest compounded annually, what first cost of the hydro plant would make the two alternatives equally attractive investments?
Ans.: $21,593,000.

3.21. A proposed investment consists of constructing a building, purchasing production machinery, and operating for 20 years. The expected life of the building is 20 years; its first cost is $250,000 with a salvage value of $50,000. Since the maximum life of the machinery is 12 years, it will be necessary to renew the machinery once during the 20 years. The first cost of the machinery is $132,000, and its salvage value is $132,000/(age, years). The annual income less the operating expense is expected to be $50,000. Annual interest is 6 percent compounded annually.
(a) When is the most favorable time to replace the machinery?
(b) Compute the present worth of the profit if the machinery is replaced at the time indicated by part (a).
Ans.: (b) $152,100.

3.22. Owners of a plant that manufactures edible oil are considering constructing a tank to store unrefined oil; this will permit buying raw oil at more favorable prices. The cost of the tank is $150,000, it has an expected life of 10 years and a salvage value at the end of its life of $20,000. The anticipated annual saving in oil cost is $25,000. What is the rate of return on the investment?
Ans.: 11.65%.

3.23. A new facility is expected to show zero profit during each of the first 5 years. For the 10 years thereafter the expected profit is to be $80,000 per year. If 12 percent interest (compounded annually) is desired as the return on the investment, what amount of investment is justified?
Ans.: $256,500.

3.24. The anticipated income from an investment is $40,000 per year for the first 5 years and $30,000 per year for the remaining 5 years of life. The desired rate of return on the investment is 12 percent. The salvage value at the end of 10 years is expected to be 20 percent of the first cost. Determine the first cost that will result in the 12 percent return.
Ans.: $219,700.

3.25. How many years will be required to double an investment if it draws interest at a rate of 8 percent, compounded semiannually?

3.26. You just received a notice of an insurance premium of $45 per month starting October 1 but were also offered the option of making one annual payment on October 1. If you want 10 percent annual rate of interest, what annual payment would you be willing to make?
Ans.: $516.12.

3.27. A sum of $10,000 is invested and draws interest at a rate of 8 percent, compounded annually. Starting at the end of the first year and each year thereafter $1000 is withdrawn. For how many years can this plan continue until the money is exhausted?

   Ans.: 21 years.

3.28. A firm borrows $200,000 at 9 percent nominal interest, compounded monthly and is to repay the loan in 12 years with regular monthly payments of $2276.06. The firm has the option of paying off in advance, and after the sixth year makes an additional $50,000 payment. If it continues the $2276.06 payments, how many additional months are required to pay off the loan?

   Ans.: 39 months

3.29. A car rental agency which leases cars to another firm buys cars for $9,000 and sells them for $6,000 two years later. It charges a monthly rate the second year of rental of 80 percent of that of the first year. The agency seeks 1 percent per month return. What are the monthly rates each year?

   Ans.: 2nd year, $177.74

3.30. A 20-year loan is to be paid off by monthly payments of M. The nominal annual interest rate is i. Develop a closed-form expression for the unpaid balance at year n.

3.31. A firm has capital on hand and is considering an investment in a plant that is expected to show a net annual return (income less expense) of $80,000 per year. The life of the facility is 10 years, and the salvage value at the end of that time is 20 percent of the first cost. If the firm wishes a 12 percent return on its investment, how much can it justify as the first cost?

   Ans.: $483,131.

3.32. The first cost of an investment is $600,000 borrowed at 11 percent interest compounded semiannually. The expected income (less operating expense) for every 6-month period is $60,000. If there is no salvage value, how long must the plant operate in order to pay off the investment?

   Ans.: $7\frac{1}{2}$ years

3.33. A 15-year mortgage of $40,000 at 10 percent interest compounded monthly is to be paid off with monthly payments. How much total interest will be paid during the first two years?

   Ans.: $7,763.57

3.34. A sum of $1,000 is invested and draws interest at the rate of 8 percent compounded annually. At the end of the first year and each year thereafter $50 is withdrawn from the invested amount. How much money is still available in the investment after the 20th annual withdrawal?

   Ans.: $2372.96

3.35. In a certain financing arrangement the sum of $100,000 is loaned at 12 percent compounded monthly as though it were to be paid off in 25 years. At the end of 5 years the agreement calls for the borrower to pay back the unpaid balance at that time. What is the unpaid balance after 5 years?

   Ans.: $95,653.

3.36. The expected annual income from a new facility that is under consideration is $120,000, and the anticipated annual operating expenditures are $60,000. The salvage value at the end of the expected life of 12 years will be 20 percent of the first cost. What first cost would result in a rate of return of 15 percent?

Ans.: $337,868.

3.37. A processing plant has a first cost of $600,000 and an expected life of 15 years with no salvage value. Money is borrowed at 8 percent compounded annually, and the first cost is paid off with 15 equal annual payments. The expected annual income is $200,000, and annual operating expenses are $40,000. Corporation income tax is 50 percent of the profits before taxes, and the SYD method of depreciation is applicable on the tax life of the facility, which is 12 years with no salvage value. Compute the income tax for (a) the first year and (b) the second year.

Ans.: (a) $9846; (b) $14,576.

3.38. A client who is constructing a warehouse instructs the contractor to omit insulation. The client explains that he will operate the building for several months and then install the insulation as a repair, so that he can deduct the expense from income tax at the end of the first year rather than spread it as straight-line depreciation over the 8-year tax life of the warehouse. The contractor points out that a later installation will cost more than the $20,000 cost of installing the insulation with the original construction. How much could the client afford to pay for the later installation for equal profit if he plans on a 15 percent return on his investment and corporation income taxes are 50 percent?

Ans.: $25,461.

3.39. A $200,000 facility has an 8-year tax life, and the firm expects a 12 percent return on its investment and pays 50 percent corporation income tax on profits. The firm is comparing the relative advantage of the SYD and straight-line methods of depreciation. If the taxes computed by the two methods are expressed as uniform annual amounts, what is the advantage of the SYD method?

Ans.: $1630.

3.40. A firm borrows $250,000 for a facility that it will pay off in 10 equal install-ments at 12 percent interest, compounded annually. In computing income tax the firm can deduct the *actual* interest paid during the year. What is the actual interest paid the second year?

Ans.: $28,290.

3.41. An investor pays $80,000 for a building and expects to sell it for twice that amount at the end of eight years. He can depreciate the building on a straight-line basis during the eight years, or he can charge off no depreciation at all. On the capital gains of the sale at the end of eight years, which is $160,000 − ($80,000 − depreciation) he pays half the income tax that he does on regular income. State which is the most profitable depreciation plan and give all the reasons why it is most profitable.

3.42. Regular payments of $1400 are to be made annually, starting at the end of the first year. These amounts will be invested at 6 percent compounded continuously. How many years will be needed for the payments plus interest to accumulate to $24,000?

Ans.: 12 years.

3.43. An investment of $300,000 yields an annual profit of $86,000 that is spread uniformly over the year and is reinvested immediately (thus continuously compounded). The life is 6 years, and there is no salvage value. What is the rate of return on the investment?

  Ans.: 20%.

ADDITIONAL READINGS

Barish, N. N.: *Economic Analysis for Engineering and Managerial Decision Making*, McGraw-Hill, New York, 1962.

DeGarmo, E. P.: *Engineering Economy*, Macmillan, New York, 1967.

Grant, E. L., and W. G. Ireson: *Principles of Engineering Economy*, 4th ed., Ronald Press, New York, 1960.

Smith, G. W.: *Engineering Economy*, Iowa State University Press, Ames, 1968.

Taylor, G. A.: *Managerial and Engineering Economy*, Van Nostrand, Princeton, N. J., 1964.

CHAPTER
4

EQUATION FITTING

4.1 MATHEMATICAL MODELING

This chapter and the next present procedures for developing equations that represent the performance characteristics of equipment, the behavior of processes, and thermodynamic properties of substances. Engineers may have a variety of reasons for wanting to develop equations, but the crucial ones in the design of thermal systems are (1) to facilitate the process of system simulation and (2) to develop a mathematical statement for optimization. Most large, realistic simulation and optimization problems must be executed on the computer, and it is usually more expedient to operate with equations than with tabular data. An emerging need for expressing equations is in *equipment selection;* some designers are automating equipment selection, storing performance data in the computer, and then automatically retrieving them when a component is being selected.

Equation development will be divided into two different categories; this chapter treats equation fitting and Chapter 5 concentrates on modeling thermal equipment. The distinction between the two is that this chapter approaches the development of equations as purely a number-processing operation, while Chapter 5 uses some physical laws to help equation development. Both approaches are appropriate. In modeling a reciprocating compressor, for example, obviously there are physical explanations for the performance, but by the time the complicated flow processes, compression, reexpansion, and valve mechanics are incorporated, the model is so complex that it is simpler to use experimental or catalog data and treat the problem as a number-processing exercise. On the other hand, heat exchangers follow certain laws that suggest a form for the equation, and this insight can be used to advantage, as shown in Chapter 5.

Where do the data come from on which equations are based? Usually the data used by a designer come from tables or graphs. Experimental data from the laboratory might provide the basis, and the techniques in this and the next chapter are applicable to processing laboratory data. But system designers are usually one step removed from the laboratory and are selecting commercially available components for which the manufacturer has provided performance data. In a few rare instances manufacturers may reserve several lines on a page of tabular data to provide the equation that represents the table. If and when that practice becomes widespread, the system designer's task will be made easier. That stage, however, has not yet been reached.

Much of this chapter presents systematic techniques for determining the constants and coefficients in equations, a process of following rules. The other facet of equation fitting is that of proposing the form of the equation, and this operation is an art. Some suggestions will be offered for the execution of this art. Methods will be presented for determining equations that fit a limited number of data points perfectly. Also explained is the method of least squares, which provides an equation of best fit to a large number of points.

4.2   MATRICES

All the operations in this chapter can be performed without using matrix terminology, but the use of matrices provides several conveniences and insights. In particular, the application of matrix terminology is applicable to the solution of sets of simultaneous equations.

A matrix is a rectangular array of numbers, for example,

$$
\begin{bmatrix} 5 & -2 & 0 \\ 3 & 1 & -1 \\ 0 & 1 & 1 \end{bmatrix}
\begin{bmatrix} 2 & 3 \\ -1 & 0 \\ 2 & 2 \end{bmatrix}
\begin{bmatrix} 7 & 3 \\ 1 & 2 \end{bmatrix}
\begin{bmatrix} a_{11} & a_{12} & \cdots & a_{1n} \\ a_{21} & a_{22} & \cdots & 1_{2n} \\ \cdots\cdots\cdots\cdots\cdots \\ a_{m1} & a_{m2} & \cdots & a_{mn} \end{bmatrix}
$$

The numbers that make up the array are called *elements*. The orders of these matrices, from left to right, are $3 \times 3, 3 \times 2, 2 \times 2$, and $m \times n$.

A transpose of a matrix $[A]$, designated $[A]^T$, is formed by interchanging rows and columns. Thus, if

$$
[A] = \begin{bmatrix} 3 & -1 \\ 2 & 0 \\ 4 & -2 \end{bmatrix} \quad \text{then} \quad [A]^T = \begin{bmatrix} 3 & 2 & 4 \\ -1 & 0 & -2 \end{bmatrix}
$$

To multiply two matrices, multiply elements of the first row of the left matrix by the corresponding elements of the first column of the right matrix; then sum the products to give the element of the first row and first column of the product matrix. For example, the multiplication of the two matrices

$$\begin{bmatrix} 1 & -1 & 0 \\ 2 & 0 & 1 \end{bmatrix} \quad \text{and} \quad \begin{bmatrix} -2 & 1 \\ 3 & 0 \\ 1 & 4 \end{bmatrix}$$

gives

$$\begin{bmatrix} (1)(-2) + (-1)(3) + (0)(1) & (1)(1) + (-1)(0) + (0)(4) \\ (2)(-2) + (0)(3) + (1)(1) & (2)(1) + (0)(0) + (1)(4) \end{bmatrix} = \begin{bmatrix} -5 & 1 \\ -3 & 6 \end{bmatrix}$$

The convention for the multiplication of two matrices offers a slightly shorter form for writing a system of simultaneous linear equations. The three equations

$$2x_1 - x_2 + 3x_3 = 6$$
$$x_1 + 3x_2 \qquad = 1$$
$$4x_1 - 2x_2 + x_3 = 0$$

can be written in matrix form as

$$\begin{bmatrix} 2 & -1 & 3 \\ 1 & 3 & 0 \\ 4 & -2 & 1 \end{bmatrix} \begin{bmatrix} x_1 \\ x_2 \\ x_3 \end{bmatrix} = \begin{bmatrix} 6 \\ 1 \\ 0 \end{bmatrix}$$

The determinant is a scalar (which is simply a number) and is written between vertical lines. For a 1×1 matrix it is the element itself; thus

$$|a_{11}| = a_{11}$$

A technique for evaluating the determinant of a 2×2 matrix is to sum the products of diagonal elements, assigning a positive sign to the diagonal moving downward to the right and a negative sign to the product moving upward to the right:

$$\begin{vmatrix} a_{11} & a_{12} \\ a_{21} & a_{22} \end{vmatrix} = + \searrow - \nearrow = a_{11}a_{22} - a_{21}a_{12}$$

An extension of this method applies to computing the determinant of a 3×3 matrix:

$$\begin{vmatrix} a_{11} & a_{12} & a_{13} \\ a_{21} & a_{22} & a_{23} \\ a_{31} & a_{32} & a_{33} \end{vmatrix} = + \searrow + \searrow + \searrow - \nearrow - \nearrow - \nearrow$$

$$= a_{11}a_{22}a_{33} + a_{12}a_{23}a_{31} + a_{13}a_{21}a_{32}$$

$$- a_{31}a_{22}a_{13} - a_{32}a_{23}a_{11} - a_{33}a_{21}a_{12}$$

Evaluation of determinants 4×4 and larger requires a more general procedure, which applies also to 2×2 and 3×3 matrices. This procedure is *row expansion* or *column expansion*. The determinant of a 3×3 matrix found by expanding about the first column is

$$\det = a_{11}A_{11} + a_{21}A_{21} + a_{31}A_{31}$$

where A_{ij} is the *cofactor* of the element a_{ij}. The cofactor is found as follows:

$$A_{ij} = [(-1)^{i+j}] \begin{vmatrix} \text{submatrix formed} \\ \text{by striking out} \\ i\text{th row and } j\text{th} \\ \text{column of } [A] \end{vmatrix}$$

For example, the cofactor of a_{21}, which is A_{21}, is

$$A_{21} = [(-1)^{2+1}] \begin{vmatrix} a_{11} & a_{12} & a_{13} \\ a_{21} & a_{22} & a_{23} \\ a_{31} & a_{32} & a_{33} \end{vmatrix} = (-1)^3 \begin{vmatrix} a_{12} & a_{13} \\ a_{32} & a_{33} \end{vmatrix}$$

$$A_{21} = -(a_{12}a_{33} - a_{32}a_{13})$$

Example 4.1. Evaluate

$$\begin{vmatrix} 1 & 2 & -1 & 0 \\ 0 & 1 & 2 & 0 \\ 3 & -1 & 1 & 2 \\ 4 & 2 & 1 & 5 \end{vmatrix}$$

Solution. Two elements of the second row are zero, so that row would be a convenient one about which to expand.

$$\det = a_{21}A_{21} + a_{22}A_{22} + a_{23}A_{23} + a_{24}A_{24}$$

$$= (0)A_{21} + (1)(-1)^{2+2} \begin{vmatrix} 1 & -1 & 0 \\ 3 & 1 & 2 \\ 4 & 1 & 5 \end{vmatrix}$$

$$+ (2)(-1)^{2+3} \begin{vmatrix} 1 & 2 & 0 \\ 3 & -1 & 2 \\ 4 & 2 & 5 \end{vmatrix} + (0)A_{24}$$

$$= 0 + 10 + 46 + 0 = 56$$

4.3 SOLUTION OF SIMULTANEOUS EQUATIONS

There are many ways of solving sets of simultaneous equations, two of which will be described in this section, Cramer's rule and gaussian elimination. For a set of linear simultaneous equations

$$a_{11}x_1 + a_{12}x_2 + a_{13}x_3 = b_1$$
$$a_{21}x_1 + a_{22}x_2 + a_{23}x_3 = b_2 \qquad (4.1)$$
$$a_{31}x_1 + a_{32}x_2 + a_{33}x_3 = b_3$$

which can be written in matrix form

$$[A][X] = \begin{bmatrix} a_{11} & a_{12} & a_{13} \\ a_{21} & a_{22} & a_{23} \\ a_{31} & a_{32} & a_{33} \end{bmatrix}\begin{bmatrix} x_1 \\ x_2 \\ x_3 \end{bmatrix} = \begin{bmatrix} b_1 \\ b_2 \\ b_3 \end{bmatrix} = [B] \qquad (4.2)$$

Cramer's rule states that

$$x_i = \frac{|\,[A] \text{ matrix with } [B] \text{ matrix substituted in } i\text{th column }|}{|A|} \qquad (4.3)$$

Example 4.2. Using Cramer's rule, solve for x_2 in this set of simultaneous linear equations:

$$\begin{bmatrix} 2 & 1 & -1 \\ 1 & -2 & 2 \\ -1 & 0 & 3 \end{bmatrix}\begin{bmatrix} x_1 \\ x_2 \\ x_3 \end{bmatrix} = \begin{bmatrix} 3 \\ 9 \\ 0 \end{bmatrix}$$

Solution

$$x_2 = \frac{\begin{vmatrix} 2 & 3 & -1 \\ 1 & 9 & 2 \\ -1 & 0 & 3 \end{vmatrix}}{\begin{vmatrix} 2 & 1 & -1 \\ 1 & -2 & 2 \\ -1 & 0 & 3 \end{vmatrix}} = \frac{30}{-15} = -2$$

Equation (4.3) suggests that none of the x's can be determined if $|A|$ is zero. The equations are dependent in this case, and there is no unique solution to the set.

Another method of solving simultaneous linear equations is gaussian elimination, which will be illustrated by solving

$$x_1 - 4x_2 + 3x_3 = -7 \qquad (4.4)$$
$$3x_1 + x_2 - 2x_3 = 14 \qquad (4.5)$$
$$2x_1 + x_2 + x_3 = 5 \qquad (4.6)$$

The two major steps in gaussian elimination are conversion of the coefficient matrix into a triangular matrix and solution for x_n to x_1 by back substitution.

In the example set of equations, the first part of step 1 is to eliminate the coefficients of x_1 in Eq. (4.5) by multiplying Eq. (4.4) by a suitable constant and adding the product to Eq. (4.5). Specifically, multiply Eq. (4.4) by -3 and add to Eq. (4.5). Similarly, multiply Eq. (4.4) by -2 and add to Eq. (4.6):

$$x_1 - 4x_2 + 3x_3 = -7 \qquad (4.7)$$

$$13x_2 - 11x_3 = 35 \qquad (4.8)$$

$$9x_2 - 5x_3 = 19 \qquad (4.9)$$

The last part of step 1 is to multiply Eq. (4.8) by $-\frac{9}{13}$ and add to Eq. (4.9), which completes the triangularization

$$x_1 - 4x_2 + 3x_3 = -7 \qquad (4.10)$$

$$13x_2 - 11x_3 = 35 \qquad (4.11)$$

$$\frac{34}{13}x_3 = -\frac{68}{13} \qquad (4.12)$$

In step 2 the value of x_3 can be determined directly from Eq. (4.12) as $x_3 = -2$. Substituting the value of x_3 into Eq. (4.11) and solving gives $x_2 = 1$. Finally, substitute the values of x_2 and x_3 into Eq. (4.10) to find that $x_1 = 3$.

If a different set of equations were being solved, and in the equation corresponding to Eq. (4.8) if the coefficient of x_2 had been zero instead of 13, it would have been necessary to exchange the positions of Eqs. (4.8) and (4.9). If both the x_2 coefficients in Eqs. (4.8) and (4.9) had been zero, this would indicate that the set of equations is dependent.

Most computer departments have in their library a routine for solving a set of simultaneous linear equations which can be called as needed. It may be convenient to write one's own subprogram using a method like gaussian elimination.[1] It will be useful for future work in this text to have access to an equation-solving routine on a digital computer.

4.4   POLYNOMIAL REPRESENTATIONS

Probably the most obvious and most useful form of equation representation is a polynomial. If y is to be represented as a function of x, the polynomial form is

$$y = a_o + a_1 x + a_2 x^2 + \cdots + a_n x^n \qquad (4.13)$$

where a_o to a_n are constants. The degree of the equation is the highest exponent of x, which in Eq. (4.13) is n.

Equation (4.13) is an expression giving the function of one variable in terms of another. In other common situations one variable is a function of two or more variables, e.g., in an axial-flow compressor

Flow rate $= f$ (inlet pressure, inlet temperature,

compressor speed, outlet pressure)

This form of equation will be presented in Sec. 4.8.

When the number of data points available is precisely the same as the degree of the equation plus 1, $n + 1$, a polynomial can be devised that exactly expresses those data points. When the number of available data points exceeds $n + 1$, it may be advisable to seek a polynomial that gives the "best fit" to the data points (see Sec. 4.10).

The first and simplest case to be considered is where one variable is a function of another variable and the number of data points equals $n + 1$.

4.5 POLYNOMIAL, ONE VARIABLE A FUNCTION OF ANOTHER VARIABLE AND $n + 1$ DATA POINTS

Two available data points are adequate to describe a first-degree, or linear, equation (Fig. 4-1). The form of this first-degree equation is

$$y = a_o + a_1 x \tag{4.14}$$

The xy pairs for the two known points (x_o, y_o) and (x_1, y_1) can be substituted into Eq. (4.14), providing two linear equations with two unknowns, a_o and a_1

$$y_o = a_o + a_1 x_o$$

$$y_1 = a_o + a_1 x_1$$

For a second-degree, or quadratic, equation, three data points are needed; for example, points 0, 1, and 2 in Fig. 4-2. The xy pairs for the three known points can be substituted into the general form for the quadratic equation

$$y = a_o + a_1 x + a_2 x^2 \tag{4.15}$$

FIGURE 4-1
Two points describing a linear equation.

FIGURE 4-2
x   Three points describing a quadratic equation.

which gives three equations

$$
\begin{bmatrix} 1 & x_0 & x_0^2 \\ 1 & x_1 & x_1^2 \\ 1 & x_2 & x_2^2 \end{bmatrix}
\begin{bmatrix} a_0 \\ a_1 \\ a_2 \end{bmatrix}
=
\begin{bmatrix} y_0 \\ y_1 \\ y_2 \end{bmatrix}
$$

The solution of these three linear simultaneous equations provides the values of a_o, a_1, and a_2.

The coefficients of the high-degree terms in a polynomial may be quite small, particularly if the independent variable is large. For example, if the enthalpy of saturated water vapor h is a function of temperature t in the equation

$$
h = a_o + a_1 t + \cdots + a_5 t^5 + a_6 t^6
$$

where the range of t extends into hundreds of degrees, the value of a_5 and a_6 may be so small that precision problems result. Sometimes this difficulty can be surmounted by defining a new independent variable, for example, $t/100$.

$$
h = a_o + a_1 \frac{t}{100} + \cdots + a_5 \left(\frac{t}{100}\right)^5 + a_6 \left(\frac{t}{100}\right)^6
$$

4.6 SIMPLIFICATIONS WHEN THE INDEPENDENT VARIABLE IS UNIFORMLY SPACED

Sometimes a polynomial is used to represent a function, say $y = f(x)$, where the values of y are known at equally spaced values of x. This situation exists, for instance, when the data points are read off a graph and the points can be chosen at equal intervals of x. The solution of simultaneous equations to determine the coefficients in the polynomial can be performed symbolically in advance,[2] and the execution of the calculations requires a relatively small effort thereafter.

Suppose that the curve in Fig. 4-3 is to be reproduced by a fourth-degree polynomial. The $n + 1$ data points (five in this case) establish a polynomial of degree n (four in this case). The spacing of the points is $x_1 - x_0 = x_2 - x_1 = x_3 - x_2 = x_4 - x_3$. The range of $x, x_4 - x_o$, is designated R, and the symbols are $\Delta y_1 = y_1 - y_0, \Delta y_2 = y_2 - y_0$, etc.

Instead of the polynomial form of Eq. (4.13), an alternate form is used

$$y - y_0 = a_1 \left[\frac{n}{R}(x - x_0) \right] + a_2 \left[\frac{n}{R}(x - x_0) \right]^2 + a_3 \left[\frac{n}{R}(x - x_0) \right]^3$$

$$+ a_4 \left[\frac{n}{R}(x - x_0) \right]^4 \tag{4.16}$$

To find a_1 to a_4, first substitute the (x_1, y_1) pair into Eq. (4.16)

$$\Delta y_1 = a_1 \frac{4(x_1 - x_0)}{R} + a_2 \left[\frac{4(x_1 - x_0)}{R} \right]^2 + a_3 \left[\frac{4(x_1 - x_0)}{R} \right]^3$$

$$+ a_4 \left[\frac{4(x_1 - x_0)}{R} \right]^4 \tag{4.17}$$

Because of the uniform spacing of the points along the x axis, $n(x_1 - x_0)/R = 1$, and so Eq. (4.17) can be rewritten as

$$\Delta y_1 = a_1 + a_2 + a_3 + a_4 \tag{4.18}$$

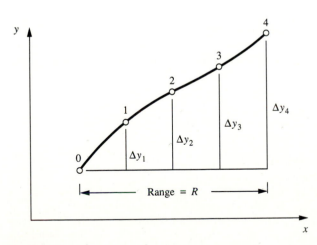

FIGURE 4-3
Polynomial representation when points are equally spaced along the x axis.

TABLE 4.1
Constants in Eq. (4.16)

Equation	a_4	a_3	a_2	a_1
Fourth degree	$\frac{1}{24}(\Delta y_4 - 4\Delta y_3 + 6\Delta y_2 - 4\Delta y_1)$	$\frac{\Delta y_3}{6} - \frac{\Delta y_2}{2} + \frac{\Delta y_1}{2} - 6a_4$	$\frac{\Delta y_2}{2} - \Delta y_1 - 3a_3 - 7a_4$	$\Delta y_1 - a_2 - a_3 - a_4$
Cubic		$\frac{1}{6}(3\Delta y_1 + \Delta y_3 - 3\Delta y_2)$	$\frac{1}{2}(\Delta y_2 - 2\Delta y_1) - 3a_3$	$\Delta y_1 - a_2 - a_3$
Quadratic			$\frac{1}{2}(\Delta y_2 - 2\Delta y_1)$	$\Delta y_1 - a_2$
Linear				Δy_1

Using the (x_2, x_2) pair and the fact that $n(x_2 - x_0)/R = 2$ gives

$$\Delta y_2 = 2a_1 + 4a_2 + 8a_3 + 16a_4 \tag{4.19}$$

Similarly, for (x_3, y_3) and (x_4, y_4)

$$\Delta y_3 = 3a_1 + 9a_2 + 27a_3 + 81a_4 \tag{4.20}$$

$$\Delta y_4 = 4a_1 + 16a_2 + 64a_3 + 256a_4 \tag{4.21}$$

The expressions for a_1 to a_4 found by solving Eqs. (4.18) to (4.21) simultaneously are shown in Table 4.1, along with the constants for the cubic, quadratic, and linear equations.

4.7 LAGRANGE INTERPOLATION

Another form of polynomial results when using Lagrange interpolation. This method is applicable, unlike the method described in Sec. 4.6, to arbitrary spacing along the x axis. It has the advantage of not requiring the simultaneous solution of equations but is cumbersome to write out. This disadvantage is not applicable if the calculation is performed on a digital computer, in which case the programming is quite compact.

With a quadratic equation as an example, the usual form for a function of one variable is

$$y = a_o + a_1 x + a_2 x^2 \tag{4.22}$$

For Lagrange interpolation, a revised form is used

$$y = c_1(x - x_2)(x - x_3) + c_2(x - x_1)(x - x_3) + c_3(x - x_1)(x - x_2) \tag{4.23}$$

The three available data points are (x_1, y_1), (x_2, y_2), and (x_3, y_3). Equation (4.23) could be multiplied out and terms collected to show the correspondence to the form in Eq. (4.22).

By setting $x = x_1$, x_2, and x_3 in turn in Eq. (4.23) the constants can be found quite simply:

$$c_1 = \frac{y_1}{(x_1 - x_2)(x_1 - x_3)}$$

$$c_2 = \frac{y_2}{(x_2 - x_1)(x_2 - x_3)}$$

$$c_3 = \frac{y_3}{(x_3 - x_1)(x_3 - x_2)}$$

The general form of the equation for finding the value of y for a given x when n data points are known is

$$y = \sum_{i=1}^{n} y_i \prod_{j=1}^{n} \frac{(x - x_j) \text{ omitting } (x - x_i)}{(x_i - x_j) \text{ omitting } (x_i - x_i)} \qquad (4.24)$$

where the pi, or product sign, indicates multiplication.

The equation represented by Eq. (4.24) is a polynomial of degree $n - 1$.

4.8   FUNCTION OF TWO VARIABLES

A performance variable of a component is often a function of two other variables,[3] not just one. For example, the pressure rise developed by the centrifugal pump shown in Fig. 4-4 is a function of both the speed S and the flow rate Q.

If a polynomial expression for the pressure rise Δp is sought in terms of a second-degree equation in S and Q, separate equations can be written for each of the three curves in Fig. 4-4. Three points on the 30 r/s curve would provide the constants in the equation

$$\Delta p_1 = a_1 + b_1 Q + c_1 Q^2 \qquad (4.25)$$

Similar equations for the curves for the 24 and 16 r/s speeds are

$$\Delta p_2 = a_2 + b_2 Q + c_2 Q^2 \qquad (4.26)$$

$$\Delta p_3 = a_3 + b_3 Q + c_3 Q^2 \qquad (4.27)$$

Next the a constants can be expressed as a second-degree equation in terms of S, using the three data points $(a_1, 30)$, $(a_2, 24)$, and $(a_3, 16)$. Such an equation would have the form

$$a = A_0 + A_1 S + A_2 S^2 \qquad (4.28)$$

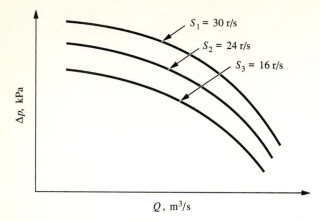

FIGURE 4-4
Performance of a centrifugal pump.

Similarly for b and c

$$b = B_0 + B_1 S + B_2 S^2 \tag{4.29}$$

$$c = C_0 + C_1 S + C_2 S^2 \tag{4.30}$$

Finally, the constants of Eqs. (4.28) to (4.30) are put into the general equation

$$\Delta p = A_0 + A_1 S + A_2 S^2 + (B_0 + B_1 S + B_2 S^2)Q \\ + (C_0 + C_1 S + C_2 S^2)Q^2 \tag{4.31}$$

The A, B, and C constants can be computed if nine data points from Fig. 4-4 are available.

Example 4.3. Manufacturers of cooling towers often present catalog data showing the outlet-water temperature as a function of the wet-bulb temperature of the ambient air and the range. The range is the difference between the inlet and outlet temperatures of the water. In Table 4.2, for example, when the wet-bulb temperature is 20°C and the range is 10°C, the temperature of leaving water is 25.9°C, and so the temperature of the entering water is 25.9 + 10 = 35.9°C. Express the outlet-water temperature t in Table 4.2 as a function of the wet-bulb temperature (WBT) and the range R.

Solution. A second-degree polynomial equation in both independent variables will be chosen as the form of the equation, and three different methods for developing the equation will be illustrated.
    Method 1. The three pairs of points for WBT $= 20°$C, (10, 25.9), (16, 27.0), and (22, 28.4), can be represented by a parabola

$$t = 24.733 + 0.075006R + 0.004146R^2$$

TABLE 4.2
Outlet-water temperature, °C, of cooling tower in Example 4.2

Range, °C	Wet-bulb temperature, °C		
	20	23	26
10	25.9	27.5	29.4
16	27.0	28.4	30.2
22	28.4	29.6	31.3

For WBT $= 23°C$

$$t = 26.667 + 0.041659R + 0.0041469R^2$$

and for WBT $= 26°C$

$$t = 28.733 + 0.024999R + 0.0041467R^2$$

Next, the constant terms 24.733, 26.667, and 28.733 can be expressed by a second-degree equation of WBT,

$$15.247 + 0.32637\text{WBT} + 0.007380\text{WBT}^2$$

The coefficients of R and R^2 can also be expressed by equations in terms of the WBT, which then provide the complete equation

$$t = (15.247 + 0.32637\text{WBT} + 0.007380\text{WBT}^2)$$

$$+ (0.72375 - 0.050978\text{WBT} + 0.000927\text{WBT}^2)R$$

$$+ (0.004147 + 0\text{WBT} + 0\text{WBT}^2)R^2 \tag{4.32}$$

Method 2. An alternate polynomial form using second-degree expressions for R and WBT is

$$t = c_1 + c_2\text{WBT} + c_3\text{WBT}^2 + c_4R + c_5R^2 + c_6(R)(\text{WBT})$$

$$+ c_7(\text{WBT})^2(R) + c_8(\text{WBT})(R)^2 + c_9(\text{WBT})^2(R)^2 \tag{4.33}$$

The nine sets of t-R-WBT combinations expressed in Table 4.2 can be substituted into Eq. (4.33) to develop nine simultaneous equations, which can be solved for the unknowns c_1 to c_9. The c values thus obtained are

$$c_1 = 15.247 \qquad c_2 = 0.32631 \qquad c_3 = 0.0073991$$
$$c_4 = 0.723753 \qquad c_5 = 0.0041474 \qquad c_6 = -0.0509782$$
$$c_7 = 0.00092704 \qquad c_8 = 0.0 \qquad c_9 = 0.0$$

It is possible to multiply and collect the terms in Eq. (4.32) to develop the equation of the form of Eq. (4.33).

Method 3. Section 4.7 described a polynomial representation of a dependent variable as a function of one independent variable by use of Lagrange interpolation. Lagrange interpolation can be extended to a function of two independent variables. For example, if $z = f(x, y)$, the form can be chosen

$$z = c_{11}(x - x_2)(x - x_3)(y - y_2)(y - y_3)$$
$$+ c_{12}(x - x_2)(x - x_3)(y - y_1)(y - y_3)$$
$$+ c_{13}(x - x_2)(x - x_3)(y - y_1)(y - y_2)$$
$$+ \cdots + c_{33}(x - x_1)(x - x_2)(y - y_1)(y - y_2) \qquad (4.34)$$

To represent the data in Table 4.2, z could refer to the outlet-water temperature, x the WBT, and y the range. In Eq. (4.34) $x_1 = 20$, $x_2 = 23$, and $x_3 = 26$, while $y_1 = 10$, $y_2 = 16$, and $y_3 = 22$.

To determine the magnitude of c_{12}, for example, values applicable when $x = x_1$ and $y = y_2$ can be substituted into Eq. (4.34).

$$c_{12} = \frac{27.0}{(20 - 23)(20 - 26)(16 - 10)(16 - 22)} = -0.04167$$

4.9 EXPONENTIAL FORMS

The dependence of one variable on a second variable raised to an exponent is a physical relation occurring frequently in engineering practice. The graphical method of determining the constants b and m in the equation

$$y = bx^m \qquad (4.35)$$

is a simple example of mathematical modeling of an exponential form. On a graph of the known values of x and y on a log-log plot (Fig. 4-5) the slope of the straight line through the points equals m, and the intercept at $x = 1$ defines b.

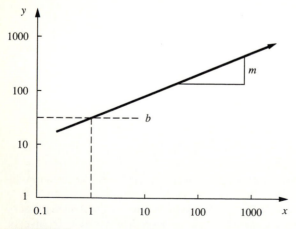

FIGURE 4-5
Graphical determination of the constant b and exponent m.

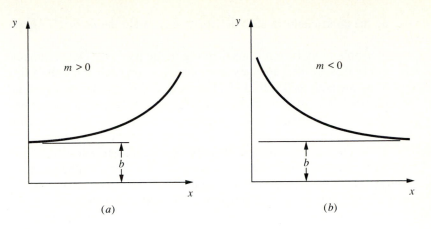

FIGURE 4-6
Curve of the equation $y = b + ax^m$.

The simple exponential form of Eq. (4.35) can be extended to include a constant

$$y = b + ax^m \tag{4.36}$$

The equation permits representations of curves similar to those shown in Fig. 4-6. The curve shown in Fig. 4-6b is especially common in engineering practice. The function y approaches some value b asymptotically as x increases.

One possible graphical method of determining a, b, and m in Eq. (4.36) when pairs of xy values are known is as follows:

1. Estimate the value of b.
2. Use the steep portion of the curve to evaluate m by a log-log plot of $y - b$ vs. x in a manner similar to that shown in Fig. 4-5.
3. With the value of m from step 2, plot a graph of y vs. x^m. The resulting curve should be a straight line with a slope of a and an intercept that indicates a more correct value of b. Iterate starting at step 2 if desired.

4.10 BEST FIT: METHOD OF LEAST SQUARES

This chapter has concentrated so far on finding equations that give a perfect fit to a limited number of points. If m coefficients are to be determined in an equation, m data points are required. If more than m points are available, it is possible to determine the m coefficients that result in the best fit of the equation to the data. One definition of a best fit is the one where the sum of the absolute values of the deviations from the data points is a minimum. In another type of best fit slightly different from the one just mentioned the sum of the squares of the deviation is a minimum. The procedure in establishing the coefficients in such an equation is called the *method of least squares*.

Some people proudly announce their use of the method of least squares in order to emphasize the care they have lavished on their data analysis. Misuses of the method, as illustrated in Fig. 4-7a and b, are not uncommon. In Fig. 4-7a, while a straight line can be found that results in the least-squares deviation, the correlation between the x and y variables seems questionable and perhaps no such device can improve the correlation. The scatter may be due to the omission of some significant variable(s). In Fig. 4-7b it would have been preferable to eyeball in the curve, rather than to fit a straight line to the data by the least-squares method. The error was not in using least squares but in applying a curve of too low a degree.

The procedure for using the least-squares method for first- and second-degree polynomials will be explained here. Consider first the linear equation of the form

$$y = a + bx \tag{4.37}$$

where m pairs of data points are available: $(x_1, y_1), (x_2, y_2), \ldots, (x_m, y_m)$. The deviation of the data point from that calculated from the equation is $a + bx_i - y_i$. We wish to choose an a and a b such that the summation

$$\sum_{i=1}^{m}(a + bx_i - y_i)^2 \rightarrow \text{minimum} \tag{4.38}$$

The minimum occurs when the partial derivatives of Eq. (4.38) with respect to a and b equal zero.

$$\frac{\partial \sum\limits_{i=1}^{m}(a + bx_i - y_i)^2}{\partial a} = \Sigma 2(a + bx_i - y_i) = 0$$

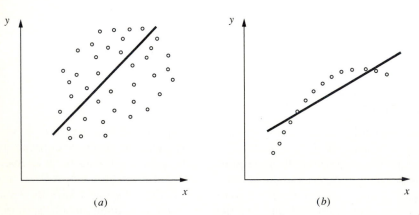

FIGURE 4-7
Misuses of the method of least squares.

and

$$\frac{\partial \sum_{i=1}^{m} (a + bx_i - y_i)^2}{\partial b} = \Sigma 2(a + bx_i - y_i)x_i = 0$$

Dividing by 2 and separating the above two equations into individual terms gives

$$ma + b\Sigma x_i = \Sigma y_i \tag{4.39}$$

$$a\Sigma x_i + b\Sigma x_i^2 = \Sigma x_i y_i \tag{4.40}$$

Example 4.4. Determine a_0 and a_1 in the equation $y = a_0 + a_1 x$ to provide a best fit in the sense of least-squares deviation to the data points $(1, 4.9)$, $(3, 11.2)$, $(4, 13.7)$, and $(6, 20.1)$.

Solution. The summations to substitute into Eqs. (4.39) and (4.40) are

	x_i	y_i	x_i^2	$x_i y_i$
	1	4.9	1	4.9
	3	11.2	9	33.6
	4	13.7	16	54.8
	6	20.1	36	120.6
Σ	$\overline{14}$	$\overline{49.9}$	$\overline{62}$	$\overline{213.9}$

and $m = 4$

The simultaneous equations to be solved are

$$4a_0 + 14a_1 = 49.9$$

$$14a_0 + 62a_1 = 213.9$$

yielding $a_0 = 1.908$ and $a_1 = 3.019$. Thus

$$y = 1.908 + 3.019x$$

A similar procedure can be followed when fitting a parabola of the form

$$y = a + bx + cx^2 \tag{4.41}$$

to m data points. The summation to be minimized is

$$\sum_{i=1}^{m} (a + bx_i + cx_i^2 - y_i)^2 \rightarrow \text{minimum}$$

Differentiating partially with respect to a, b, and c, in turn, results in three linear simultaneous equations expressed in matrix form

$$\begin{bmatrix} m & \Sigma x_i & \Sigma x_i^2 \\ \Sigma x_i & \Sigma x_i^2 & \Sigma x_i^3 \\ \Sigma x_i^2 & \Sigma x_i^3 & \Sigma x_i^4 \end{bmatrix} \begin{bmatrix} a \\ b \\ c \end{bmatrix} = \begin{bmatrix} \Sigma y_i \\ \Sigma x_i y_i \\ \Sigma x_i^2 y_i \end{bmatrix} \tag{4.42}$$

A comparison of the matrix equation (4.42) with Eqs. (4.39) and (4.40) shows a pattern evolving which by analogy permits developing the equations for higher degree polynomials without even differentiating the summation of the squared deviation.

4.11   METHOD OF LEAST SQUARES APPLIED TO NONPOLYNOMIAL FORMS

The explanation of the method of least squares was applied to polynomial forms in Sec. 4.10, but it should not be suggested that the method is limited to those forms. The method is applicable to any form which contains constant coefficients. For example, if the form of the equation is

$$y = a \sin 2x + b \ln x^2$$

the summation comparable to Eq. (4.38) is

$$\sum_{i=1}^{m} (y_i - a \sin 2x_i - b \ln x_i^2)^2 \tag{4.43}$$

Partial differentiation with respect to a and b yields

$$a \sum (\sin 2x_i)^2 + b \sum (\sin 2x_i)(\ln x_1^2) = \sum y_i \sin 2x_i$$

$$a \sum (\sin 2x_i)(\ln x_i^2) + b \sum (\ln x_i^2)^2 = \sum y_i \ln x_i^2$$

which can be solved for a and b.

A crucial characteristic of the equation form that makes it tractable to the method of least squares is that the equation have constant coefficients. In an equation of the form

$$y = \sin 2ax + bx^c$$

the terms a and c do not appear as coefficients, and this equation cannot be handled in a straightforward manner by least squares.

4.12   THE ART OF EQUATION FITTING

While there are methodical procedures for fitting equations to data, the process is also an art. The art of intuition is particularly needed in deciding upon the form of the equation, namely, the choice of independent variables to be included and the form in which these variables should appear. There are no fixed rules for knowing what variables to include or what their form should be in the equation, but making at least a rough plot of the data will often provide some insight. If the dependent variable is a function of two independent variables, as in $z = f(x, y)$, two plots might be made, as illustrated in Fig. 4-8.

FIGURE 4-8
Cross plots to aid in developing the form of the equation.

The insight provided by Fig. 4-8a is that z bears a linear relation to y, and the fact that the straight lines are parallel shows no influence of x on the slope. Figure 4-8b suggests at least a second-degree representation of z as a function of x. A reasonable form to propose, then, is

$$z = a_0 + a_1x + a_2y + a_3x^2$$

Several frequently used forms merit further discussion.

Polynomials

If there is a lack of special indicators that other forms are more applicable, a polynomial would probably be explored. When the curve has a reverse curvature (inflection point), as shown in Fig. 4-9, at least a third-degree polynomial must be chosen. Extrapolation of a polynomial beyond the

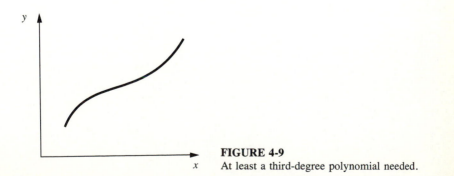

FIGURE 4-9
At least a third-degree polynomial needed.

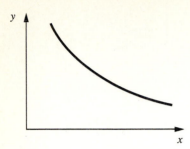

FIGURE 4-10
Negative exponents of polynomials for a curve that flattens out.

borders of the data used to develop the equation often results in serious error.

Polynomials with Negative Exponents

When curves approach a constant value at large magnitudes of the independent variable, polynomials with negative exponents

$$y = a_0 + a_1 x^{-1} + a_2 x^{-2}$$

may provide a good representation; see Fig. 4-10.

Exponential Equations

Section 4.9 has described several examples of exponential forms. The shape of the curve in Fig. 4-10 might also include a c^{-x} term. Plots on log-log paper would be a routine procedure, although a simple plot of log y vs. log x yields a straight line only with equations in the form of Eq. (4.35).

Gompertz Equation

The Gompertz equation,[4] or S curve (Fig. 4-11), appears frequently in engineering practice. The Gompertz curve, for example, represents the

FIGURE 4-11
Gompertz, or S curve.

FIGURE 4-12
Combination of two forms.

sales volume vs. years for many products which have low sales when first introduced, experience a period of rapid increase, then reach saturation. The personnel required in many projects also often follows the curve. The form that represents Fig. 4-11 is

$$y = ab^{c^x} \tag{4.44}$$

where a, b, and c are constants and b and c have magnitudes less than unity.

Combination of Forms

It may be possible to fit a curve by combining two or more forms. For example, in Fig. 4-12, suppose that the value of y approaches asymptotically a straight line as x increases. A reasonable way to attack this modeling task would be to propose that

$$y = y_1 + y_2 = (a + bx) + (c + dx^m)$$

where m is a negative exponent.

4.13 AN OVERVIEW OF EQUATION FITTING

The task of finding suitable equations to represent the performance of components or thermodynamic properties is a common preliminary step to simulating and optimizing complex systems. Data may be available in tabular or graphic form, and we seek to represent the data with an equation that is both simple and faithful. A requirement for keeping the equation simple is to choose the proper terms (exponential, polynomial, etc.) to include in the equation. It is possible, of course, to include all the terms that could possibly be imagined, evaluate the coefficients by the method of least squares, and then eliminate terms that provide little contribution. This process is essentially one of *regression analysis*,[5] which also is used to assess which variables are important in representing the dependent variable.

The field of statistical analysis of data is an extensive one, and this chapter has only scratched the surface. On the other hand, much of the effort in the statistical analysis of data is directed toward fitting experimental data to equations where random experimental error occurs. In equation fitting for the design of thermal systems, since catalog tables and charts are the most frequent source of data, there usually has already been a process of smoothing of the experimental data coming from the laboratory. Because of the growing need for fitting catalog data to equations, many designers hope that manufacturers will present the equation that represents the table or graph to save each engineer the effort of developing the equation again when needed.

This chapter presented one approach to mathematical modeling where the relationship of dependent and independent variables was developed without the help of physical laws. Chapter 5 explores some special important cases where physical insight into some thermal equipment can be used to advantage in fitting equations to performance data. Chapter 13 extends the experience on mathematical modeling and also concentrates on the important topic of thermodynamic properties.

PROBLEMS

4.1. Compute

$$\begin{vmatrix} 2 & -1 & 0 & 3 \\ 1 & -2 & 2 & 4 \\ -3 & 1 & 0 & -1 \\ 4 & 2 & 0 & 3 \end{vmatrix}$$

Ans.: 50.

4.2. Test the coefficient matrix in the set of linear equations

$$\begin{bmatrix} 1 & 2 & -2 & 3 \\ 2 & -1 & 3 & -2 \\ -1 & 3 & 1 & -4 \\ 1 & -3 & 5 & -5 \end{bmatrix} \begin{bmatrix} x_1 \\ x_2 \\ x_3 \\ x_4 \end{bmatrix} = \begin{bmatrix} 5 \\ 18 \\ -6 \\ 13 \end{bmatrix}$$

and determine whether the set of equations is dependent or independent.

4.3. Using a computer program (gaussian elimination or any other that is available for solving a set of linear simultaneous equations), solve for the x's:

$$2x_1 + x_2 - 4x_3 + 6x_4 + 3x_5 - 2x_6 = 16$$
$$-x_1 + 2x_2 + 3x_3 + 5x_4 - 2x_5 \qquad = -7$$
$$x_1 - 2x_2 - 5x_3 + 3x_4 + 2x_5 + x_6 = 1$$
$$4x_1 + 3x_2 - 2x_3 + 2x_4 \qquad + x_6 = -1$$
$$3x_1 + x_2 - x_3 + 4x_4 + 3x_5 + 6x_6 = -11$$
$$5x_1 + 2x_2 - 2x_3 + 3x_4 + x_5 + x_6 = 5$$

Ans.: 2, −1, 1, 0, 3, −4.

4.4. A second-degree equation of the form

$$y = a + bx + cx^2$$

has been proposed to pass through the three (x, y) points (1,3), (2,4), and (2, 6). Proceed with the solution for a, b, and c.
(*a*) Describe any unusual problems encountered.
(*b*) Propose an alternate second-degree relation between x and y that will successfully represent these three points.

4.5. Use data from Table 4.3 at $t = 0$, 50, and 100°C to establish a second-degree polynomial that fits h_g to t. Using the equation, compute h_g at 80°C.
Ans.: 2643.3 kJ/kg.

4.6. Using the data from Table 4.3 for v_g at $t = 40$, 60, 80 and 100°C, develop a third-degree equation similar in form to Eq. (4.16). Compute v_g at 70°C using this equation.
Ans.: 4.91 m³/kg.

4.7. Lagrange interpolation is to be used to represent the enthalpy of saturated air, h_s kJ/kg, as a function of the temperature t°C. The pairs of (h_s, t) values to be used as the basis are (9.470, 0), (29.34, 10), (57.53, 20), and (99.96, 30).
(*a*) Determine the values of the coefficients c_1 to c_4 in the equation for h_s.
(*b*) Calculate h_s at 15°C.
Ans.: (b) From tables 42.09 kJ/kg.

TABLE 4.3
Properties of saturated water

Temperature		Pressure	Specific volume	Enthalpy	
t, °C	*T, K*	*p*, kPa	v_g, m³ /kg	h_f, kJ/kg	h_g, kJ/kg
0	273.15	0.6108	206.3	−0.04	2501.6
10	283.15	1.227	106.4	41.99	2519.9
20	293.15	2.337	57.84	83.86	2538.2
30	303.15	4.241	32.93	125.66	2556.4
40	313.15	7.375	19.55	167.45	2574.4
50	323.15	12.335	12.05	209.26	2592.2
60	333.15	19.92	7.679	251.09	2609.7
70	343.15	31.16	5.046	292.97	2626.9
80	353.15	47.36	3.409	334.92	2643.8
90	363.15	70.11	2.361	376.94	2660.1
100	373.15	101.33	1.673	419.06	2676.0

4.8. An equation of the form

$$y - y_0 = a_1(x - 1) + a_2(x - 1)^2$$

is to fit the following three (x, y) points: (1, 4), (2, 8), and (3, 10). What are the values of y_0, a_1, and a_2?
   Ans.: $a_1 = 5$.

4.9. The pumping capacity of a refrigerating compressor (and thus the capability for developing refrigerating capacity) is a function of the evaporating and condensing pressures. The refrigerating capacities in kilowatts of a certain reciprocating compressor at combinations of three different evaporating and condensing temperatures are shown in Table 4.4. Develop an equation similar to the form of Eq. (4.33), namely,

$$q_e = c_1 + c_2 t_e + c_3 t_e^2 + c_4 t_c + \cdots + c_9 t_e^2 t_c^2$$

   Ans.: c_1 to c_9 are 239.51, 10.073, −0.10901, −3.4100, −0.0025000, −0.20300, 0.0082004, 0.0013000, −0.000080005.

4.10. The data in Table 4.4 are to be fit to an equation using Lagrange interpolation with a form similar to Eq. (4.34). The variable x corresponds to t_e, y corresponds to t_c, and z to q_e. Compute the coefficient c_{23}.
   Ans.: −0.02026.

4.11. The values of c_1 and c_2 are to be determined so that the curve represented by the equation $y = c_1/(c_2 + x)^2$ passes through the (x, y) points (2, 4) and (3, 1). Find the *two* $c_1 - c_2$ combinations.
   Ans.: One value of c_1 is $\frac{4}{9}$.

4.12. Using the graphical method for the form $y = b + ax^m$ described in Sec. 4.9, determine the equation that represents the following pairs of (x, y) points: (0.2, 26), (0.5, 7), (1, 2.8), (2, 1.3), (4, 0.79), (6, 0.65), (10, 0.58), (15, 0.54).
   Ans.: $y = 0.5 + 2.3x^{-1.5}$.

4.13. A function y is expected to be of the form $y = cx^m$ and the xy data develop a straight line on log-log paper. The line passes through the (x, y) points (100, 50) and (1000, 10). What are the values of c and m?
   Ans.: $c = 1250$.

4.14. Compute the constants in the equation $y = a_0 + a_1x + a_2x^2$ to provide a best fit in the sense of least squares for the following (x, y) points: (1, 9.8), (3, 13.0), (6, 9.1), and (8, 0.6).
   Ans.: 6.424, 3.953, −0.585.

TABLE 4.4
Refrigerating capacity q_e kW

Evaporating temperature t_e, °C	Condensing temperature, t_c, °C		
	25	35	45
0	152.7	117.1	81.0
5	182.9	141.9	101.3
10	215.4	170.7	126.5

4.15. An equation of the form $y = ax + b/x$ has been chosen to fit the following (x, y) pairs of points: (1, 10.5), (3, 8), and (8,18). Choose a and b to give the best fit to the points in the sense of least sum of the deviations squared.

Ans.: $b = 8.14$.

4.16. The proposed form of the equation to represent z as a function of x and y is $z = ax + b[\ln(xy)]$, where a and b are constants. Some data relating these variables are

z	x	y
2	1	2
5	2	1
4	2	2

Determine the values of a and b that give the best fit of the equation to the data in the sense of least square deviation.

Ans.: $b = -0.35$.

4.17. With the method of least squares, fit the enthalpy of saturated liquid h_f by means of a cubic equation to the temperature t in degrees Celsius using the 11 points on Table 4.3. Then compute the values of h_f at the 11 points with the equation just developed.

Ans.: $h_f = -0.0037 + 4.2000t - 0.000505t^2 + 0.000003935t^3$.

4.18. A frequently used form of equation to relate saturation pressures to temperatures is

$$\ln p = A + \frac{B}{T}$$

where $p =$ saturation pressure, kPa
$T =$ absolute temperature, K

With the method of least squares and the 11 points for Table 4.3, determine the values of A and B that give the best fit. Then compute the values of p at the 11 points using the equation just developed.

Ans.: $\ln p = 18.60 - 5206.9/T$.

4.19. The variable z is to be expressed in an equation of the form

$$z = ax + by + cxy$$

The following data points are available, and a least-squares fit is desired:

z	x	y
0.1	1	1
-0.9	1	2
2.0	2	2
-1.8	3	1

Determine the values of a, b, and c.

Ans.: $-2.0467, -0.9167$, and 1.8833.

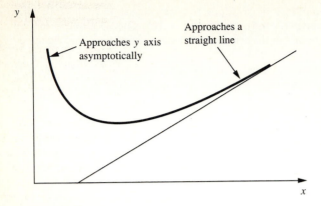

FIGURE 4-13
Function in Prob. 4.19.

4.20. Three points, (x_1, y_1), (x_2, y_2), and (x_3, y_3), lie precisely on the straight line $y = a + bx$. If a least-squares best fit were applied to these three points to determine the values of A and B in the equation $y = A + Bx$, show that the process would indeed give $A = a$, and $B = b$.

4.21. An equation is to be found that represents the function shown in Fig. 4-13. Since one single simple expression seems inadequate, propose that $y = f_1(x) + f_2(x)$. Suggest appropriate forms for f_1 and f_2 and sketch these functions.

4.22. The enthalpy of a solution is a function of the temperature t and the concentration x and consists of straight lines at a constant temperature, as shown in Fig. 4-14. Develop an equation that accurately represents h as a function of x and t.

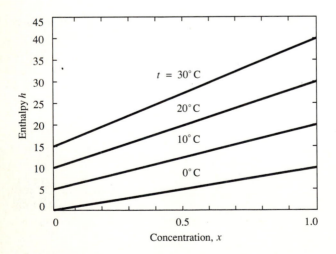

FIGURE 4-14
Enthalpy as a function of temperature and concentration in Prob. 4.20.

FIGURE 4-15
Gompertz equation in Prob. 4.23.

4.23. In a certain Gompertz equation which is $y = ab^{c^x}$ and represented by Fig. 4-15, $c = 0.5$, $y_0 = 2$ and the asymptote has a value of 6. Determine the values of a and b.

        Ans.: $a = 6$.

REFERENCES

1. *Procedures for Simulating the Performance of Components and Systems for Energy Calculations,* American Society of Heating, Refrigerating, and Air-Conditioning Engineers, New York, 1975.
2. M. W. Wambsganss, Jr., "Curve Fitting with Polynomials," *Mach. Des.,* vol. 35, no. 10, p. 167, Apr. 25, 1963.
3. C. Daniel and F. S. Wood, *Fitting Equations to Data,* Wiley-Interscience, New York, 1971.
4. D. S. Davis, *Nomography and Empirical Equations,* Reinhold, New York, 1955.
5. N. R. Draper and H. Smith, *Applied Regression Analysis,* Wiley, New York, 1966.

CHAPTER
5

MODELING THERMAL EQUIPMENT

5.1 USING PHYSICAL INSIGHT

This chapter continues the objective of the previous chapter, of fitting equations to the performance of components. Chapter 4, however, assumed that physical insight into the equation either did not exist or offered no particular advantage. In contrast, this chapter concentrates on three classes of components that appear almost universally in thermal systems, heat exchangers, distillation separators, and turbomachinery, where the knowledge of the physical relationships helps structure the equations.

The reasons for singling out these three components are different for each component. The preface of this book indicates that a background in heat transfer is assumed. In the study of thermal systems a segment of that knowledge is particularly important, namely, predicting the performance of an existing heat exchanger. Not only is the selection of a heat exchanger important, but it is also crucial to be able to calculate how a certain heat exchanger will perform when operating at off-design conditions. A useful tool to be stressed in this chapter is the *effectiveness* of heat exchangers.

The techniques this book explains are particularly applicable to the thermal processing industry, which includes petroleum refining, and other process industries where the separation of mixtures of several substances by means of distillation is an integral process. Even an understanding of the separation of binary mixtures, which is explained in this chapter, expands the horizons of applications of the future topics of simulation and optimization.

The third class of thermal equipment treated in this chapter is turbomachinery, the performance of which can often be expressed in terms of dimensionless groups. This chapter will show how the use of dimensionless groups can sometimes simplify the equation representation of a turbomachine.

5.2 SELECTING VS. SIMULATING A HEAT EXCHANGER

A common engineering task is to *select,* design, or specify a heat exchanger to perform a certain heat-transfer duty. The engineer then decides on the type of heat exchanger and its details. Three of the several dozen types of heat exchangers available are shown in Fig. 5-1. Figure 5-1*a* shows a shell-and-tube heat exchanger, commonly used to transfer heat between two liquids. One of the fluids flows inside the tubes and is called the tube-side fluid, while the other flows over the outside of the tubes and is called the shell-side fluid. The heat exchanger in Fig. 5-1*a* has two tube passes, which means that the tube-side fluid flows through half the tubes in one direction and back through the other half. The head of the heat exchanger on the left end is equipped with a divider that separates the incoming from the outgoing tube-side fluid. Baffles are placed in the shell so that the shell-side fluid flows across the tubes a number of times before leaving the heat exchanger, instead of short-circuiting to the outlet.

The finned-coil heat exchanger in Fig. 5-1*b* is the type often chosen to transfer heat between a gas and a liquid. Since the resistance to heat transfer on the gas side is usually high because of the low heat-transfer coefficient of a gas, fins are installed on the gas side to increase the heat-transfer area.

The third type of heat exchanger, shown in Fig. 5-1*c,* is a compact heat exchanger; it usually consists of a stack of metal plates that are often

(*a*) Shell-and-tube   (*b*) Finned-coil   (*c*) Compact

FIGURE 5-1
Several types of heat exchangers.

corrugated and arranged so that the two fluids flow through alternate spaces between the plates.

We now return to the distinction between selecting and simulating a heat exchanger. To select a shell-and-tube heat exchanger, for example, the flow rates, entering temperatures, and leaving temperatures of both fluids would be known. The task of the designer is to select the combination of shell diameter, tube length, number of tubes, number of tube passes, and the baffle spacing that will accomplish the specified heat-transfer duty. The design must also ensure that certain pressure-drop limitations of the fluids flowing through the heat exchanger are not exceeded.

In simulation, on the other hand, the heat exchanger already exists, either in actual hardware or as a specific design. Furthermore the performance characteristics of the heat exchanger are available, such as the area and overall heat-transfer coefficients. Simulation of a heat exchanger consists of predicting outlet conditions, such as temperatures, for various inlet temperatures and flow rates. The emphasis of the next several sections will be on predicting outlet conditions of a given heat exchanger when the inlet conditions are known.

5.3 COUNTERFLOW HEAT EXCHANGER

For a heat exchanger between two fluids with given inlet and outlet temperatures, the most favorable difference in temperature between these two fluids is achieved with a counterflow arrangement. A counterflow heat exchanger is therefore a good choice for a standard of comparison. Figure 5-2 shows a

FIGURE 5-2
A counterflow heat exchanger.

counterflow heat exchanger with the symbols that will be used in developing equations.

Three equations for the rate of heat transfer q in watts are

$$q = W_1(t_{1,i} - t_{1,o}) \tag{5.1}$$

$$q = W_2(t_{2,o} - t_{2,i}) \tag{5.2}$$

$$q = UA \frac{(t_{1,i} - t_{2,o}) - t_{1,o} - t_{2,i})}{\ln[(t_{1,i} - t_{2,o})/(t_{1,o} - t_{2,i})]} \tag{5.3}$$

where
q = rate of heat transfer, W
w_1, w_2 = flow rates of respective fluids, kg/s
c_{p1}, c_{p2} = specific heats of respective fluids, J/(kg · K)
t = temperature, °C
U = overall heat-transfer coefficient, W/(m² · K)
A = heat-transfer area, m²
$W_1 = w_1, c_{p1}$
$W_2 = w_2 c_{p2}$

Equations (5.2) to (5.3) have made no assumption about which is the hot fluid and which the cold. If q is positive, fluid 1 is hotter than fluid 2, but if q is negative, the opposite is true. If the W's, UA, and the entering temperatures are known, the three equations contain three unknowns, q, $t_{1,o}$, and $t_{2,o}$. The number of equations can be reduced to two by eliminating q, to give

$$W_1(t_{1,i} - t_{1,o}) = W_2(t_{2,o} - t_{2,i}) \tag{5.4}$$

$$W_1(t_{1,i} - t_{1,o}) = UA \frac{(t_{1,i} - t_{2,o}) - (t_{1,o} - t_{2,i})}{\ln[(t_{1,i} - t_{2,o})/(t_{1,o} - t_{2,i})]} \tag{5.5}$$

Solving for $t_{2,o}$ in Eq. (5.4) and substituting into Eq. (5.5) gives

$$\ln \frac{t_{1,i} - [t_{2,i} + (W_1/W_2)(t_{1,i} - t_{1,o})]}{t_{1,o} - t_{2,i}} = UA \left(\frac{1}{W_1} - \frac{1}{W_2} \right)$$

Define D as

$$D = UA \left(\frac{1}{W_1} - \frac{1}{W_2} \right)$$

Then

$$\frac{t_{1,i} - t_{2,i} - (W_1/W_2)(t_{1,i} - t_{1,o})}{t_{1,o} - t_{2,i}} = e^D$$

Solving for $t_{1,o}$ gives

$$t_{1,o} = \frac{t_{1,i}(W_1/W_2 - 1) + t_{2,i}(1 - e^D)}{W_1/W_2 - e^D}$$

or, in alternate form,

$$t_{1,o} = t_{1,i} - (t_{1,i} - t_{2,i})\frac{1 - e^D}{W_1/W_2 - e^D} \qquad (5.6)$$

Equation (5.6) permits computation of one outlet temperature of a heat exchanger of known characteristics when both entering temperatures are known. The other outlet temperature can be computed by application of Eq. (5.4), and q can be computed from Eq. (5.1). If the outlet temperature of fluid 2 is the one sought directly, subscripts 1 and 2 can be interchanged in Eq. (5.6) and in the equation for D.

5.4 SPECIAL CASE OF COUNTERFLOW HEAT EXCHANGER WHERE PRODUCTS OF FLOW RATES AND SPECIFIC HEATS ARE EQUAL

The direct application of Eq. (5.6) is unsuccessful when

$$w_1 c_{p1} = w_2 c_{p2} \qquad \text{thus} \qquad W_1 = W_2$$

The value of D is zero, and Eq. (5.6) is indeterminate. There are two ways to develop an alternate expression for the outlet temperature, one mathematical and the other by physical reasoning.

The mathematical solution uses the expression for e^x as a series.

$$1 + x + \frac{x^2}{2} + \cdots$$

The indeterminate part of Eq. (5.6) can be written

$$\frac{1 - e^D}{W_1/W_2 - e^D} = \frac{1 - \left\{1 + \dfrac{UA}{W_1}\left(1 - \dfrac{W_1}{W_2}\right) + \dfrac{1}{2}\left[\dfrac{UA}{W_1}\left(1 - \dfrac{W_1}{W_2}\right)\right]^2 + \cdots\right\}}{\dfrac{W_1}{W_2} - \left\{1 + \dfrac{UA}{W_1}\left(1 - \dfrac{W_1}{W_2}\right) + \dfrac{1}{2}\left[\dfrac{UA}{W_1}\left(1 - \dfrac{W_1}{W_2}\right)\right]^2 + \cdots\right\}}$$

Canceling where possible and dividing both the numerator and denominator by $1 - W_1/W_2$ results in

$$\frac{1 - e^D}{(W_1/W_2) - e^D} = \frac{-(UA/W_1) - \dfrac{1}{2}(UA/W_1)^2(1 - W_1/W_2) + \cdots}{-1 - UA/W_1 - \dfrac{1}{2}(UA/W_1)^2(1 - W_1/W_2) + \cdots}$$

Finally let $W_1 \rightarrow W_2$ and call this common value W. Then

$$\frac{1 - e^D}{W/W - e^D} = \frac{UA/W}{1 + UA/W} = \frac{1}{W/UA + 1}$$

Substituting back into Eq. (5.6) gives

$$t_{1,o} = t_{1,i} - \frac{t_{1,i} - t_{2,i}}{W/UA + 1} \tag{5.7}$$

which is the equation for computing one outlet temperature.

The physical analysis that leads to Eq. (5.7) is to recognize that the change of temperature of one fluid while flowing past a differential area dA in Fig. 5-3 equals the change in temperature of the other fluid. The slopes of the two temperature lines are then the same at all positions along the area. (This condition does not yet establish that the lines are straight.) A further stipulation demanded by the heat-transfer rate equation is that

$$dq = W \, dt_2 = W \, dt_1 = U \, dA(t_1 - t_2) \tag{5.8}$$

Starting at the left edge of the graph in Fig. 5-3, the inlet temperature $t_{2,i}$ of fluid is specified, and there is some outlet temperature of fluid 1, designated $t_{1,o}$, yet to be determined. Regardless of the value of $t_{1,o}$, the slopes of the temperature lines are fixed for the first increment of area dA

$$\frac{dt_1}{dA} = \frac{dt_2}{dA} = \frac{U}{W}(t_{1,o} - t_{2,i})$$

Since the slopes through dA are identical, the temperature difference after dA is still $t_{1,o} - t_{2,i}$, and by similar reasoning the temperature difference remains constant through the entire heat exchanger. Thus the temperature

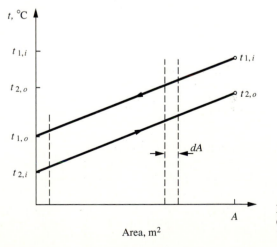

t, °C

$t_{1,i}$

$t_{2,o}$

$t_{1,o}$

$t_{2,i}$

$t_{1,i}$

$t_{2,o}$

dA

A

Area, m²

FIGURE 5-3
Counterflow heat exchanger where $W_1 = W_2$.

lines are straight, parallel lines. The choice of $t_{1,o}$ was arbitrary, but now we see that it must be chosen so that the t_1 line terminates at the specified inlet temperature of $t_{1,i}$. Since the mean temperature difference in the heat exchanger is $t_{1,o} - t_{2,i}$, the rate equation can be written

$$q = UA(t_{1,o} - t_{2,i}) = W(t_{1,i} - t_{1,o})$$

Then

$$t_{1,o}(UA + W) = Wt_{1,i} + UAt_{2,i}$$

and

$$t_{1,o} = \frac{Wt_{1,i} + UAt_{2,i}}{UA + W}$$

Add zero in the form of $(UAt_{1,i} - UAt_{1,i})/(UA + W)$ to the right side; then

$$t_{1,o} = t_{1,i} - \frac{t_{1,i} - t_{2,i}}{W/UA + 1}$$

which checks the mathematical derivation, Eq. (5.7).

Example 5.1. In the counterflow heat exchanger shown in Fig. 5-4, a flow rate of 0.5 kg/s of water enters one circuit of the heat exchanger at a temperature of 30°C, and the same flow rate of water enters the other circuit at a temperature of 65°C. The UA of the heat exchanger is 4 kW/K. What is the mean temperature difference between the two streams?

Solution. The products of the flow rates and specific heats of the two streams are the same, (0.5 kg/s) [4.19 kJ/(kg · K)] $=2.095$ kW/K. For this special case where the wc_p products are equal, Eq. (5.7) gives the outlet temperature for the hot stream $t_{1,o}$

$$t_{1,o} = 65 - \frac{65 - 30}{2.095/4 + 1} = 42°C$$

The temperature difference at either end of the heat exchanger prevails throughout, and so the mean temperature difference is $42 - 30 = 12°C$.

FIGURE 5-4
Heat exchanger in Example 5.1.

5.5   EVAPORATORS AND CONDENSERS

A special set of equations is possible—and indeed necessary—when one of the fluids flowing through a heat exchanger changes phase. In an evaporator or condenser, as shown in Fig. 5-5, assume that there is no superheating or subcooling of the fluid that changes phase. That fluid will then remain at a constant temperature, provided that its pressure does not change.

The log-mean temperature difference still applies and in combination with a heat balance gives

$$q = UA\frac{(t_c - t_i) - (t_c - t_o)}{\ln\left[(t_c - t_i)/(t_c - t_o)\right]} = wc_p(t_o - t_i) \tag{5.9}$$

Equation (5.9) can be converted into the form

$$\frac{UA}{wc_p} = \ln\frac{t_c - t_i}{t_c - t_o} = -\ln\frac{t_c - t_o}{t_c - t_i}$$

Taking the antilog gives

$$e^{-UA/wc_p} = \frac{t_c - t_o}{t_c - t_i} = \frac{t_c - t_o + t_i - t_i}{t_c - t_i}$$

Then

$$t_o = t_i + (t_c - t_i)(1 - e^{-UA/wc_p}) \tag{5.10}$$

For a heat exchanger of known characteristics Eq. (5.10) can be used to compute the outlet temperature of the fluid that does not change phase when its entering temperature and the temperature of the boiling or condensing fluid t_c are known.

The characteristic shape of the temperature curves of the two fluids is shown in Fig. 5-6, applicable to a condenser.

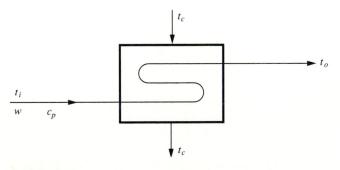

FIGURE 5-5
An evaporator or condenser where one fluid remains at a constant temperature.

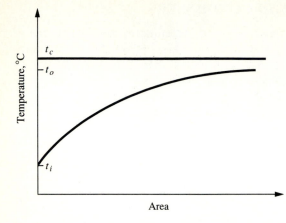

FIGURE 5-6
Temperature distribution in fluids in a condenser.

Example 5.2. Water is continuously heated from 25 to 50°C by steam condensing at 110°C. If the water flow rate remains constant but its inlet temperature drops to 15°C, what will its new outlet temperature be?

Solution. The terms U, A, w, c_p, and t_c all remain constant, so

Original	New

$$\frac{t_o - t_i}{t_c - t_i} = 1 - e^{-UA/wc_p} = \frac{t_o - t_i}{t_c - t_i}$$

$$\frac{50 - 25}{110 - 25} \qquad = \frac{t_o - 15}{110 - 15}$$

New $t_o = 42.9°C$.

5.6 HEAT-EXCHANGER EFFECTIVENESS

The effectiveness ϵ of a heat exchanger is defined as

$$\epsilon = \frac{q_{actual}}{q_{max}} \tag{5.11}$$

where q_{actual} = actual rate of heat transfer, kW
$\quad q_{max}$ = maximum possible rate of heat transfer, kW, with same inlet temperatures, flow rates, and specific heats as actual case

Another approach to defining q_{max} is to designate it as the rate of heat transfer that a heat exchanger of infinite area would transfer with given inlet temperatures, flow rates, and specific heats.

Example 5.3. What is the maximum rate of heat transfer possible in a counterflow heat exchanger shown in Fig. 5-7 if water enters at 30°C and cools oil entering at 60°C?

FIGURE 5-7
Oil cooler in Example 5.3.

$$
\text{Flow rate} = \begin{cases} 2.6 \text{ kg/s} & \text{oil} \\ 1.5 \text{ kg/s} & \text{water} \end{cases}
$$

$$
\text{Specific heat} = \begin{cases} 2.2 \text{ kJ/(kg} \cdot \text{K)} & \text{oil} \\ 4.19 \text{ kJ/(kg} \cdot \text{K)} & \text{water} \end{cases}
$$

Solution. The break in the heat exchanger indicates that to achieve the maximum rate of heat transfer the area must be made infinite. The next question, then, is: What are the outlet temperatures? Does the oil leave at 30°C, or does the water leave at 60°C?

From energy balances those two options give the following consequences:

1. Oil leaves at 30°C

$$q = (2.6 \text{ kg/s})[2.2 \text{ kJ/(kg} \cdot \text{K)}](60 - 30°C) = 171.6 \text{ kW}$$

   and water leaves at

$$30°C + \frac{171.6 \text{ kW}}{(1.5 \text{ kg/s})[4.19 \text{ kJ/(kg} \cdot \text{K)}]} = 57.3°C$$

2. Water leaves at 60° C

$$q = (1.5)(4.19)(60 - 30) = 188.6 \text{ kW}$$

   and oil leaves at

$$60°C - \frac{188.6}{(2.6)(2.2)} = 27°C$$

The second case is clearly impossible because the oil temperature would drop below that of the entering water, which would violate the second law of thermodynamics. Thus, $q_{max} = 171.6 \text{ kW}$.

The concept that Example 5.3 has exposed is that the maximum rate of heat transfer occurs when the fluid with the minimum product of flow rate and specific heat changes temperature to the entering temperature of the other fluid. Equation (5.11) can be rewritten

$$\epsilon = \frac{q_{\text{actual}}}{(wc_p)_{\min}(t_{\text{hot,in}} - t_{\text{cold,in}})} \tag{5.12}$$

where $(wc_p)_{\min}$ is the smaller wc_p of the two fluids.

5.7 TEMPERATURE PROFILES

The shapes of the temperature curves in the counterflow heat exchanger are shown by Fig. 5-8*a* or *b*. The minimum product of wc_p is possessed by fluid 1 in Fig. 5-8*a* and by fluid 2 in Fig. 5-8*b*, which is indicated by the fluid whose temperature changes most. The curves are steepest in the portion of the heat exchanger where the rate of heat transfer is highest, and this region occurs where the temperature differences are largest.

5.8 EFFECTIVENESS OF A COUNTERFLOW HEAT EXCHANGER

The form of Eq. (5.12) can be applied to the equation for a counterflow heat exchanger, Eq. (5.6), to develop an expression for the effectiveness of the counterflow heat exchanger. Denote fluid 1 as the fluid with the lesser value of wc_p. Note that

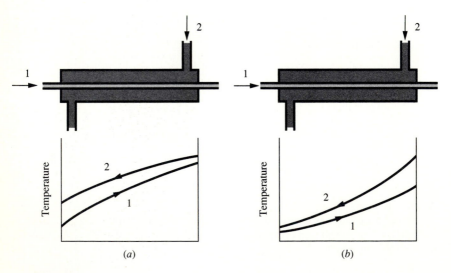

FIGURE 5-8
Temperature profiles in a counterflow heat exchanger.

$$\frac{t_{1,o} - t_{1,i}}{t_{2,i} - t_{1,i}} = \frac{(wc_p)_{\min}(t_{1,o} - t_{1,i})}{(wc_p)_{\min}(t_{2,i} - t_{1,i})} = \frac{q_{actual}}{(wc_p)_{\min}(t_{hot,in} - t_{cold,in})} = \epsilon$$

Combining with Eq. (5.6) gives

$$\epsilon = \frac{t_{1,o} - t_{1,i}}{t_{2,i} - t_{1,i}} = \frac{1 - e^D}{W_{\min}/W_2 - e^D} \tag{5.13}$$

where

$$D = UA\left(\frac{1}{W_{\min}} - \frac{1}{W_2}\right) = \frac{UA}{W_{\min}}\left(1 - \frac{W_{\min}}{W_2}\right) \tag{5.14}$$

Examination of Eqs. (5.13) and (5.14) shows that the effectiveness of the counterflow heat exchanger can be expressed as a function of two dimensionless groups, $UA/W_{\min}$ and $W_{\min}/W_2$.

5.9 NUMBER OF TRANSFER UNITS, NTU'S

The ability to express the effectiveness as a function of UA/W_{min} and W_{min}/W_2 leads to graphic presentations of effectiveness of heat exchangers of various configurations. The group UA/W_{min} is called the *number of transfer units* (NTU). Typical graphic presentations are shown in Figs. 5-9 and 5-10 for two different configurations of heat exchangers. Figure 5-9 is the graphic presentation of Eqs. (5.13) and (5.14).

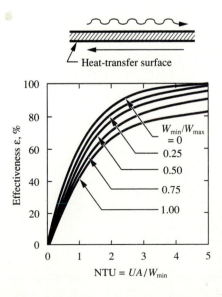

FIGURE 5-9
Effectiveness of a counterflow heat exchanger. (From W. M. Kays and A. L. London, *Compact Heat Exchangers*, 2d ed., McGraw-Hill Book Company, New York, 1964, p.50, used by permission.)

FIGURE 5-10
Effectiveness of a parallel counterflow heat exchanger. (From W. M. Kays and A. L. London, *Compact Heat Exchangers*, 2d ed., McGraw-Hill Book Company, New York, 1964, p.54, used by permission.)

Example 5.4. Compute the effectiveness of a counterflow heat exchanger having a U value of 1.1 kW/(m² · K) and an area of 16 m² when one fluid has a flow rate of 6 kg/s and a specific heat of 4.1 kJ/(kg · K) and the other fluid a flow rate of 3.8 kg/s and a specific heat of 3.3 kJ/(kg · K).

Solution

$$W_a = (6 \text{ kg/s})[4.1 \text{ kJ/(kg} \cdot \text{K)}] = 24.6 \text{ kW/K}$$

$$W_b = (3.8)(3.3) = 12.54 \text{ kW/K} \quad \text{designate } W_{min}$$

$$\text{NTU} = \frac{UA}{W_{min}} = \frac{(1.1)(16)}{12.54} = 1.40$$

$$\frac{W_{min}}{W_2} = \frac{12.54}{24.6} = 0.51$$

From Fig. 5-9

$$\epsilon = 0.67$$

or from Eq. (5.14)

$$D = 1.40(1 - 0.51) = 0.686$$

and from Eq. (5.13)

$$\epsilon = \frac{1 - e^{0.686}}{0.51 - e^{0.686}} = 0.668$$

It is now appropriate to put the previous nine sections on modeling of heat exchangers into perspective. The direction from which the modeling was approached was to predict the performance of an existing heat exchanger, which is a common assignment in systems simulation. The outlet temperatures can be computed if the inlet temperatures and effectiveness are known. The effectiveness can be computed or determined from a graph when the product of UA and the products of mass rate of flow and specific heat are known. Equations for the effectiveness of a counterflow heat exchanger have been presented, and graphs are available[1] for various configurations of heat exchangers. Worsøe-Schmidt[2] and Høgaard Knudsen have provided coefficients for equations that fit the curves of Fig. 5-10 as well as other heat exchangers.

5.10 BINARY SOLUTIONS

Figure 5-11 illustrates a binary solution where equilibrium prevails between liquid and vapor. The liquid is a solution of substances A and B, both exerting a vapor pressure such that both substances A and B exist in the vapor. The concentration of substance A in the vapor phase is not necessarily the same as its concentration in the liquid phase. A frequent assignment in petroleum, cryogenic, petrochemical, food, and other thermal processing industries is to separate substance A from substance B. System simulation computer programs (Chapter 6) are heavily used in these industries, and the separation process (usually by distillation) of binary solutions is often an integral function of these programs. We wish to be able to incorporate distillation equipment into systems along with heat exchangers, fans, pumps, compressors, expanders, and other thermal components.

5.11 TEMPERATURE-CONCENTRATION-PRESSURE CHARACTERISTICS

The temperature-concentration, t-vs.-x, graph of a binary mixture for a given pressure is shown in Fig. 5-12. The abscissa is the fraction of material A and a complementary scale is the fraction of material B. The fractions can be expressed either as mass fractions, for example,

$$\text{Mass fraction of A} = \frac{\text{mass of A, kg}}{\text{mass of A, kg} + \text{mass of B, kg}}$$

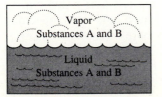

FIGURE 5-11
A binary solution.

FIGURE 5-12
Temperature-concentration diagram of a binary solution at a constant pressure.

or as mole fractions

$$\text{mole fraction of A} = \frac{\text{mol A}}{\text{mol A} + \text{mol B}}$$

$$= \frac{(\text{mass of A, kg})/(MW_a)}{(\text{mass of A, kg})/(MW_a) + (\text{mass of B, kg})/(MW_b)}$$

where MW_a and MW_b are the molecular weights of substances A and B, respectively.

The curves on Fig. 5-12 mark off three regions on the graph. The upper region is vapor (actually superheated vapor). If the temperature of vapor is reduced at a constant concentration, the state reaches the dew-point line, where some vapor begins to condense. The region between the two curves is the liquid-vapor region, where both vapor and liquid exist in equilibrium, as represented by the vessel in Fig. 5-11. The lowest region in Fig. 5-12 is the liquid region (actually subcooled liquid). If the temperature of subcooled liquid is increased at a constant concentration, the state reaches the bubble-point line, where some liquid begins to vaporize.

The temperature where the curves join at the left axis is the boiling temperature of substance B at the pressure in question, and the temperature intercepted at the right axis is the boiling temperature of substance A.

The pair of curves shown in Fig. 5-12 applies to a given pressure, and there will be a pair of curves for each different pressure, as shown in Fig. 5-13.

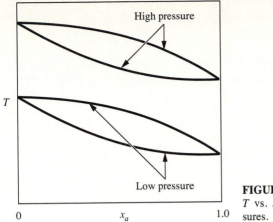

FIGURE 5-13
T vs. x diagrams for two different pressures.

An alternate form of expressing the properties of a binary solution is to present a p-vs.-x diagram, as in Fig. 5-14, where the pair of curves apply to a given temperature.

5.12 DEVELOPING A T-VS.-X DIAGRAM

Several ideal relationships can be assembled to compute the data for a graph like Fig. 5-12. In some cases the results will be very close to the actual binary properties, while with other combinations of substances the real properties deviate from the ideal. In most cases the idealization will give a fair approximation of the properties of the real mixture.

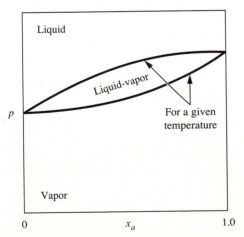

FIGURE 5-14
Pressure-concentration diagram.

The three tools used to develop the binary properties are (1) the saturation pressure-temperature relationships of the two substances, (2) Raoult's law, and (3) Dalton's law.

Saturation pressure-temperature relation. A simple equation form that relates pressure and temperature at saturated conditions (see Prob. 4.18) is

$$\ln P = C + \frac{D}{T}$$

where P = saturation pressure, kPa
$\quad\quad T$ = absolute temperature, K
$\quad C, D$ = constants

The values of C and D are unique to each substance and must be developed from experimental data. It is presumed that C_a and D_a for substance A and C_b and D_b for substance B are known in the equations

$$\ln P_a = C_a + \frac{D_a}{T} \tag{5.15}$$

$$\ln P_b = C_b + \frac{D_b}{T} \tag{5.16}$$

Raoult's law. Raoult's law states that the vapor pressure of one component in a mixture is

$$p_a = x_{a,l}P_a \tag{5.17}$$

where p_a = vapor pressure of substance A in mixture, kPa
$\quad x_{a,1}$ = mole fraction of substance A in liquid, dimensionless
$\quad\quad P_a$ = saturation pressure of pure A at existing temperature, kPa

Thus, as for the binary mixture of A and B in Fig. 5-15, if at the temperature T the saturation pressure of pure A is 500 kPa and the mole fraction in the liquid $x_{a,l} = 0.3$, the partial pressure of substance A in the vapor is $(0.3)(500) = 150$ kPa.

Dalton's law. Dalton's law states that the total pressure of the vapor mixture is the sum of the partial pressures of the constituents

$$p_t = p_a + p_b \tag{5.18}$$

Furthermore,
$$p_a = x_{a,v}p_t \tag{5.19}$$

and
$$p_b = x_{b,v}p_t \tag{5.20}$$

where p_t = total pressure, kPa
$\quad x_{a,v}$ = mole fraction of A in vapor
$\quad x_{b,v}$ = mole fraction of B in vapor

P_a at T = 500 kPa

FIGURE 5-15
Raoult's law.

Example 5.5. A binary solution of n-butane and n-heptane exists in liquid-vapor equilibrium at a pressure of 700 kPa. The saturation pressure-temperature relationships are

Butane: $\ln P = 21.77 - \dfrac{2795}{T}$

Heptane: $\ln P = 22.16 - \dfrac{3949}{T}$

where P is in pascals. Compute the mole fraction of butane in the liquid and in the vapor at a temperature of 120°C.

Solution. At 120°C the saturation pressures of pure substances are

Butane: $P_{but} = \exp\left(21.77 - \dfrac{2795}{393}\right) = 2322$ kPa

Heptane: $P_{hep} = \exp\left(22.16 - \dfrac{3949}{393}\right) = 182$ kPa

The combination of Eqs. (5.17) and (5.18) along with the recognition that $x_{hep,\,l} + x_{but,\,l} = 1.0$ yields

$$P_{hep} + P_{but} = (1 - x_{but,\,l})(182) + (x_{but,\,l})(2322) = 700$$

and so

$$x_{but,\,l} = 0.242$$

To find the concentration of butane in the vapor, use Eqs. (5.18) to (5.20)

$$x_{but,v} = \frac{x_{but,\,l}P_{but}}{P_t} = \frac{(0.242)(2322)}{700} = 0.803$$

The results may be compared with the experimentally determined properties shown in Fig. 5-16.

5.13 CONDENSATION OF A BINARY MIXTURE

When a pure substance condenses at constant pressure, the temperature remains constant but in a binary mixture the temperature progressively changes, even though the pressure remains constant. The condensation process of a binary system taking place in a tube (Fig. 5-17) can be represented

FIGURE 5-16
Binary system of n-butane and n-heptane.

for a specific binary mixture as in Fig. 5-16. Assume that the condensation takes place at a constant pressure of 2800 kPa and vapor enters with a mole fraction of 0.6 n-butane. Point 1 is superheated vapor at a temperature of 223°C, which is first cooled until point 2, where the temperature is 218°C and condensation begins. Removal of heat from the mixture results in continued condensation. At point 3 the temperature has dropped to 190°C, and here the liquid has a concentration of 0.33 and the vapor 0.80. The system is a mixture of liquid and vapor at point 3, and a mass balance of the butane can indicate how much liquid and how much vapor exists at 3. For 1 mol of combination

0.33 mol in liquid state $+ (0.80)(1 - $ mol in liquid state$) = (1$ mol$)(0.60)$

Therefore at point 3 in 1 mol of combination there is 0.427 mol of liquid and 0.573 mol of vapor.

FIGURE 5-17
Condensation of a binary mixture.

Condensation continues to point 4, where the temperature is 158°C; here all the vapor has condensed. Further removal of heat results in subcooling the liquid. During the condensation process the temperature drops from 218 to 158°C.

5.14 SINGLE-STAGE DISTILLATION

A simple distillation unit is the single-stage still shown in Fig. 5-18. If the still operates at a pressure for which the bubble-point and dew-point curves are shown in Fig. 5-19, and if the entering liquid to the partial vaporizer is at point 1, various outlet conditions of the vapor and liquid are possible. The limiting cases are combination 2-3 and combination 2″-2″. If the liquid is heated to point 3, only liquid would leave the still and no vapor. If the vaporizer carries on the process to 2″, only vapor leaves the still and no liquid. The desired operating condition will be 2′-3′, for example, where there is some separation in the sense that the vapor leaves with a high

FIGURE 5-18
Single-stage still.

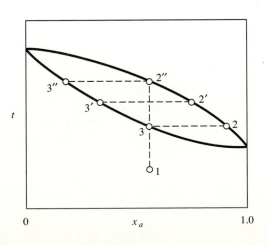

FIGURE 5-19
Some possible outlet conditions from the still in Fig. 5-18.

concentration of A and the liquid leaves with a lower concentration than entered at 1.

Example 5.6. A single-stage distillation tower receives 3.2 mol/s of butane-heptane (Fig. 5-16). Liquid enters with a mole fraction of 0.4 butane, the still operates at 700 kPa, and the mixture leaves the partial vaporizer at a temperature of 120°C. What are the flow rates of liquid and vapor leaving the separator?

Solution. At 120°C and a pressure of 700 kPa, the mole fraction of butane in the liquid is 0.242 and in the vapor is 0.803. A material balance on the butane states that

$$(3.2 \text{ mol/s})(0.4) = (w_l \text{ mol/s})(0.242) + (3.2 - w_l)(0.803)$$

$$w_l = 2.30 \text{ mol/s} \quad w_v = 3.2 - 2.30 = 0.90 \text{ mol/s}$$

5.15 RECTIFICATION

The single-stage still described in Sec. 5.14 performs a separation but has rather poor performance. The other end of the performance spectrum is

FIGURE 5-20
A rectification column.

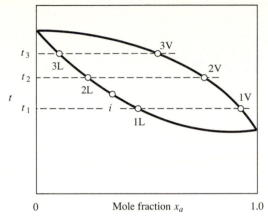

FIGURE 5-21
States of binary system in rectification column.

0   Mole fraction x_a 1.0

an ideal rectification tower, shown in Fig. 5-20. The t-vs.-x properties corresponding to the key locations in the rectification column are shown in Fig. 5-21. Liquid feed, assumed saturated, enters at point i, passes through a partial vaporizer, and is heated to temperature 2. In the column, vapor flows upward and liquid downward. The liquid is heated at the bottom to drive off some vapor, and a heat exchanger at the top of the tower condenses some of the vapor, providing a source of liquid to drain down the column.

Figure 5-21 shows that the separation of the two components is quite effective, since the liquid leaves the bottom of the tower at condition 3L and vapor leaves the top at condition 1V. There is a continuous transfer of heat and mass between the rising vapor and descending liquid. Furthermore in the ideal rectification tower there is equilibrium of temperature and vapor pressure between the liquid and vapor along the column. For accurate simulation of towers there must be a temperature difference between the vapor and liquid to transfer heat and vapor-pressure difference between the vapor and liquid in order to transfer mass.[3–6]

5.16 ENTHALPY

Enthalpy values of binary solutions and mixtures of vapor are necessary when making energy calculations. For system simulation (Chapter 6) the enthalpy data would be most convenient in equation form. More frequently the enthalpy data appear in graphic form, as shown in the skeleton diagram of Fig. 5-22. Figure 5-22 is an enthalpy-concentration, h-vs.-x, diagram for solutions and vapor. Since pressure has a negligible effect on the enthalpy of the liquid, the chart is applicable to subcooled as well as saturated liquid, but because the enthalpy of the vapor is somewhat sensitive to the pressure, the enthalpy curves for vapor apply only to saturated conditions.

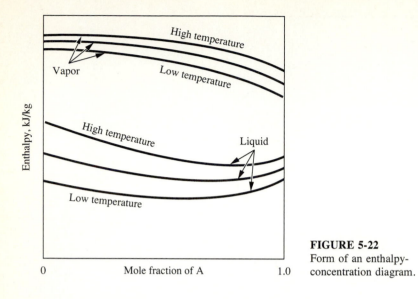

FIGURE 5-22
Form of an enthalpy-concentration diagram.

5.17 PRESSURE DROP AND PUMPING POWER

A cost that appears in most economic analyses of thermal systems is the pumping cost. The size of a heat exchanger transferring heat to a liquid can be reduced, for example, if the flow rate of liquid or the velocity for a given flow rate is increased. The cost whose increase eventually overtakes the reduction in the cost of the heat exchanger as the velocity or flow increase is the pumping cost. Another example of the emergence of pumping cost is in the selection of optimum pipe size. The smaller the pipe the less the first cost but (for a given flow) the higher the pumping cost for the life of the system.

Since the pumping-cost term appears so frequently, it is appropriate to review the expression for pumping power. The pressure drop of an incompressible fluid flowing turbulently through pipes, fittings, heat exchangers, and almost any confining conduit varies as w^n

$$\Delta p = C(w^n)$$

where C is a constant, w is the mass rate of flow, and the exponent n varies between about 1.8 and 2.0. Generally the value of n is close to 2.0, except for flow in straight pipes at Reynolds numbers in the low turbulent range.

The ideal work per unit mass required for pumping fluid in steady flow is $\int v\,dp$, and for an incompressible fluid the power required is

$$\text{Power} = \frac{w}{\rho}\Delta p$$

where ρ is the density.

Thus the equation for the pumping power is

$$\text{Power} = \frac{C}{\rho} w^{n+1} \approx \frac{C}{\rho} w^3 \tag{5.21}$$

which is further modified by dividing by the pump, fan, or compressor efficiency.

5.18 TURBOMACHINERY

The methods of mathematical modeling explained in this chapter have generally been limited to expressing one variable as a function of one or two other variables. In principle it is possible to extend these methods to functions of three variables, but the execution might be formidable. Turbomachines, such as fans, pumps, compressors, and turbines, are used in practically all thermal systems, and in these components the dependent variable may be a function of three of more independent variables. Fortunately, the tool of dimensional analysis frequently permits reducing the number of independent variables to a smaller number by treating groups of terms as individual variables. The performance of a centrifugal compressor, for example, will typically appear as in Fig. 5-23. Instead of attempting to express p_2 as a function of six variables, an equation could be developed to express p_2/p_1 as a function of the two other dimensionless groups. Then in calculating

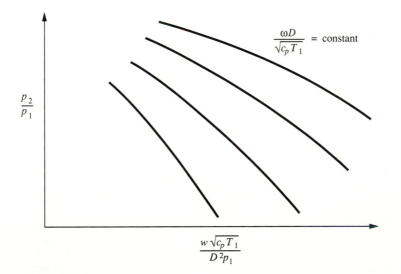

FIGURE 5-23
Performance of a centrifugal compressor expressed in dimensionless groups to reduce the number of independent variables.

p_2 the two independent variables $\omega D / \sqrt{c_p T_1}$ and $w \sqrt{c_p T_1}/D^2 p_1$ would be calculated first and next p_2/p_1; finally, p_2 would be computed from p_2/p_1 and p_1.

PROBLEMS

5.1. What is the effectiveness of a counterflow heat exchanger that has a UA value of 24 kW/K if the respective mass rates of flow and specific heats of the two fluids are 10 kg/s, 2 kJ/(kg · K) and 4 kg/s, 4kJ/(kg · K)?
   Ans.: 0.636.

5.2. Water flows through one side of a heat exchanger with a flow rate of 0.2 kg/s rising in temperature from 20 to 50°C. The specific heat of water is 4.19 kJ/(kg · K). The fluid on the other side of the heat exchanger enters at 80°C and leaves at 40°C. What is the effectiveness of the heat exchanger?
   Ans.: 0.667.

5.3. A flow rate of 2 kg/s of water, $c_p = 4.19$ kJ/(kg · K), enters one end of a counterflow heat exchanger at a temperature of 20°C and leaves at 40°C. Oil enters the other side of the heat exchanger at 60°C and leaves at 30°C. If the heat exchanger were made infinitely large while the entering temperatures and flow rates of the water and oil remained constant, what would the rate of heat transfer in the exchanger be?
   Ans.: 223.5 kW.

5.4. The evaporating fluid in an evaporator has a temperature of 5°C and the heat exchanger transfers 50 kW. The UA of the exchanger is 12 kW/K, and the wc_p of the fluid being cooled is 8 kw/K. What are the entering and leaving temperatures of the fluid being chilled?

5.5. Stream 1 enters a multipass heat exchanger at a temperature of 82°C with a flow rate of 4.1 kg/s; the fluid has a specific heat of 4.19 kJ/(kg · K). Stream 2 enters at a temperature of 18°C, with a flow rate of 4.5 kg/s; the fluid has a specific heat of 3.2 kJ/(kg · K). The effectiveness of the heat exchanger is 0.46. What is the rate of heat transfer in kW in the heat exchanger?

5.6. A counterflow heat exchanger having a UA value of 2 kW/K receives stream 1 at a temperature of 60°C and stream 2 at a temperature of 20°C. The flow rates are 0.3 and 0.4 kg/s, and the c_p in kJ/(kg · K) are 3.2 and 4.19 for streams 1 and 2, respectively. What is the outlet temperature of stream 2?

5.7. Fluid 1 enters a counterflow heat exchanger at a temperature of 60°C having a flow rate of 1.0 kg/s and a c_p of 3.2 kJ/(kg · K). Fluid 2 enters with a temperature of 20°C having a flow rate of 0.8 kg/s and a c_p of 4.0 kJ/(kg · K). The UA value of the heat exchanger is 3.0 kW/K. What is the outlet temperature of Fluid 1?

5.8. A flow rate of 0.8 kg/s of water is heated in a heat exchanger by condensing steam at 100°C. When water enters at 15°C, it leaves the heat exchanger at 62°C. If the inlet water temperature were changed to 20°C while its flow rate and the condensing temperature remained constant, what would its outlet temperature be?
   Ans.: 64.2°C.

5.9. A counterflow heat exchanger cools 5 kg/s of oil, $c_p = 2.4$ kJ/(kg · K), with water that has a flow rate of 7.5 kg/s. The specific heat of water is 4.19 kJ/(kg · K). Under the original operating conditions the oil is cooled from 75 to 40°C when water enters at 25°C. To what temperature will the oil be cooled if it enters at 65°C and if there is no change in the entering water temperature, the flow rates of either fluid, or the heat-transfer coefficients?
> **Ans.:** 37°C.

5.10. In a processing plant a material must be heated from 20 to 80°C in order for the desired reaction to proceed, whereupon the material is cooled in a regenerative heat exchanger, as shown in Fig. 5-24. The specific heat of the material before and after the reaction is 3.0 kJ/(kg · K). If the *UA* of this counterflow regenerative heat exchanger is 2.1 kW/K and the flow rate is 1.2 kg/s, what is the temperature *t* leaving the heat exchanger?
> **Ans.:** 57.9°C.

5.11. A condenser having a *UA* value of 480 kW/K condenses steam at a temperature of 40°C. The cooling water enters at 20°C with a flow rate of 160 kg/s. What is the outlet temperature of the cooling water? The specific heat of water is 4.19 kJ/(kg · K).
> **Ans.:** 30.2°C

5.12. A heat exchanger with one shell pass and two tube passes (Fig. 5-10) uses seawater at 15°C, $c_p = 3.8$ kJ/(kg · K), to cool a flow rate of fresh water of 1.6 kg/s entering at 40°C. The specific heat of the fresh water is 4.19 kJ/(kg · K). If the *UA* of the heat exchanger is 10 kW/K, what must the flow rate of seawater be in order to cool the fresh water to 22.5°C?
> **Ans.:** 7kg/s.

5.13. A double-pipe heat exchanger serves as an oil cooler with oil flowing in one direction through the inner tube and cooling water in the opposite direction through the annulus. The oil flow rate is 0.63 kg/s, the oil has a specific heat of 1.68 kJ/(kg · K), the water flow rate is 0.5 kg/s, and its specific heat is 4.19 kJ/(kg · K). In a test of a prototype, oil entering at 78°C was cooled to 54°C when the entering water temperature was 30°C. The possibility of increasing the area of the heat exchanger by increasing the length of the double pipe is to be considered. If the flow rates, fluid properties, and entering temperatures remain unchanged, what will the expected outlet temperature of the oil be if the area is increased by 20 percent?
> **Ans.:** 51.3°C.

FIGURE 5-24
Regenerative heat exchanger in Prob. 5.10.

5.14. To ventilate a factory building, 5 kg/s of factory air at a temperature of 27°C is exhausted, and an identical flow rate of outdoor air at a temperature of -12°C is introduced to take its place. To recover some of the heat of the exhaust air, heat exchangers are placed in the exhaust and ventilation air ducts, as shown in Fig. 5-25, and 2 kg/s of water is pumped between the two heat exchangers. The UA value of both of these counterflow heat exchangers is 6.33 kW/K. What is the temperature of air entering the factory?

  Ans.: 2.9°C.

5.15. A solar air heater consists of a flat air duct composed on one side of an absorbing sheet backed by insulation and on the other side by a transparent sheet, as shown in Fig. 5-26. The absorbing sheet absorbs 500 W/m² and delivers all this heat to the air being heated, which loses some to the atmosphere through the transparent sheet. The convection heat-transfer coefficient

FIGURE 5-25
Heat-recovery system in Prob. 5.14.

FIGURE 5-26
Solar air heater in Prob. 5.15.

from the transparent sheet to the ambient air is 12 W/(m² · K) and from the air being heated to the transparent sheet is 45 W/(m² · K). The air enters with a temperature that is the same as the ambient, namely 15°C, and the flow rate of air is 0.02 kg/s per meter width. Develop the equation for the temperature of heated air t as a function of length along the collector x assuming no conduction in the sheet in the direction of airflow.

Ans.: $t = 15 + 52.8(1 - e^{-x/2.11})$.

5.16. The chain of heat exchangers shown in Fig. 5-27 has the purpose of elevating the temperature of a fluid to 390 K, at which temperature the desired chemical reaction takes place. The fluid has a specific heat of 3.2 kJ/(kg · K) both before and after the reaction, and its flow rate is 1.5 kg/s. The entering temperature of the fluid to heat exchanger I is 290 K and the UA of this heat exchanger is 2.88 kW/K. Steam is supplied to heat exchanger II at 375 K, and condensate leaves at the same temperature. The UA values of heat exchangers II and III are 4.7 and 9.6 kW/K, respectively. What are the values of temperatures t_1 to t_4?

Ans.: $t_4 = 365$ K.

5.17. A binary solution of liquid and vapor exists at equilibrium. In the liquid the molal fraction of material A is 0.4 and of material B is 0.6. At the existing temperature the saturation pressure of pure A is 530 kPa and of pure B is 225 kPa. (a) What is the pressure in the vessel, and (b) what is the fraction of A in the vapor?

Ans.: (b) 0.61.

5.18. A single-stage still, as in Fig. 5-18, is supplied with a feed of 0.6 mole fraction of n-butane and 0.4 mole fraction of n-heptane with properties as shown in Fig. 5-16. The still operates at a pressure of 700 kPa. How many moles of vapor are derived from 1 mol of feed if the vapor is to leave the still with a mole fraction of butane of 0.8?

Ans.: 0.51.

5.19. A vapor mixture of n-butane and n-heptane, with properties as shown in Fig. 5-16, at a pressure of 700 kPa and a temperature of 170°C, and a mole fraction of butane of 0.4 enters a condenser.

FIGURE 5-27
Chain of heat exchangers in Prob. 5.16.

(a) At what temperature does condensation begin?
(b) At what temperature is condensation complete?
(c) When the temperature is 120°C, what is the fraction in liquid form?
  Ans.: (c) 0.78.

5.20. A mixture of butane and propane is often sold as a fuel. We are interested in determining the T-vs.-x relationship of a binary mixture of butane and propane at standard atmospheric pressure of 101.3 kPa. The pressure-temperature relationships at saturated conditions for the pure substances are

$$\ln P = \begin{cases} 21.40 - \dfrac{2286}{T} & \text{propane} \\[2mm] 21.77 - \dfrac{2795}{T} & \text{butane} \end{cases}$$

where P = pressure, Pa
  T = absolute temperature, K

Present the T-vs.-x curves for vapor and liquid neatly on a full-size sheet of graph paper, where x represents the mole fraction of propane.
  Ans.: One point on vapor curve, $x = 0.5$ when $T = 260\ K$.

5.21. A distillation tower (Fig. 5-28) receives a two-component solution in liquid form. The two components are designated as A and B, and x indicates the mass fraction of material A. The concentration of the feed $x_1 = 0.46$ and the enthalpies entering and leaving the still are $h_1 = 80$ kJ/kg, $h_3 = 360$ kJ/kg, and $h_5 = 97$ kJ/kg. The condenser operates at $t = 30°C$, at which state $x_2 = 0.92$, $h_2 = 320$ kJ/kg, $x_4 = 0.82$, $h_4 = 23$ kJ/kg, and the condenser rejects 550 kW to the cooling water. The reboiler operates at $t = 210°C$, at which temperature $x_6 = 0.08$, $h_6 = 108$ kJ/kg, $x_7 = 0.13$, $h_7 = 415$ kJ/kg, and the reboiler receives 820 kW from high-pressure steam. Complete Table 5.1.
  Ans.: Flow rate at $1 = 2.18$ kg/s.

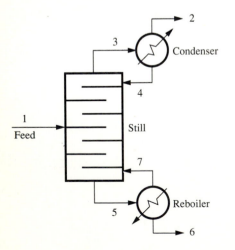

FIGURE 5-28
Distillation tower in Prob. 5.21

TABLE 5.1
Operating conditions of distillation tower

Position	Flow rate, kg/s	x	h, kJ/kg
1		0.46	80
2		0.92	320
3	.		360
4		0.82	23
5			97
6		0.08	108
7		0.13	415

5.22. Dimensional analysis suggests that the performance of a centrifugal fan can be expressed as a function of two dimensionless groups:

$$\frac{SP}{D^2\omega^2\rho} \quad \text{and} \quad \frac{Q}{D^3\omega}$$

where SP = static pressure, Pa
$\quad\quad D$ = diameter of wheel, m
$\quad\quad \omega$ = rotative speed, rad/s
$\quad\quad \rho$ = density, kg/m^3
$\quad\quad Q$ = volume rate of airflow, m^3/s

For a given fan operating with air at a constant density, it should be possible to plot one curve, as in Fig. 5-29, that represents the performance at all speeds. The performance of a certain 0.3-m-diameter fan of Lau Blower Company is presented in Table 5.2.
(a) Plot neatly on graph paper the above performance data in the form of Fig. 5-29.

Q/ω

FIGURE 5-29
Performance of a centrifugal fan.

TABLE 5.2
Performance of fan in Prob. 5.22

rad/s	Q, m³/s	SP, Pa	rad/s	Q, m³/s	SP, Pa
157	1.42	861	126	3.30	114
	1.89	861	94	0.94	304
	2.36	796		1.27	299
	2.83	694		1.89	219
	3.02	635		2.22	134
	3.30	525		2.36	100
126	1.42	548	63	0.80	134
	1.79	530		1.04	122
	2.17	473		1.42	70
	2.36	428		1.51	55
	2.60	351			

(*b*) If the SP is to be computed as a function of Q and ω, propose a convenient form of the equation (just use symbols for the coefficients; do not evaluate them numerically).

REFERENCES

1. W. M. Kays and A. L. London, *Compact Heat Exchangers,* 2d ed., McGraw-Hill, New York, 1964.
2. P. Worsøe-Schmidt and H. J. Høgaard Knudsen, "Thermal Modelling of Heat Exchangers for Simulation Purposes," *25th Heat Transfer Fluid Mech. Inst.,* Davis, Calif., June 1976.
3. W. Badger and J. T. Banchero, *Introduction to Chemical Engineering,* McGraw-Hill, New York, 1955.
4. W. L. McCabe and J. C. Smith, *Unit Operations of Chemical Engineering,* 3d ed., McGraw-Hill, New York, 1976.
5. C. M. Thatcher, *Fundamentals of Chemical Engineering,* Merrill, Columbus, Ohio, 1962.
6. C. G. Kirkbride, *Chemical Engineering Fundamentals,* McGraw-Hill, New York, 1947.

CHAPTER
6

SYSTEM SIMULATION

6.1 DESCRIPTION OF SYSTEM SIMULATION

System simulation, as practiced in this chapter, is the calculation of operating variables (such as pressures, temperatures, and flow rates of energy and fluids) in a thermal system operating in a steady state. System simulation presumes knowledge of the performance characteristics of all components as well as equations for thermodynamic properties of the working substances. The equations for performance characteristics of the components and thermodynamic properties, along with energy and mass balances, form a set of simultaneous equations relating the operating variables. The mathematical description of system simulation is that of solving these simultaneous equations, many of which may be nonlinear.

A *system* is a collection of components whose performance parameters are interrelated. *System simulation* means observing a synthetic system that imitates the performance of a real system. The type of simulation studied in this chapter can be accomplished by calculation procedures, in contrast to simulating one physical system by observing the performance of another physical system. An example of two corresponding physical systems is when an electrical system of resistors and capacitors represents the heat-flow system in a solid wall.

6.2 SOME USES OF SIMULATION

System simulation may be used in the design stage to help achieve an improved design, or it may be applied to an existing system to explore prospective modifications. Simulation is not needed at the *design conditions* because in the design process the engineer probably chooses reasonable values of the operating variables (pressures, temperatures, flow rates, etc.) and selects the components (pumps, compressors, heat exchangers, etc.) that correspond to the operating variables. It would be for the nondesign conditions that system simulation would be applied, e.g., as at part-load or overload conditions. The designer may wish to investigate off-design operation to be sure that pressures, temperatures, or flow rates will not be too high or too low.

The steep increase in the cost of energy has probably been responsible for the blossoming of system simulation during recent years. Thermal systems (power generation, thermal processing, heating, and refrigeration) operate most of the time at off-design conditions. To perform energy studies in the design stage the operation of the system must be simulated throughout the range of operation the system will experience.

System simulation is sometimes applied to existing systems when there is an operating problem or a possible improvement is being considered. The effect on the system of changing a component can be examined before the actual change to ensure that the operating problem will be corrected and to find the cheapest means of achieving the desired improvement.

After listing some of the classes of system simulation, this chapter concentrates on just one class for the remainder of the chapter. Next the use of information-flow diagrams and the application to sequential and simultaneous calculations are discussed. The process of simulating thermal systems operating at steady state usually simmers down to the solution of simultaneous nonlinear algebraic equations, and procedures for their solution are examined.

6.3 CLASSES OF SIMULATION

System simulation is a popular term and is used in different senses by various workers. We shall first list some of the classes of system simulation and then designate the type to which our attention will be confined.

Systems may be classified as *continuous* or *discrete*. In a continuous system, the flow through the system is that of a continuum, e.g., a fluid or even solid particles, flowing at such rates relative to particle sizes that the stream can be considered as a continuum. In discrete systems, the flow is treated as a certain number of integers. The analysis of the flow of people through a supermarket involving the time spent at various shopping areas and the checkout counter is a discrete system. Another example of a discrete-system analysis is that performed in traffic control on expressways and city streets. Our concern, since it is primarily directed toward fluid and energy systems, is continuous systems.

Another classification is *deterministic* v. *stochastic*. In the deterministic analysis the input variables are precisely specified. In stochastic analysis the input conditions are uncertain, either being completely random or (more commonly) following some probability distribution. In simulating the performance of a steam-electric generating plant that supplies both process steam and electric power to a facility, for example, a deterministic analysis starts with one specified value of the steam demand along with one specified value of the power demand. A stochastic analysis might begin with some probability description of the steam and power demands. We shall concentrate in this chapter on deterministic analysis, and reserve for Chapter 19 the study of probabilistic influences.

Finally, system simulation may be classed as *steady-state* or *dynamic,* where in a dynamic simulation there are changes of operating variables with respect to time. Dynamic analyses are used for such purposes as the study of a control system in order to achieve greater precision of control and to avoid unstable operating conditions (Chapter 15). The dynamic simulation of a given system is more difficult than the steady-state simulation, since the steady state falls out as one special case of the transient analysis. On the other hand, steady-state simulations are required much more often than dynamic simulations and are normally applied to much larger systems.

The simulation to be practiced here will be that of continuous, deterministic steady-state systems.

6.4 INFORMATION-FLOW DIAGRAMS

Fluid- and energy-flow diagrams are standard engineering tools. In system simulation, an equally useful tool is the information-flow diagram. A block diagram of a control system is an information-flow diagram in which a block signifies that an output can be calculated when the input is known. In the block diagram used in automatic-control work the blocks represent *transfer functions,* which could be considered differential equations. In steady-state system simulation the block represents an algebraic equation. A centrifugal pump might appear in a fluid-flow diagram like that shown in Fig. 6-1a, while in the information-flow diagram the blocks (Fig. 6-1b) represent

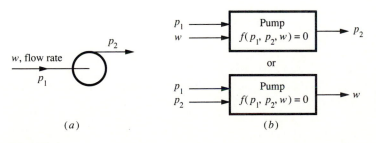

FIGURE 6-1
(a) Centrifugal pump in fluid-flow diagram (b) possible information-flow blocks representing pump.

functions or expressions that permit calculation of the outlet pressure for the one block and the flow rate for the other. A block, as in Fig. 6-1b, is usually an equation, here designated as $f(p_1, p_2, w) = 0$, or it may be tabular data to which interpolation would be applicable.

Figure 6-1 shows only one component. To illustrate how these individual blocks can build the information-flow diagram for a system, consider the fire-water facility shown in Fig. 6-2. A pump having pressure-flow characteristics shown in Fig. 6-2 draws water from an open reservoir and delivers it through a length of pipe to hydrant A, with some water continuing through additional pipe to hydrant B. The water flow rates in the pipe sections are designated w_1 and w_2, and the flow rates passing out the hydrants are w_A and w_B. The equations for the water flow rate through open hydrants are $w_A = C_A \sqrt{p_3 - p_{at}}$ and $w_B = C_B \sqrt{p_4 - p_{at}}$, where C_A and C_B are constants and p_{at} is the atmospheric pressure. The equation for the pipe section 0–1 is $p_{at} - p_1 = C_1 w_1^2 + h\rho g$, where $C_1 w_1^2$ accounts for friction and $h\rho g$ is the pressure drop due to the change in elevation h. In pipe sections 2–3 and 3–4

$$p_2 - p_3 = C_2 w_1^2 \quad \text{and} \quad p_3 - p_4 = C_3 w_2^2$$

These five equations can be written in functional form

$$f_1(w_A, p_3) = 0 \tag{6.1}$$

$$f_2(w_B, p_4) = 0 \tag{6.2}$$

$$f_3(w_1, p_1) = 0 \tag{6.3}$$

$$f_4(w_1, p_2, p_3) = 0 \tag{6.4}$$

$$f_5(w_2, p_3, p_4) = 0 \tag{6.5}$$

The atmospheric pressure p_{at} is not listed as a variable since it will have a

FIGURE 6-2
Fire-water system and pump characteristics.

FIGURE 6-3
Information-flow diagram for fire-water system.

known value. An additional function is provided by the pump characteristics

$$f_6(w_1, p_1, p_2) = 0 \qquad (6.6)$$

The preceding six equations can all be designated as *component performance characteristics*. There are eight unknown variables, $w_1, w_2,$ $w_A, w_B, p_1, p_2, p_3,$ and p_4, but only six equations so far. Mass balances provide the other two equations

$$w_1 = w_A + w_2 \qquad \text{or} \qquad f_7(w_1, w_A, w_2) = 0 \qquad (6.7)$$

and

$$w_2 = w_B \qquad \text{or} \qquad f_8(w_2, w_B) = 0 \qquad (6.8)$$

Several correct flow diagrams can be developed to express this system, one of which is shown in Fig. 6-3. Each block is arranged so that there is only one output, which indicates that the equation represented by that block is solved for the output variable.

6.5 SEQUENTIAL AND SIMULTANEOUS CALCULATIONS

Sometimes it is possible to start with the input information and immediately calculate the output of a component. The output information from this first component is all that is needed to calculate the output information of the next component, and so on to the final component of the system, whose output is the output information of the system. Such a system simulation consists of *sequential calculations*. An example of a sequential calculation might occur in an on-site power-generating plant using heat recovery to generate steam for heating or refrigeration, as shown schematically in Fig. 6-4. The exhaust gas from the engine flows through the boiler, which generates steam

FIGURE 6-4
On-site power generation with heat recovery to develop steam for refrigeration.

to operate an absorption refrigeration unit. If the output information is the refrigeration capacity that would be available when the unit generates a given electric-power requirement, a possible information-flow diagram for this simulation is shown in Fig. 6-5.

Starting with the knowledge of the engine-generator speed and electric-power demand, we can solve the equations representing performance characteristics of the components in sequence to arrive at the output information, the refrigeration capacity.

The *sequential* simulation shown by the information-flow diagram of

FIGURE 6-5
Information-flow diagram for an on-site power-generating plant of Fig. 6-4.

Fig. 6-5 is in contrast to the *simultaneous* simulation required for the information-flow diagram of Fig. 6-3. Sequential simulations are straightforward, but simultaneous simulations are the challenges on which the remainder of the chapter concentrates.

6.6 TWO METHODS OF SIMULATION: SUCCESSIVE SUBSTITUTION AND NEWTON-RAPHSON

The task of simulating a system, after the functional relationships and interconnections have been established, is one of solving a set of simultaneous algebraic equations, some or all of which may be nonlinear. Two of the methods available for this simultaneous solution are *successive substitution* and *Newton-Raphson*. Each method has advantages and disadvantages which will be pointed out.

6.7 SUCCESSIVE SUBSTITUTION

The method of successive substitution is closely associated with the information-flow diagram of the system (Fig. 6-3). There seems to be no way to find a toe-hold to begin the calculations. The problem is circumvented by assuming a value of one or more variables, beginning the calculation, and proceeding through the system until the originally-assumed variables have been recalculated. The recalculated values are substituted successively (which is the basis for the name of the method), and the calculation loop is repeated until satisfactory convergence is achieved.

Example 6.1 A water-pumping system consists of two parallel pumps drawing water from a lower reservoir and delivering it to another that is 40 m higher, as illustrated in Fig. 6-6. In addition to overcoming the pressure difference due to the elevation, the friction in the pipe is $7.2w^2$ kPa, where w is the combined flow rate in kilograms per second. The pressure-flow-rate characteristics of the pumps are

Pump 1: $\quad\quad\quad \Delta p, \text{kPa} = 810 - 25w_1 - 3.75w_1^2$

Pump 2: $\quad\quad\quad \Delta p, \text{kPa} = 900 - 65w_2 - 30w_2^2$

where w_1 and w_2 are the flow rates through pump 1 and pump 2, respectively.
Use successive substitution to simulate this system and determine the values of Δp, w_1, w_2, and w.

Solution. The system can be represented by four simultaneous equations. The pressure difference due to elevation and friction is

FIGURE 6-6
Water-pumping system in Example 6.1.

$$\Delta p = 7.2w^2 + \frac{(40 \text{ m})(1000 \text{ kg/m}^3)(9.807 \text{ m/s}^2)}{1000 \text{ Pa/kPa}} \tag{6.9}$$

Pump 1: $\quad\quad \Delta p = 810 - 25w_1 - 3.75w_1^2 \tag{6.10}$

Pump 2: $\quad\quad \Delta p = 900 - 65w_2 - 30w_2^2 \tag{6.11}$

Mass balance: $\quad w = w_1 + w_2 \tag{6.12}$

One possible information-flow diagram that represents this system is shown in Fig. 6-7. If a trial value of 4.2 is chosen for w_1, the value of Δp can be computed from Eq. (6.10), and so on about the loop. The values of the variables resulting from these iterations are shown in Table 6.1. The calculation appears to be converging slowly to the values $w_1 = 3.991, w_2 = 1.99, w = 5.988$, and $\Delta p = 650.5$.

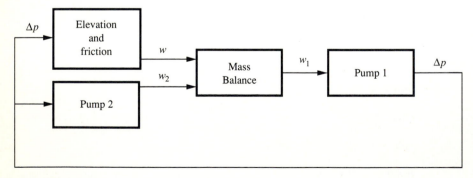

FIGURE 6-7
Information-flow diagram 1 for Example 6.1.

TABLE 6.1
Successive substitution on information-flow diagram of Fig. 6-7

Iteration	Δp	w_2	w	w_1
1	638.85	2.060	5.852	3.792
2	661.26	1.939	6.112	4.174
3	640.34	2.052	5.870	3.818
4	659.90	1.946	6.097	4.151
. . .	. . .	. . .	. . .	. . .
47	649.98	2.000	5.983	3.983
48	650.96	1.995	5.994	3.999
49	650.04	2.000	5.983	3.984
50	650.90	1.995	5.993	3.998

6.8 PITFALLS IN THE METHOD OF SUCCESSIVE SUBSTITUTION

Figure 6-7 is only one of the possible information-flow diagrams that can be generated from the set of equations (6.9) to (6.12). Two additional flow diagrams are shown in Figs. 6-8 and 6-9.

The trial value of $w_2 = 2.0$ was chosen for the successive substitution method on information-flow diagram 2, and the results of the iterations are shown in Table 6.2. A trial value of $w = 6.0$ was chosen for the solution of information-flow diagram 3, and the results are shown in Table 6.3.

Information-flow diagram 1 converged to the solution, while diagrams 2 and 3 diverged. This experience is typical of successive substitution. It should be observed that the divergence in diagrams 2 and 3 is attributable to the calculation sequence and not a faulty choice of the trial value. In both cases the trial value was essentially the correct solution: $w_2 = 2.0$ and $w = 6.0$.

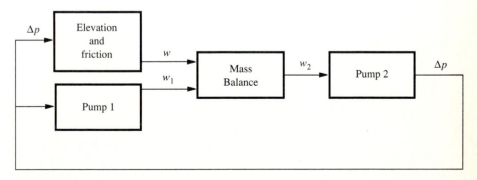

FIGURE 6-8
Information-flow diagram 2 for Example 6.1.

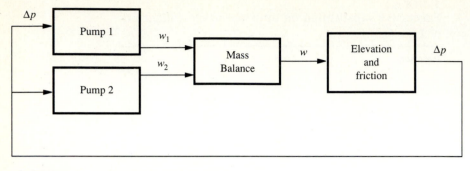

FIGURE 6-9
Information-flow diagram 3 for Example 6.1.

TABLE 6.2
Iterations of information-flow diagram 2

Iteration	Δp	w	w_1	w_2
1	650.0	5.983	4.000	1.983
2	653.2	6.019	3.942	2.077
3	635.5	5.812	4.258	1.554
4	726.5	6.814	2.443	4.371
5	42.8	†		

† Value of w became imaginary.

TABLE 6.3
Iterations of information-flow diagram 3

Iteration	Δp	w_1	w_2	w
1	651.5	3.973	1.992	5.965
2	648.5	4.028	2.008	6.036
3	654.6	3.916	1.975	5.891
4	642.1	4.142	2.042	6.184
5	667.6	3.672	1.903	5.575
6	616.1	4.593	2.178	6.771
7	722.3	2.539	1.580	4.120
8	514.5	6.149	2.662	8.811
9	951.2	†		

† Value of w_1 became imaginary.

Are there means of checking a flow diagram in advance to determine whether the calculations will converge or diverge? Yes, and a technique will be explained in Chapter 14, but the effort of such a check is probably greater than simply experimenting with various diagrams until one is found that converges.

In the method of successive substitution each equation is solved for one variable, and the equation may be nonlinear in that variable, as was true, for example, for the calculations of w and w_2 in diagram 1. No particular problem resulted in computing w and w_2 here because the equations were quadratic. An iterative technique, which may be required in some cases, is described in Sec. 6.10.

6.9 TAYLOR-SERIES EXPANSION

The second technique of system simulation, presented in Secs. 6.10 and 6.11, the Newton-Raphson method, is based on a Taylor-series expansion. It is therefore appropriate to review the Taylor-series expansion. If a function z, which is dependent upon two variables x and y, is to be expanded about the point $(x = a, y = b)$, the form of the series expansion is

$$z = \text{const} + \text{first-degree terms}$$
$$+ \text{second-degree terms} + \text{higher-degree terms}$$

or, more specifically,

$$z = c_o + [c_1(x - a) + c_2(y - b)] + [c_3(x - a)^2$$
$$+ c_4(x - a)(y - b) + c_5(y - b)^2] + \cdots \quad (6.13)$$

Now determine the values of the constants in Eq. (6.13). If x is set equal to a and y is set equal to b, all the terms on the right side of the equation reduce to zero except c_o, so that the value of the function at (a, b) is

$$c_o = z(a, b) \quad (6.14)$$

To find c_1, partially differentiate Eq. (6.13) with respect to x; then set $x = a$ and $y = b$. The only term remaining on the right side of Eq. (6.13) is c_1, so

$$c_1 = \frac{\partial z(a, b)}{\partial x} \quad (6.15)$$

In a similar manner,

$$c_2 = \frac{\partial z(a, b)}{\partial y} \quad (6.16)$$

The constants c_3, c_4, and c_5 are found by partial differentiation twice followed by substitution of $x = a$ and $y = b$ to yield

$$c_3 = \frac{1}{2}\frac{\partial^2 z(a, b)}{\partial x^2} \qquad c_4 = \frac{\partial^2 z(a, b)}{\partial x \, \partial y} \qquad c_5 = \frac{1}{2}\frac{\partial^2 z(a, b)}{\partial y^2} \quad (6.17)$$

For the special case where y is a function of one independent variable x, the Taylor-series expansion about the point $x = a$ is

$$y = y(a) + \frac{dy(a)}{dx}(x - a) + \left[\frac{1}{2}\frac{d^2y(a)}{dx^2}\right](x - a)^2 + \cdots \qquad (6.18)$$

The general expression for the Taylor-series expansion if y is a function of n variables $x_1, x_2, \ldots, x_n$ around the point $(x_1 = a_1, x_2 = a_2, \ldots, x_n = a_n)$ is

$$y(x_1, x_2, \ldots, x_n) = y(a_1, a_2, \ldots, a_n)$$

$$+ \sum_{j=1}^{n} \frac{\partial y(a_1, \ldots, a_n)}{\partial x_j}(x_j - a_j)$$

$$+ \frac{1}{2}\sum_{i=1}^{n}\sum_{j=1}^{n} \frac{\partial^2 y(a_1, \ldots, a_n)}{\partial x_i \, \partial x_j}(x_i - a_i)(x_j - a_j) + \cdots \qquad (6.19)$$

Example 6.2. Express the function $\ln (x^2/y)$ as a Taylor-series expansion about the point $(x = 2, y = 1)$.

Solution

$$z = \ln \frac{x^2}{y} = c_o + c_1(x - 2) + c_2(y - 1) + c_3(x - 2)^2$$

$$+ c_4(x - 2)(y - 1) + c_5(y - 1)^2 + \cdots$$

Evaluating the constants, we get

$$c_0 = \ln \frac{2^2}{1} = \ln 4 = 1.39$$

$$c_1 = \frac{\partial z(2, 1)}{\partial x} = \frac{2x/y}{x^2/y} = \frac{2}{x} = 1$$

$$c_2 = \frac{\partial z(2, 1)}{\partial y} = -\frac{x^2/y^2}{x^2/y} = -\frac{1}{y} = -1$$

$$c_3 = \frac{1}{2}\frac{\partial^2 z(2, 1)}{\partial x^2} = \frac{1}{2}\left(-\frac{2}{x^2}\right) = -\frac{1}{4}$$

$$c_4 = \frac{\partial^2 z(2, 1)}{\partial x \, \partial y} = 0$$

$$c_5 = \frac{1}{2}\frac{\partial^2 z(2, 1)}{\partial y^2} = \frac{1}{2}\frac{1}{y^2} = \frac{1}{2}$$

The first several terms of the expansion are

$$z = 1.39 + (x - 2) - (y - 1) - \left(\frac{1}{4}\right)(x - 2)^2 + \left(\frac{1}{2}\right)(y - 1)^2 + \cdots$$

6.10 NEWTON-RAPHSON WITH ONE EQUATION AND ONE UNKNOWN

In the Taylor-series expansion of Eq. (6.18) when x is close to a, the higher-order terms become negligible. The equation then reduces approximately to

$$y \approx y(a) + [y'(a)](x - a) \tag{6.20}$$

Equation (6.20) is the basis of the Newton-Raphson iterative technique for solving a nonlinear algebraic equation. Suppose that the value of x is sought that satisfies the equation

$$x + 2 = e^x \tag{6.21}$$

Define y as

$$y(x) = x + 2 - e^x \tag{6.22}$$

and denote x_c as the correct value of x that solves Eq. (6.21) and makes $y = 0$

$$y(x_c) = 0 \tag{6.23}$$

The Newton-Raphson process requires an initial assumption of the value of x. Denote as x_t this temporary value of x. Substituting x_t into Eq. (6.22) gives a value of y which almost certainly does not provide the desired value of $y = 0$. Specifically, if $x_t = 2$,

$$y(x_t) = x_t + 2 - e^{x_t} = 2 + 2 - 7.39 = -3.39$$

Our trial value of x is incorrect, but now the question is how the value of x should be changed in order to bring y closer to zero.

Returning to the Taylor expansion of Eq. (6.20), express y in terms of x by expanding about x_c

$$y(x) \approx y(x_c) + [y'(x_c)](x - x_c) \tag{6.24}$$

For $x = x_t$, Eq. (6.24) becomes

$$y(x_t) \approx y(x_c) + [y'(x_t)](x_t - x_c) \tag{6.25}$$

Equation (6.25) contains the further approximation of evaluating the derivative at x_t rather than at x_c, because the value of x_c is still unknown. From Eq. (6.23) $y(x_c) = 0$, and so Eq. (6.25) can be solved approximately for the unknown value of x_c

$$x_c \approx x_t - \frac{y(x_t)}{y'(x_t)} \tag{6.26}$$

In the numerical example

$$x_c \approx 2 - \frac{-3.39}{1 - e^2} = 1.469$$

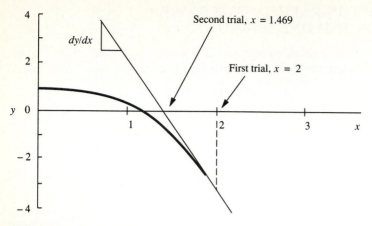

FIGURE 6-10
Newton-Raphson iteration.

The value of $x = 1.469$ is a more correct value and should be used for the next iteration. The results of the next several iterations are

x	1.469	1.208	1.152
$y(x)$	-0.876	-0.132	-0.018

The graphic visualization of the iteration is shown in Fig. 6-10, where we seek the root of the equation $y = x + 2 - e^x$. The first trial is at $x = 2$, and the deviation of y from zero divided by the slope of the curve there suggests a new trial of 1.469.

The Newton-Raphson method, while it is a powerful iteration technique, should be used carefully because if the initial trial is too far off from the correct result, the solution may not converge. Some insight into the nature of the function being solved is therefore always helpful.

6.11 NEWTON-RAPHSON METHOD WITH MULTIPLE EQUATIONS AND UNKNOWNS

The solution of a nonlinear equation for the unknown variable discussed in Sec. 6.10 is only a special case of the solution of a set of multiple nonlinear equations. Suppose that three nonlinear equations are to be solved for the three unknown variables x_1, x_2, and x_3

$$f_1(x_1, x_2, x_3) = 0 \qquad (6.27)$$

$$f_2(x_1, x_2, x_3) = 0 \qquad (6.28)$$

$$f_3(x_1, x_2, x_3) = 0 \qquad (6.29)$$

The procedure for solving the equations is an iterative one in which the following steps are followed:

1. Rewrite the equations so that all terms are on one side of the equality sign [Eqs. (6.27) to (6.29) already exist in this form].
2. Assume temporary values for the variables, denoted $x_{1,t}$, $x_{2,t}$, and $x_{3,t}$.
3. Calculate the values of f_1, f_2, and f_3 at the temporary values of x_1, x_2, and x_3.
4. Compute the partial derivatives of all functions with respect to all variables.
5. Use the Taylor-series expansion of the form of Eq. (6.19) to establish a set of simultaneous equations. The Taylor-series expansion for Eq. (6.27), for example, is

$$f_1(x_{1,t}, x_{2,t}, x_{3,t}) \approx f_1(x_{1,c}, x_{2,c}, x_{3,c})$$

$$+ \frac{\partial f_1(x_{1,t}, x_{2,t}, x_{3,t})}{\partial x_1}(x_{1,t} - x_{1,c})$$

$$+ \frac{\partial f_1(x_{1,t}, x_{2,t}, x_{3,t})}{\partial x_2}(x_{2,t} - x_{2,c})$$

$$+ \frac{\partial f_1(x_{1,t}, x_{2,t}, x_{3,t})}{\partial x_3}(x_{3,t} - x_{3,c}) \qquad (6.30)$$

The set of equations is

$$\begin{bmatrix} \dfrac{\partial f_1}{\partial x_1} & \dfrac{\partial f_1}{\partial x_2} & \dfrac{\partial f_1}{\partial x_3} \\[2mm] \dfrac{\partial f_2}{\partial x_1} & \dfrac{\partial f_2}{\partial x_2} & \dfrac{\partial f_2}{\partial x_3} \\[2mm] \dfrac{\partial f_3}{\partial x_1} & \dfrac{\partial f_3}{\partial x_2} & \dfrac{\partial f_3}{\partial x_3} \end{bmatrix} \begin{bmatrix} x_{1,t} - x_{1,c} \\[2mm] x_{2,t} - x_{2,c} \\[2mm] x_{3,t} - x_{3,c} \end{bmatrix} = \begin{bmatrix} f_1 \\[2mm] f_2 \\[2mm] f_3 \end{bmatrix} \qquad (6.31)$$

6. Solve the set of linear simultaneous equations (6.31) to determine $x_{i,t} - x_{i,c}$.
7. Correct the x's

$$x_{1,\,new} = x_{1,\,old} - (x_{1,t} - x_{1,c})$$
$$\dotfill$$
$$x_{3,\,new} = x_{3,\,old} - (x_{3,t} - x_{3,c})$$

8. Test for convergence. If the absolute magnitudes of all the f's or all the Δx's are satisfactorily small, terminate; otherwise return to step 3.

Example 6.3. Solve Example 6.1 by the Newton-Raphson method.

Solution

   Step 1.

$$f_1 = \Delta p - 7.2w^2 - 392.28 = 0$$

$$f_2 = \Delta p - 810 + 25w_1 + 3.75w_1^2 = 0$$

$$f_3 = \Delta p - 900 + 65w_2 + 30w_2^2 = 0$$

$$f_4 = w - w_1 - w_2 = 0$$

   Step 2. Choose trial values of the variables, which are here selected as $\Delta p = 750$, $w_1 = 3$, $w_2 = 1.5$, and $w = 5$.

   Step 3. Calculate the magnitudes of the f's at the temporary values of the variables, $f_1 = 177.7$, $f_2 = 48.75$, $f_3 = 15.0$, and $f_4 = 0.50$.

   Step 4. The partial derivatives are shown in Table 6.4.

   Step 5. Substituting the temporary values of the variables into the equations for the partial derivatives forms a set of linear simultaneous equations to be solved for the corrections to x:

$$\begin{bmatrix} 1.0 & 0.0 & 0.0 & -72.0 \\ 1.0 & 47.5 & 0.0 & 0.0 \\ 1.0 & 0.0 & 155.0 & 0.0 \\ 0.0 & -1.0 & -1.0 & 1.0 \end{bmatrix} \begin{bmatrix} \Delta x_1 \\ \Delta x_2 \\ \Delta x_3 \\ \Delta x_4 \end{bmatrix} = \begin{bmatrix} 177.7 \\ 48.75 \\ 15.0 \\ 0.50 \end{bmatrix}$$

where $\Delta x_i = x_{i,t} - x_{i,c}$

   Step 6. Solution of the simultaneous equations

$$\Delta x_1 = 98.84 \quad \Delta x_2 = -1.055 \quad \Delta x_3 = -0.541 \quad \Delta x_4 = -1.096$$

   Step 7. The corrected values of the variables are

$$\Delta p = 750.0 - 98.84 = 651.16 \quad w_1 = 4.055 \quad w_2 = 2.041 \quad w = 6.096$$

These values of the variables are returned to step 3 for the next iteration.

   The values of the f's and the variables resulting from continued iterations are shown in Table 6.5.

   The calculations converged satisfactorily after three iterations.

TABLE 6.4

	$\partial/\partial \Delta p$	$\partial/\partial w_1$	$\partial/\partial w_2$	$\partial/\partial w$
$\partial f_1/\partial$	1	0	0	$-14.4w$
$\partial f_2/\partial$	1	$25 + 7.5w_1$	0	0
$\partial f_3/\partial$	1	0	$65 + 60w_2$	0
$\partial f_4/\partial$	0	-1	-1	1

TABLE 6.5

After iteration	f_1	f_2	f_3	f_4	Δp	w_1	w_2	w
1	−8.681	4.170	8.778	0.000	651.16	4.055	2.041	6.096
2	−0.081	0.0148	0.056	0.000	650.48	3.992	1.998	5.989
3	0.000	0.000	0.000	0.000	650.49	3.991	1.997	5.988

6.12 SIMULATION OF A GAS TURBINE SYSTEM

A simulation of a more extensive thermal system will be given for a nonregenerative gas-turbine cycle. This cycle, shown in Fig. 6-11, consists of a compressor, combustor, and turbine whose performance characteristics are known. The turbine-compressor combination operates at 120 r/s.

The objective of the simulation is to determine the power output at the shaft, E_s kW, if 8000 kW of energy is added at the combustor by burning fuel. The turbine draws air and rejects the turbine exhaust to atmospheric pressure of 101 kPa. The entering air temperature is 25°C. Certain simplifications will be introduced in the solution, but it is understood that the simulation method can be extended to more refined calculations. The simplifications are

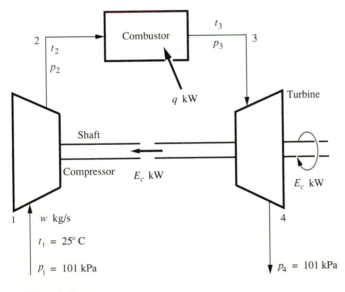

FIGURE 6-11
Gas-turbine cycle.

1. Assume perfect-gas properties throughout the cycle and a c_p constant at 1.03 kJ/(kg · K)
2. Neglect the mass added in the form of fuel in the combustor so that the mass rate of flow w is constant throughout the cycle
3. Neglect the pressure drop in the combustor so that $p_2 = p_3$ and the high pressure in the system can be designated simply as p
4. Neglect heat transfer to the environment

The performance characteristics of the axial-flow compressor and the gas turbine[1] operating at 120 r/s with an atmospheric pressure of 101 kPa that will be used in the simulation are shown in Figs. 6-12 and 6-13, respectively.

With the techniques presented in Chap. 4 equations can be developed for the curves in Fig. 6-12,

$$p = 331 + 45.6w - 4.03w^2 \tag{6.32}$$

and

$$E_c = 1020 - 0.383p + 0.00513p^2 \tag{6.33}$$

where p = discharge pressure of compressor, kPa
w = mass rate of flow, kg/s
E_c = power required by compressor, kW

When operating at a given speed and discharge pressure, the characteristics of the turbine take the form shown in Fig. 6-13. With the techniques

FIGURE 6-12
Performance of axial-flow compressor operating at 120 r/s with 101 kPa inlet pressure.

FIGURE 6-13
Performance of gas turbine operating at 120 r/s and 101 kPa discharge pressure.

of Sec. 4.8 equations can be developed for the curves in Fig. 6-13

$$w = 8.5019 + 0.02332p + 0.48 \times 10^{-4}p^2 - 0.02644t$$
$$+ 0.1849 \times 10^{-4}t^2 + 0.000121pt - 0.2736 \times 10^{-6}p^2t$$
$$- 0.1137 \times 10^{-6}pt^2 + 0.2124 \times 10^{-9}p^2t^2 \qquad (6.34)$$

and

$$E_t = 1727.5 - 10.06p + 0.033033p^2 - 7.4709t + 0.003919t^2$$
$$+ 0.050921pt - 0.8525 \times 10^{-4}p^2t - 0.2356 \times 10^{-4}pt^2$$
$$+ 0.4473 \times 10^{-7}p^2t^2 \qquad (6.35)$$

where t = entering temperature = t_3, °C
E_t = power delivered by turbine, kW

To achieve the simulation the values of the following unknown variables must be determined: w, p, E_c, t_2, E_s, t_3, and E_t. Seven independent equations must be found to solve for this set of unknowns. Four equations are available from the performance characteristics of the compressor and turbine. The three other equations come from energy balances:

Compressor: $\qquad\qquad E_c = wc_p(t_2 - 25) \qquad\qquad (6.36)$

Combustor: $\qquad\qquad 8000 = wc_p(t_3 - t_2) \qquad\qquad (6.37)$

Turbine power: $\qquad\qquad E_t = E_c + E_s \qquad\qquad (6.38)$

TABLE 6.6
Newton-Raphson solution of gas-turbine simulation

	w	p	E_c	t_2	E_s	t_3	E_t
Trial value	10.00	300.0	1000.0	250.0	1500.0	800.0	2400.0
After iteration:							
1	10.91	352.2	1507.8	150.8	1569.3	877.5	3077.0
2	10.77	354.8	1530.0	162.8	1597.7	884.2	3127.7
3	10.77	354.9	1530.1	173.0	1598.5	884.5	3128.6
4	10.77	354.9	1530.1	173.0	1598.5	884.5	3128.6

The execution of the solution follows the steps outlined in Sec. 6.11. A summary of the trial values and results after the Newton-Raphson iterations is presented in Table 6.6.

The shaft power delivered by this system is 1598.5 kW.

6.13   OVERVIEW OF SYSTEM SIMULATION

Steady-state simulation of thermal systems is fast increasing in applicability. The uses of simulation include evaluation of part-load operation directed toward identifying potential operating problems and also predicting annual energy requirements of systems.[2] System simulation can also be one step in an optimization process. For example, the effect on the output of the system of making a small change in one component, e.g., the size of a heat exchanger, is essentially a partial derivative of the type that will be needed in certain of the optimization techniques to be explained in later chapters.

If the exposure to system simulation in this chapter was the reader's first experience with it, wrestling with the techniques may be the major preoccupation. After the methods have been mastered, setting up the equations becomes the major challenge. In large systems it may not be simple to choose the proper combination of equations that precisely specifies the system while avoiding combinations of dependent equations. Unfortunately, no methodical procedure has yet been developed for choosing the equations; a thorough grounding in thermal principles and a bit of intuition are still the necessary tools.

The mathematical description of steady-state system simulation is that of solving a simultaneous set of algebraic equations, some of which are nonlinear. One impulse might be first to eliminate equations and variables by substitution. This strategy is not normally recommended when using a computer to perform the successive substitution or Newton-Raphson solution. Working with the full set of equations provides the solution to a larger number of variables directly, some of which may be of interest. Performing the substitution always presents the hazard of making an algebraic error, and the equations in combined form are more difficult to check than their simpler basic arrangement.

Two methods of system simulation have been presented in this chapter, successive substitution and Newton-Raphson. Successive substitution is a straight-forward technique and is usually easy to program. It uses computer memory sparingly. The disadvantages are that sometimes the sequence may either converge very slowly or diverge. As far as can be determined through the web of commercial secrecy, many of the large simulation programs used in the petroleum, chemical, and thermal processing industries rely heavily on the successive-substitution method.[3,4] The experienced programmer will enhance his chances of a convergent sequence by choosing the blocks in the information-flow diagram in such a way that the output is only moderately affected by large changes in the input. The Newton-Raphson technique, while a bit more complex, is powerful.

This chapter serves as an introduction to system simulation and provides the tools for solving useful engineering problems. Chapter 14 continues the study of steady-state system simulation in greater depth. In that chapter the successive-substitution method is explored further by identifying the nature of calculation sequences that result in convergence. Methods are developed to accelerate convergence or damp divergence. The Newton-Raphson technique as explained in the foregoing chapter required extracting partial derivatives by hand—a process that is tedious. Chapter 14 explains the structure of a generalized program[5] that extracts the partial derivatives numerically. A challenge also addressed by Chapter 14 is the simulation of large systems where the number of equations and unknowns, n, becomes very large. The Newton-Raphson technique requires the solution of an $n \times n$ matrix, so acceleration techniques become valuable.

PROBLEMS

6.1. The operating point of a fan-and-duct system is to be determined. The equations for the two components are

Duct: $\qquad SP = 80 + 10.73Q^{1.8}$

Fan: $\qquad Q = 15 - (73.5 \times 10^{-6})SP^2$

where $\qquad SP =$ static pressure, Pa

$\qquad\qquad\quad Q =$ airflow rate, m^3/s

Use successive substitution to solve for the operating point, choosing as trial values $SP = 200$ Pa or $Q = 10$ m^3/s.

Ans.: 6 m^3/s and 350 Pa.

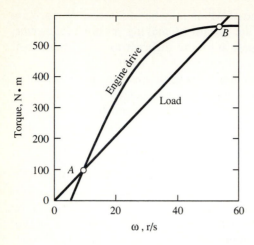

FIGURE 6-14
Torque-rotative-speed curves of engine drive and load on a truck.

6.2. The torque-rotative-speed curve of the engine-and-drive train of a truck operating at a certain transmission setting is shown in Fig. 6-14. The T vs. ω curve for the load on the truck is also shown and is appropriate for the truck moving slowly uphill. The equations for the two curves are

Engine drive: $T = -170 + 29.4\omega - 0.284\omega^2$

Load: $T = 10.5\omega$

where T = torque, $N \cdot m$
$\qquad \omega$ = rotative speed, r/s

(a) Determination of the operating condition of the truck is a simulation of a two-component system. Perform this simulation with both flow diagrams shown in Fig. 6-15. Use an initial value for both simulations of $\omega = 40$ r/s and show the results in the form of Table 6.7.
(b) From a physical standpoint, explain the behavior of the system when operating in the immediate vicinity (on either side) of A.

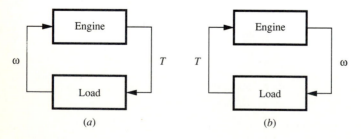

FIGURE 6-15
Flow diagrams in Prob. 6.2.

TABLE 6.7
Results of iterations in Prob. 6.2

Flow diagram Fig. 6-15a			Flow diagram Fig. 6-15b		
Iteration	T	ω	Iteration	T	ω
0	✕	40	0	✕	40
1			1		
2			2		
3			3		

☐ Converging to point _____ ☐ Converging to point _____
☐ Diverging ☐ Diverging

6.3. A seawater desalination plant operates on the cycle shown in Fig. 6-16. Seawater is pressurized, flows through a heat exchanger, where its temperature is elevated by the condensation of what becomes the desalted water, and flows next through a steam heat exchanger, where it is heated but is still in a liquid state at point 3. In passing through the float valve the pressure drops and some of the liquid flashes into vapor, which is the vapor that condenses as fresh water. The portion at point 4 that remains liquid flows out as waste at point 6. The following conditions and relationships are known:

> Temperature and flow rate of entering seawater.
> UA_1 and UA_2 of the heat exchangers.
> Enthalpies of saturated liquid and saturated vapor of seawater and the fresh water as functions of temperature:

$$h_f = f_1(t) \qquad \text{and} \qquad h_g = f_2(t)$$

> For heat exchangers with one fluid condensing, use Eq. (5.10).
> The system operates so that essentially $t_4 = t_5 = t_6$.

FIGURE 6-16
Desalination plant in Prob. 6.3.

Set up an information-flow diagram that would be used for a successive-substitution system simulation, indicating which equations apply to each block. For convenience in checking, use these variables: t_2, t_3, h_3, h_4, $h_{f,4}$, $h_{g,4}$, t_4, w_5, x_4, and q where x_4 is the fraction of vapor at point 4 and q is the rate of heat transfer at the fresh-water condenser.

6.4. On some high-speed aircraft an air-cycle refrigeration unit is used for cabin cooling, one concept of which is shown in Fig. 6-17. Equations available for the compressor are: power $= f_1(t_1, p_1, p_2, \dot{m})$ and $\dot{m} = f_2(t_1, p_1, \omega, p_2)$. For the turbine the equations are: power $= f_3(t_3, p_3, \dot{m}, p_4)$ and $\dot{m} = f_4(t_3, p_3, \omega, p_4)$ where $f(\ )$ indicates a function or equation in terms of the variables in the parentheses. Assume no pressure drop through the heat exchanger. The compressor-turbine combination operates when p_1 is greater than p_4. The following data are imposed and known: p_1, t_1, p_4, the UA of the heat exchanger, and the temperature and flow rate of ambient air through the heat exchanger. Construct an information-flow diagram to simulate the system using the equations and variables previously listed as well as others that are necessary.

6.5. Steam boilers sometimes use a continuous blowdown of water to control the amount of impurities in the water. This high-temperature water is capable of heating feedwater, as shown in Fig. 6-18. A flow rate of 0.2 kg/s at a temperature of 340°C is blown down from the boiler. The flow rate of the feed water to the heater is 3 kg/s and its entering temperature is 80°C. The UA value of the feed-water heater is 10 kW/K. Equations for the enthalpy of saturated liquid and vapor are, respectively, $h_f = 4.19t$ and $h_g = 2530 + 0.4t$, where t is the temperature in °C. The system is to be simulated and the following variables computed: t_2, t_o, w_A and w_B.
(a) Construct an information-flow diagram.
(b) Using successive substitution, compute the values of the variables.
    Ans.: $t_2 = 108.2°C$.

6.6. In a synthetic-ammonia plant (Fig. 6-19) a 1:3 mixture, on a molar basis, of N_2 and H_2 along with an impurity, argon, passes through a reactor

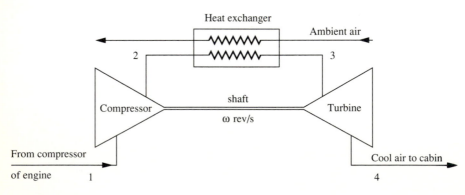

FIGURE 6-17
Air-cycle refrigeration unit in Prob. 6.4.

FIGURE 6-18
Blowdown from a boiler in Prob. 6.5.

where some of the nitrogen and hydrogen combine to form ammonia. The ammonia product formed leaves the system at the condenser and the remaining H_2, N_2, and Ar recycle to the reactor.

The presence of the inert gas argon is detrimental to the reaction. If no argon is present, the reactor converts 60 percent of the incoming N_2 and H_2 into ammonia, but as the flow rate of argon through the reactor increases, the

FIGURE 6-19
A synthetic-ammonia plant.

percent conversion decreases. The conversion efficiency follows the equation

$$\text{Conversion, \%} = 60e^{-0.016w}$$

where w is the flow rate of argon through the reactor in moles per second.

To prevent the reaction from coming to a standstill, a continuous bleed of 25 mol/s of mixture of N_2, H_2, and Ar is provided. If the incoming feed consists of 100 mol/s of N_2, 300 mol/s of H_2, and 1 mol/s of Ar, simulate this system by successive substitution to determine the flow rate of mixture through the reactor and the rate of liquid ammonia production in moles per second.

Ans.: 893.2 and 188 mol/s.

6.7. For $x(\tan x) = 2.0$, where x is in radians, use the Newton-Raphson method to determine the value of x.

Ans.: 1.0769.

6.8. The heat exchanger in Fig. 6-20 heats water entering at 30°C with steam entering as saturated vapor at 50°C and leaving as condensate at 50°C. The flow of water is to be chosen so that the heat exchanger transfers 50 kW. The area of the heat exchanger is 1.4 m², and the U value of the heat exchanger based on this area is given by

$$\frac{1}{U}(\text{m}^2 \cdot \text{K})/\text{kW} = \frac{0.0445}{w^{0.8}} + 0.185$$

where w is the flow rate of water in kilograms per second. Use the Newton-Raphson method for one equation and one unknown to determine the value of w that results in the transfer of 50 kW.

Ans.: 0.6934 kg/s.

6.9. Solve Prob. 6.1 using the multiple-equation Newton-Raphson method with trial values of SP $= 200$ Pa and $Q = 10$ m³/s.

6.10. An oil pipeline has ten pumping stations, each station having the pressure-flow characteristic of

$$\Delta p = 2100 - 20w - 0.5w^2$$

where $\Delta p =$ pressure rise in the station, kPa
$\quad\quad\quad w =$ flow rate of oil, kg/s

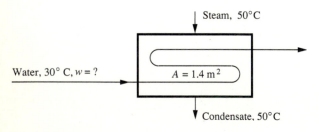

FIGURE 6-20
Heat exchanger in Prob. 6.8.

In normal operation the flow rate is 25 kg/s. The pressure drop in the pipe is proportional to the square of the flow rate. If one pumping station fails and that station bypassed, what will be the flow rate provided by the remaining nine stations?

6.11. In a closed loop a centrifugal and gear pump operate in series to deliver fluid through a long pipe. The equations relating Δp and the flow rate for the three components are:

centrifugal pump: $\qquad \Delta p = 6 + 2Q - 0.5Q^2$

gear pump: $\qquad\qquad \Delta p = 40 - 5Q$

pipe: $\qquad\qquad\qquad \Delta p = 0.1Q^2$

where Δp = pressure rise (or drop in the pipe), kPa
$\qquad\quad Q$ = flow rate, m³/s

(a) Plot on a $\Delta p - Q$ graph the performance of all components.
(b) If a system simulation were performed (no need to perform this simulation), what would be the approximate solution? Discuss the physical implications of the solution.

6.12. In some cryogenic liquefaction systems the temperature of a stream of liquid is reduced by flashing off some of the liquid into vapor through a throttling valve and heat exchanger as shown in Fig. 6-21. With the values shown in Fig. 6-21, use a Newton-Raphson simulation to determine the flow rate of liquid leaving the heat exchanger and its temperature.
   Ans.: outlet temperature of liquid is 157.3 K.

6.13. A centrifugal pump operates with a bypass as shown in Fig. 6-22. The pressure drop through the bypass line is given by the equation

$$\Delta p = 1.2(w_{bp})^2$$

the characteristics of the pump are expressed by the equation

$$\Delta p = 50 + 5w_p - 0.1w_p^2$$

FIGURE 6-21
Cryogenic liquid cooler in Prob. 6.12.

FIGURE 6-22
Centrifugal pump with bypass line
in Prob. 6.13.

and the equation for the pressure drop in the piping system is

$$p_2 - p_1 = 0.018w^2$$

where the flow rates are in kg/s and the pressures in kPa. Use the Newton-Raphson technique to determine w, w_p, w_{bp} and p_2.

6.14. Air at 28°C with a flow rate of 4 kg/s flows through a cooling coil counterflow to cold water that enters at 6°C, as shown in Fig. 6-23. Air has a specific heat, $c_p = 1.0$ kJ/(kg · K). No dehumidification of the air occurs as it passes through the coil. The product of the area and heat-transfer coefficient for the heat exchanger is 7 kW/K. The pump just overcomes the pressure drop through the control valve and coil, such that $p_1 = p_4$. The pressure-flow characteristics of the pump are

$$p_2 - p_1, \text{Pa} = 120,000 - 15,400w_w^2$$

where w_w is the flow rate of water in kilograms per second. The specific heat of the water is $c_p = 4.19$ kJ/(kg · K). The pressure drop through the coil is $p_3 - p_4 = 9260w_w^2$. The outlet-air temperature regulates the control

FIGURE 6-23
Cooling coil in Prob. 6.14.

valve to maintain an outlet-air temperature somewhere between 10 and 12°C. The flow-pressure-drop relation for the valve is $w_w = C_v \sqrt{p_2 - p_3}$, where C_v is a function of the degree of valve opening, a linear relation, as shown in Fig. 6-24. The fully open value of C_v is 0.012.

    Use the Newton-Raphson method to simulate this system, determining at least the following variables: w_w, t_4, $t_{\text{air out}}$, p_2, p_3, and C_v. Use as the test for convergence that the absolute values of pressures change less than 1.0, absolute values of temperatures less than 0.001, and absolute value of C_v less than 0.000001. Limit the number of iterations to 10.

    Ans.: $p_2 = 64,355$ (based on $p_1 = 0$), $t_4 = 14.14$, $C_v = 0.0108$.

6.15. A two-stage air compressor with intercooler shown in Fig. 6-25 compresses air (which is assumed dry) from 100 to 1200 kPa absolute. The following data apply to the components:

$$\text{Displacement rate } = \begin{cases} 0.2 \text{ m}^3/\text{s low-stage compressor} \\ 0.05 \text{ m}^3/\text{s high-stage compressor} \end{cases}$$

Volumetric efficiency η, % $=$

$$\frac{\text{flow rate measured at compressor suction, m}^3/\text{s}}{\text{displacement rate, m}^3/\text{s}} (100)$$

and for both compressors

$$\eta, \% = 104 - 4.0 \left(\frac{p_{\text{disch}}}{p_{\text{suction}}} \right)^{1.4}$$

The polytropic exponent n in the equation $p_1 v_1^n = p_2 v_2^n$ is 1.2. The intercooler is a counterflow heat exchanger receiving 0.09 kg/s of water at 22°C. The

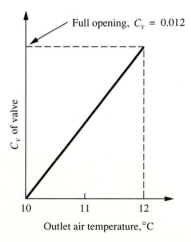

Full opening, $C_v = 0.012$

C_v of valve

10   11   12

Outlet air temperature, °C

FIGURE 6-24
Characteristics of valve in Prob. 6.14.

FIGURE 6-25
Two-stage air compression in Prob. 6.15.

product of the overall heat-transfer coefficient and the area of this heat exchanger is $UA = 0.3$ kW/K. Assume that the air is a perfect gas.

Use the Newton-Raphson method to simulate this system, determining at least the values of w, p_i, t_2 and t_3. Use as a test for convergence that all variables change less than 0.001 during an iteration. Limit the number of iterations to 10.

Ans.: $w = 0.18$ kg/s, $t_2 = 101.8°C$, $p_i = 387.7$ kPa, $t_3 = 43.84°C$.

6.16. A helium liquefier operating according to the flow diagram shown in Fig. 6-26 receives high-pressure helium vapor, liquefies a fraction of the vapor, and returns the remainder to be recycled. The following operating conditions prevail:

Point 1 (vapor entering warm side of heat exchanger), $T = 15$ K, $h = 78.3$ kJ/kg, $w = 5$ g/s, $p = 2000$ kPa
Point 5 (vapor leaving turbine), $T = 8$ K, $h = 53$ kJ/kg, $w = 4$ g/s

FIGURE 6-26
Helium liquefier in Prob. 6.16.

Separator, $p = 100$ kPa, saturation temperature at 100 kPa $= 4.2$ K, $h_f = 10$ kJ/kg, $h_g = 31$ kJ/kg

Heat exchanger, $UA = 100$ W/K

Specific heat of helium vapor:

$$c_p = \begin{cases} 6.4 \text{ kJ/(kg} \cdot \text{K)} & \text{at 2000 kPa} \\ 5.8 \text{ kJ/(kg} \cdot \text{K)} & \text{at 100 kPa} \end{cases}$$

Using the Newton-Raphson method, simulate this system, determining the values of w_4, T_2, T_7, and T_8. Use as the test for convergence that all variables change less than 0.001 during an iteration. Limit the number of iterations to 10.

Ans.: $w_4 = 0.447$ g/s, $T_2 = 7.316$, $T_7 = 5.97$, and $T_8 = 10.93$ K.

6.17. A refrigeration plant that operates on the cycle shown in Fig. 6-27 serves as a water chiller. Data on the individual components are as follows:

$$UA = \begin{cases} 30,600 \text{ W/K} & \text{evaporator} \\ 26,500 \text{ W/K} & \text{condenser} \end{cases}$$

$$\text{Rate of water flow} = \begin{cases} 6.8 \text{ kg/s} & \text{evaporator} \\ 7.6 \text{ kg/s} & \text{condenser} \end{cases}$$

The refrigeration capacity of the compressor as a function of the evaporating and condensing temperatures t_e and t_c, respectively, is given by the equation developed in Prob. 4.9.

$$q_e, \text{ kW} = 239.5 + 10.073t_e - 0.109t_e^2 - 3.41t_c - 0.00250t_c^2$$

$$-0.2030t_e t_c + 0.00820t_e^2 t_c + 0.0013t_e t_c^2 - 0.000080005t_e^2 t_c^2$$

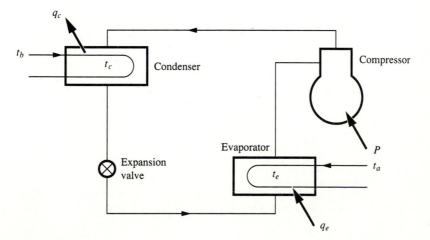

FIGURE 6-27
Refrigeration plant in Prob. 6.17.

The compression power is expressed by the equation

$$P, \text{kW} = -2.634 - 0.3081t_e - 0.00301t_e^2 + 1.066t_c - 0.00528t_c^2$$
$$-0.0011t_et_c - 0.000306t_e^2t_c + 0.000567t_et_c^2 + 0.0000031t_e^2t_c^2$$

The condenser must reject the energy added in both the evaporator and the compressor. Determine the values of t_e, t_c, q_e, and P for the following combinations of inlet water temperatures:

Evaporator inlet water t_a, °C	10	15	10	15
Condenser inlet water t_b, °C	25	25	30	30

Continue iterations until all variables change by an absolute value less than 0.1 percent during an iteration.

   Ans.: For $t_a = 10°C$ and $t_b = 25°C$, $t_e = 2.84°C$, $t_c = 34.05°C$, $q_e = 134.39$ kW and $p = 28.34$ kW.

REFERENCES

1. G. M. Dusinberre and J. C. Lester, *Gas Turbine Power,* International Textbook, Scranton, Pa., 1958.
2. *Procedures for Simulating the Performance of Components and Systems for Energy Calculations,* American Society of Heating, Refrigerating, and Air-Conditioning Engineers, New York, 1975.
3. H. A. Mosler, "PACER—A Digital Computer Executive Routine for Process Simulation and Design," M.S. thesis, Purdue University, Lafayette, Ind., January 1964.
4. C. M. Crowe, A. E. Hamielec, T. W. Hoffman, A. I. Johnson, D. R. Woods, and P. T. Shannon, *Chemical Plant Simulation; an Introduction to Computer-Aided Steady-State Process Analysis,* Prentice-Hall, Englewood Cliffs, N. J., 1971.
5. W. F. Stoecker, "A Generalized Program for Steady-State System Simulation," *ASHRAE Trans.,* vol. 77, pt. I, pp. 140–148, 1971.

ADDITIONAL READINGS

Chen, C-C., and L. B. Evans: "More Computer Programs for Chemical Engineers," *Chem. Eng.,* vol. 86, no. 11, pp. 167–173, May 21, 1979.
Henley, E. J., and E. M. Rosen: *Material and Energy Balance Computations,* Wiley, New York, 1969.
Naphtali, L. M., "Process Heat and Material Balances," *Chem. Eng. Prog.,* vol. 60, no. 9, pp. 70–74, September 1964.
Peterson, J. N., C. Chen, and L. B. Evans: "Computer Programs for Chemical Engineers: 1978," pt. 1, *Chem. Eng.,* vol. 85, no. 13, pp. 145–154, June 5, 1978.

CHAPTER
7

OPTIMIZATION

7.1 INTRODUCTION

Optimization is the process of finding the conditions that give maximum or minimum values of a function. Optimization has always been an expected role of engineers, although sometimes on small projects the cost of engineering time may not justify an optimization effort. Often a design is difficult to optimize because of its complexity. In such cases, it may be possible to optimize subsystems and then choose the optimum combination of them. There is no assurance, however, that this procedure will lead to the true optimum.

Chapter 1 pointed out that in designing a workable system the process often consists of arbitrarily assuming certain parameters and selecting individual components around these assumptions. In contrast, when optimization is an integral part of the design, the parameters are free to float until the combination of parameters is reached which optimizes the design.

Basic to any optimization process is the decision regarding which criterion is to be optimized. In an aircraft or space vehicle, minimum weight may be the criterion. In an automobile, the size of a system may be the criterion. Minimum cost is probably the most common criterion. On the other hand, the minimum owning and operating cost, even including such factors as those studied in Chapter 3 on economics, may not always be followed strictly. A manufacturer of domestic refrigerators, for example, does not try to design his system to provide minimum total cost to the consumer during the life of the equipment. The achievement of minimum first cost, which enhances sales, is more important than operating cost, although the operating cost cannot be completely out of bounds. Industrial organizations often turn aside from the most economical solution by introducing human, social, and aesthetic concerns. What is happening is that their criterion function includes not only monetary factors but also some other factors that may admittedly be only vaguely defined.

Optimization activities are often practiced under the name of *operations research*. Many developments in operations research emerged from attempts to optimize mathematical models of economic systems. It is only recently that mechanical and chemical engineers have used certain of the disciplines to optimize fluid- and energy-flow systems.

Component simulation and system simulation are often preliminary steps to optimizing thermal systems, since it may be necessary to simulate the performance over a wide range of operating conditions. A system that may be optimum for design loads may not be optimum over the entire range of its expected operation.

7.2 LEVELS OF OPTIMIZATION

Sometimes a design engineer will say: I have optimized the design by examining four alternate concepts to do the job, which probably means that the engineer has compared *workable* systems of four different concepts. The statement does emphasize the two levels of optimization, comparison of alternate concepts and optimization within a concept. All the optimization methods presented in the following chapters are optimizations within a concept. The flow diagram and mathematical representation of the system must be available at the beginning, and the optimization process consists of a give and take of sizes of individual components. All this optimization is done within a given concept. There is nothing in the upcoming procedures that will jump from one model to a better one. No optimization procedure will automatically shift the system under consideration from a steam-electric generating plant to a fuel-cell concept, for example.

A complete optimization procedure, then, consists of proposing all reasonable alternate concepts, optimizing the design of each concept, and then choosing the best of the optimized designs.

7.3 MATHEMATICAL REPRESENTATION OF OPTIMIZATION PROBLEMS

The elements of the mathematical statement of optimization include specification of the function and the constraints. Let y represent the function that is to be optimized, called the *objective function*; y is a function of

$x_1, x_2, \ldots, x_n$, which are called the *independent variables*. The objective function, then, is

$$y = y(x_1, x_2, \ldots, x_n) \rightarrow \text{optimize} \qquad (7.1)$$

In many physical situations there are constraints, some of which may be equality constraints

$$\phi_1 = \phi_1(x_1, x_2, \ldots, x_n) = 0 \qquad (7.2)$$
$$\cdots \cdots \cdots \cdots \cdots \cdots$$
$$\phi_m = \phi_m(x_1, x_2, \ldots, x_n) = 0 \qquad (7.3)$$

as well as inequality constraints

$$\Psi_1 = \Psi_1(x_1, x_2, \ldots, x_n) \le L_1 \qquad (7.4)$$
$$\cdots \cdots \cdots \cdots \cdots \cdots$$
$$\Psi_j = \Psi_j(x_1, x_2, \ldots, x_n) \le L_j \qquad (7.5)$$

The physical conditions dictate the sense of the inequalities in Eqs. (7.4) to (7.5).

An additive constant appearing in the objective function does not affect the values of the independent variables at which the optimum occurs. Thus, if

$$y = a + Y(x_1, \ldots, x_n)$$

where a is a constant, the minimum of y can be written

$$\min \, [a + Y(x_1, \ldots, x_n)] = a + \min \, [Y(x_1, \ldots, x_n)] \qquad (7.6)$$

A further property of the optimum is that the maximum of a function occurs at the same state point at which the minimum of the negative of the function occurs, thus

$$\max \, [y(x_1, \ldots, x_n)] = [-y(x_1, \ldots, x_n)] \qquad (7.7)$$

7.4   A WATER-CHILLING SYSTEM

A water-chilling system, shown schematically in Fig. 7-1, will be used to illustrate the mathematical statement. The requirement of the system is that it cool 20 kg/s of water from 13 to 8°C, rejecting the heat to the atmosphere through a cooling tower. We seek a system with a minimum first cost to perform this duty.

Designate the sizes of the components in the system by x_{CP}, x_{EV}, x_{CD}, x_P, and x_{CT}, which represent the sizes of the compressor, evaporator, condenser, pump, and cooling tower, respectively. The total cost y is the sum of the individual first plus installation costs, and this is the quantity that we wish to minimize.

$$y(x_{CP}, x_{EV}, x_{CD}, x_P, x_{CT}) \rightarrow \text{minimize} \qquad (7.8)$$

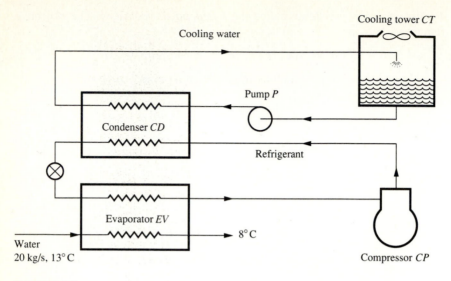

FIGURE 7-1
Water-chilling unit being optimized for minimum first cost.

With only the statement of Eq. (7.8), the minimum could be achieved by shrinking the sizes of all components to zero. Overlooked is the requirement that the combination of sizes be such that the water-chilling assignment of providing

$$(20 \text{ kg/s})(13 - 8°\text{C})[4.19 \text{ kJ/(kg} \cdot \text{K)}] = 419 \text{ kW}$$

of refrigeration is accomplished. Equation (7.9) expresses this constraint

$$\phi(x_{CP}, x_{EV}, x_{CD}, x_P, x_{CT}) = 419 \text{ kW} \tag{7.9}$$

where ϕ is understood to mean the cooling capacity as a function of component sizes when 20 kg/s of water enters at 13°C. Actually, Eq. (7.9) could be an inequality constraint, because probably no one would object to a larger capacity than the requirement of 419 kW if the cost were not increased.

Some practical considerations impose certain inequality constraints. The system should be designed so that the evaporating temperature t_{ev} is above 0°C or, at the lowest, -2 to -1°C to prevent water from freezing on the tube surfaces. This constraint is

$$t_{ev}(x_{CP}, x_{EV}, x_{CD}, x_P, x_{CT}) \geq 0°\text{C} \tag{7.10}$$

An extremely high discharge temperature t_d of the refrigerant leaving the compressor may impair the lubrication

$$t_d(x_{CP}, x_{EV}, x_{CD}, x_P, x_{CT}) \leq 110°\text{C} \tag{7.11}$$

There may be other inequality constraints, such as limiting the condenser cooling-water flow in relation to the size of the cooling tower to prevent it from splashing out.

The elements of the optimization problem are all present here, the objective function, equality constraints, and inequality constraints, all in terms of the independent variables, which are the sizes of components.

7.5 OPTIMIZATION PROCEDURES

In the next few sections several optimization methods will be listed. Although this list includes most of the frequently used methods in engineering practice, it is nowhere near exhaustive. In the optimization of systems, it is almost axiomatic that the objective function is dependent upon more than one variable. In fact, some thermal systems may have dozens or even hundreds of variables which demand sophisticated optimization techniques. While considerable effort may be required in the optimization process, developing mathematical relationships for the function to be optimized and the constraints may also require considerable effort.

7.6 CALCULUS METHODS: LAGRANGE MULTIPLIERS

The basis of optimization by calculus, presented in Chapter 8 and continued in Chapter 16, is to use derivatives to indicate the optimum. The method of Lagrange multipliers performs an optimization where equality constraints exist but the method cannot directly accommodate inequality constraints. A necessary requirement for using calculus methods is the ability to extract derivatives of the objective function and constraints.

7.7 SEARCH METHODS

These methods, covered in Chapters 9 and 17, involve examining a number of combinations of values of the independent variables and drawing conclusions from the magnitude of the objective function at these combinations. An obvious possibility is to calculate the value of the function at all possible combinations of, for example, 20 values distributed through the range of interest of one parameter, each in combination with 20 of the second parameter, and so on. Such a search method is not very imaginative and is also inefficient. The search methods of interest are those which are efficient, particularly when applied to multivariable optimization.

When applying search methods to continuous functions, since only discrete points are examined, the exact optimum can only be approached, not reached, by a finite number of trials. On the other hand, when optimizing systems where the components are available only in finite steps of sizes, search methods are often superior to calculus methods, which assume an infinite gradation of sizes.

7.8   DYNAMIC PROGRAMMING

The word "programming" here and in the next several sections means optimization and has no direct relationship with computer programming, for example. This method of optimization, introduced in Chapter 10 and amplified in Chapter 18, is unique in that the result is an optimum function rather than an optimum state point. The result of the optimization of all the other methods mentioned here is a set of values of the independent variables x_1 to x_n that result in the optimal value of the objective function y. The problem attacked by dynamic programming is one where the desired result is a *path*, e.g., the best route of a gas pipeline. The result is therefore a function relating several variables. Dynamic programming is related to the calculus of variations, and it does in a series of discrete processes what the calculus of variations does continuously.

7.9   GEOMETRIC PROGRAMMING

Probably the youngest member of the programming family is geometric programming, discussed in Chapter 11. Geometric programming optimizes a function that consists of a sum of polynomials wherein the variables may appear raised to integer and noninteger exponents. When the usefulness of polynomial expressions in Chapter 4 is recalled, it is clear that the form of the function to which geometric programming is applicable is one that occurs frequently in thermal systems.

7.10   LINEAR PROGRAMMING

Chapter 12 presents an introduction to linear programming, which is a widely used and well-developed discipline applicable when all the equations (7.1) to (7.5) are linear. The magnitude of problems now being solved by linear programming is enormous, occasionally extending into optimizations which contain several thousand variables.

7.11   SETTING UP THE MATHEMATICAL STATEMENT OF THE OPTIMIZATION PROBLEM

One of the first steps in performing an optimization is to translate the physical situation into a mathematical statement. The desired form is that comparable to Eqs. (7.1) to (7.5) because the optimization techniques can pick up the problem when the objective function and the constraints are specified. In the optimization of thermal systems, establishing the objective function is often simple and sometimes even a trivial task. The challenge usually arises in writing the constraints. A strategy that is often successful is to (1) specify all the direct constraints, such as requirements of capacity, limitations of temperature and pressure, etc., (2) describe in equation form the component characteristics and properties of working substances, and (3) write mass and energy balances. Operations (1) and (3) usually provide a set of equations containing more variables than exist in the objective function. The set of constraint equations is then reduced in number by eliminating variables that do not exist in the objective function. It should be pointed out, however, that some optimization techniques permit additional variables in the constraints that do not exist in the objective function. Steps (2) and (3) in setting up the constraints are similar to the process used in structuring a system-simulation problem.

Example 7.1. Between two stages of air compression, the air is to be cooled from 95 to 10°C. The facility to perform this cooling, shown in Fig. 7-2, first cools the air in a precooler and then in a refrigeration unit. Water passes through the condenser of the refrigeration unit, then into the precooler, and finally to a cooling tower, where heat is rejected to the atmosphere.

The flow rate of compressed air is 1.2 kg/s, and the specific heat is 1.0 kJ/(kg · K). The flow rate of water is 2.3 kg/s, and its specific heat is 4.19 kJ/(kg · K). The water leaves the cooling tower at a temperature 24°C. The system is to be designed for minimum first cost, where this first cost comprises the cost of the refrigeration unit, precooler, and cooling tower, designated x_1, x_2, and x_3, respectively, in dollars. The expressions for these costs are

Refrigeration unit: $$x_1 = 48q_1 \qquad\qquad (7.12)$$

FIGURE 7-2
Air-cooling system in Example 7.1.

Precooler:
$$x_2 = \frac{50q_2}{t_3 - t_1}$$
(7.13)

where the equation is applicable when $t_3 > t_1$

Cooling tower:
$$x_3 = 25q_3$$
(7.14)

where the q's are rates of heat transfer in kilowatts, as designated in Fig. 7-2. The compression power P kW required by the refrigeration unit in $0.25q_1$, and both q_1 and the compression power must be absorbed by the condenser cooling water passing through the refrigeration unit.

Develop (a) the objective function and (b) the constraint equations for an optimization to provide minimum first cost.

Solution. The goal of this example is only to set up the optimization problem in the form of Eqs. (7.1) to (7.5) and not to perform the actual optimization. Before proceeding, however, it would be instructive to examine qualitatively the optimization features of this system. Since the precooler is a simple heat exchanger, under most operating conditions it is less costly for a given heat-transfer rate than the refrigeration unit. It would appear preferable, then, to do as much cooling of the air as possible with the precooler. However, as the temperature t_3 approaches the value of t_1, the size of the precooler becomes very large. Some capacity is required of the refrigeration unit in order to cool the air below 24°C. The cooling tower must reject all the heat from the system, which includes the heat from the air as well as the compression power to drive the refrigeration unit. Shifting more cooling load to the refrigeration unit increases the size and cost of the cooling tower moderately.

(a) The first assignment is to develop the expression for the objective function. Since the total first cost is to be minimized, the objective function will be the first cost in terms of the variables of optimization. A choice must be made of these variables; the objective function could conceivably be written in terms of the costs of the individual components (the x's), the energy flow rates (the q's), or even the temperatures (t_1, t_2, and t_3). The most straightforward choice is to use the component costs

$$\text{Total cost} = y = x_1 + x_2 + x_3$$
(7.15)

If the q's are chosen as the variables of optimization, it is necessary to start with Eq. (7.15) and express the x's in terms of the q's.

(b) The next task is to write the constraints, and this means developing the set of equations in terms of the variables used in the objective function. Establishing the objective function is usually a simple process; the major challenge is setting up the constraints. The advice in the early part of this section was to specify the direct constraints, the component characteristics, and finally the energy and mass balances. This expanded set of equations is then reduced by eliminating the variables that do not appear in the objective function.

A direct constraint is the requirement that the airflow rate of 1.2 kg/s be cooled from 95 to 10°C. This requirement can be expressed in two equations

$$q_1 = (1.2 \text{ kg/s})[1.0 \text{ kJ/(kg} \cdot \text{K)}](t_3 - 10)$$
(7.16)

$$q_2 = (1.2 \text{ kg/s})[1.0 \text{ kJ/(kg} \cdot \text{K)}](95 - t_3)$$
(7.17)

Under the heading of component characteristics fall the expression for the compression power equaling $0.25q_1$ and the relationships of the sizes (costs) to the capacity, Eqs. (7.12) to (7.14).

The final category includes energy and mass balances

Refrigeration unit: $q_1 + P = (2.3 \text{ kg/s})[4.19 \text{ kJ/(kg} \cdot \text{K)}](t_1 - 24)$

Precooler: $(1.2)(1.0)(95 - t_3) = (2.3)(4.19)(t_2 - t_1)$

Cooling tower: $(2.3)(4.19)(t_2 - 24) = q_3$

The complete set of constraint equations is

$$q_1 = (1.2)(1.0)(t_3 - 10) \tag{7.18}$$

$$q_2 = (1.2)(1.0)(95 - t_3) \tag{7.19}$$

$$P = 0.25q_1 \tag{7.20}$$

$$x_1 = 48q_1 \tag{7.21}$$

$$x_2 = \frac{50q_2}{t_3 - t_1} \tag{7.22}$$

$$x_3 = 25q_3 \tag{7.23}$$

$$q_1 + P = (2.3)(4.19)(t_1 - 24) \tag{7.24}$$

$$(1.2)(1.0)(95 - t_3) = (2.3)(4.19)(t_2 - t_1) \tag{7.25}$$

$$(2.3)(4.19)(t_2 - 24) = q_3 \tag{7.26}$$

There are nine equations in the set, Eqs. (7.18) to (7.26) and ten unknowns, q_1, q_2, q_3, P, x_1, x_2, x_3, t_1, t_2, and t_3. The next operation is to eliminate in this set of equations all but the variables of optimization, $x_1, x_2,$ and x_3. As the elimination of variables and equations proceeds, there will always be one more unknown than the number of equations, so when all but the three x's are eliminated, there should be two equations remaining. These two constraint equations are Eqs. (7.28) and (7.29), so the complete mathematical statement of this optimization problem is as follows:

Minimize $$y = x_1 + x_2 + x_3 \tag{7.27}$$

subject to $$0.01466x_1x_2 - 14x_2 + 1.042x_1 = 5100 \tag{7.28}$$

$$7.69x_3 - x_1 = 19,615 \tag{7.29}$$

7.12   DISCUSSION OF EXAMPLE 7.1

The objective of this chapter is to introduce procedures for setting up the mathematical statement of the optimization problem. With one of the methods in the subsequent chapters the execution of the optimization process would show that the optimal values of x_1, x_2, and x_3 are $1450, $496, and $2738, respectively. Equation (7.13) [or Eq. (7.22)] was presented as valid if $t_3 > t_1$, and this condition could legitimately be listed as one of the constraints. If $t_3 < t_1$, x_2 becomes negative, which is physically impossible.

The constraints are an integral part of the statement of the optimization problem. The objective function without the constraints is meaningless because the x's could all shrink to zero and there would be no cost for the system. The constraint in Eq. (7.28) requires a positive value of x_1 which is the same as requiring the existence of a refrigeration unit. From heat-transfer considerations the precooler can cool the air no lower than a temperature of 24°C. Substituting $x_1 = 0$ in Eq. (7.28) makes x_2 negative, which is physically impossible. Equation (7.28) does permit x_2 to be zero, in which case all cooling is performed by the refrigeration unit.

The constraint equation (7.29) imposes a minimum value of the cooling-tower size and cost x_3. As the size of the refrigeration unit and x_1 increases, x_3 also increases because of the compression power associated with the refrigeration unit.

7.13   SUMMARY

While it is true that engineers have always sought to optimize their designs, it has been only since the widespread application of the digital computer that sophisticated methods of optimization have become practical for complex systems. The application of optimization techniques to large-scale thermal systems is still in its infancy, but experience so far indicates that setting up the problem to the point where an optimization method can take over represents perhaps 70 percent of the total effort. The emphasis on optimization techniques in the next five chapters may suggest that engineers are home free once they know several methods. Realistically, however, the execution of the optimization can only begin when the characteristics of the physical system have been converted into the equations for the objective function and constraints.

PROBLEMS

7.1. A pair of pumps is available to fill a tank from a constant-level reservoir, as shown in Fig. 7-3. The pumps may be operated individually or together, and the objective is to fill the tank using a minimum total amount of energy. The pump characteristics (head-flow rate) are shown on the graph, and both pumps have the same efficiency as a function of flow rate. State clearly the mode of operation of the pumps over the enter filling process that results in minimum total energy.

7.2. Two heat exchangers in a circulating water loop, as shown in Fig. 7-4, transfer heat from a fluid condensing at 80°C to a fluid boiling at 20°C. The required rate of heat transfer is 65 kW, the U value of both heat exchangers is 0.03 kW/(m$^2 \cdot$ K), the first cost of the heat exchangers is $80/m^2 of heat-transfer area, and the present worth of the lifetime pumping cost in dollars is 12,000w. Develop the objective function for the total present cost of the system along with any constraint equation(s).

$$\text{Power} = \frac{(\text{Head})(Q)}{\text{Efficiency}/100}$$

FIGURE 7-3
Combination of pumps in Prob. 7.1.

FIGURE 7-4
Heat-exchange loop in Prob. 7.2.

FIGURE 7-5
Carbon dioxide extraction in Prob. 7.3.

7.3. A plant produces CO_2 by extracting it from a mixture of CO_2 and air, as shown in Fig. 7-5. The feed rate of the CO_2-air mixture is 5×10^9 m^3/year. The system consists of two extractors—Extractor I of low cost and low efficiency, and Extractor II of high cost and high efficiency. The CO_2 in m^3/yr removed by Extractor I is (total m^3/yr entering)$(x)(1 - e^{-A/2000})$, and that removed by Extractor II is (total m^3/yr entering)$(x)(1 - e^{B/1000})$, where the x's are the fraction of CO_2 in the feed stream. The annual capital costs of the extractors in dollars/yr are $3000A$ and $5000B$, respectively. The value of the CO_2 product is $0.01Q$ dollars/yr. Develop the objective function and constraint(s) in terms of A, B, and Q.

7.4. An optimum shell-and-tube heat exchanger, as shown in Fig. 7-6 has a UA value of 1500 kW/K and has a pressure drop of the tube fluid of 300 kPa. Applicable equations are

$$A = 0.2NL$$

$$V = 5M/N$$

$$1/U = 0.08 + 1/h = 0.08 + 1/(0.4 \, V)$$

$$N = (\pi D^2/4)/160 = 0.005 \, D^2$$

where N = number of tubes
$\quad M$ = number of tube passes
$\quad A$ = heat-transfer area in m^2
$\quad V$ = velocity in m/s
$\quad U$ = overall heat-transfer coefficient in kW/(m$^2 \cdot$ K)
$\quad h$ = convection coefficient on the tube side in kW/(m$^2 \cdot$ K)
$\quad D$ = shell diameter in m
$\quad L$ = tube length in m

The pressure drop of the tube fluid per unit length of tube, kPa/m, is $0.1V^2$. The optimum heat exchanger is one of least cost that meets the other

FIGURE 7-6
Shell-and-tube heat exchanger in Prob. 7.4.

requirements, and the cost is a function of L and D. Develop the constraint(s) in terms of L and D.

7.5. The flow rate of raw material to the processing plant shown in Fig. 7-7 is 10 kg/s of mixture consisting of 50 percent A and 50 percent B. The separator can remove some of material A in pure form, and this product has a selling price of $10 per kilogram. The other product from the separator sells for $2 per kilogram, and some of this stream may be recycled. The cost of operating the separator in dollars per second is

$$\frac{0.2w_1}{1 - (w_2/w_1x_1)}$$

where w = mass rate of flow, kg/s
x = fraction of A in the mixture

(This separator cost equation indicates that the cost becomes infinite for perfect separation.)

(*a*) Set up the objective function in terms of w_1, w_2, and x_1 to maximize the profit. The cost of the raw material is constant.

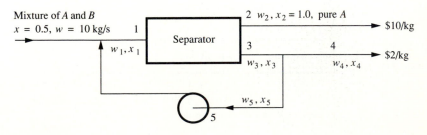

FIGURE 7-7
Processing plant in Prob. 7.5.

(b) Set up the constraint(s) in terms of w_1, w_2, and x_1.

Ans.: (b) $(5 - w_1 x_1)(w_1 - w_2) = (10 - w_1)(w_1 x_1 - w_2)$.

7.6. The plant shown schematically in Fig. 7-8 produces chlorine by the reaction of HCl and O_2 according to the equation

$$4HCl + O_2 \rightarrow 2H_2O + 2Cl_2$$

The plant receives 80 mol/s of HCl and 20 mol/s of O_2. The converter is capable of achieving only a partial conversion, and the unreacted HCl and O_2 are recycled following removal of the water and the chlorine.

The first cost of the system is to be minimized, and the major variable affecting the cost is that of the converter whose cost is represented by

$$\text{Cost per mol/s of reactants at } A, \text{ dollars} = 800 + \frac{24,000}{100 - x}$$

where x = conversion efficiency, %

or x = (fraction of entering HCl and O_2 that reacts) $\times$ 100

Determine (a) the objective function of w and x and (b) the constraint(s) that permit optimization for minimum cost.

Ans.: (b) $wx = 10,000$.

7.7. A supersonic wind-tunnel facility is being designed in which the air will flow in series through a compressor, storage tank, pressure-control valve, the wind tunnel, and thence to exhaust. During tests a flow rate of 5 kg/s must be available to the wind tunnel at a pressure of 400 kPa. The tests are intended to study heat transfer, and before each test 120 s of stabilization time is required, during which 5 kg/s must also flow. A total of 3600 s of useful test time is required during an 8-hour period, and this 3600 s can be subdivided into any number of equal-length tests.

The mode of operation is to start the compressor and allow it to run continuously at full capacity during the 8-hour period. Each cycle consists of the following stages: (1) buildup of pressure in the storage tank from 400 to 530 kPa, during which time there is no flow through the wind tunnel, (2) 120 s, during which the flow is 5 kg/s and the storage tank pressure begins to drop, and (3) the useful test, during which flow is 5 kg/s and at the end of which the storage-tank pressure has dropped to 400 kPa.

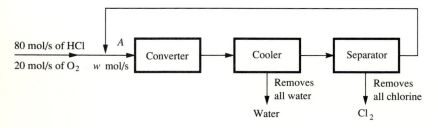

FIGURE 7-8
Chlorine plant in Prob. 7.6.

A pressure 400 kPa is available in the storage tank at the start of the day.

The compressor-storage-tank combination is to be selected for minimum total first cost. The compressor cost in dollars is given by

$$\text{Cost} = 800 + 2400S$$

where S is the capacity of the compressor in kilograms per second. The storage tank is to be a cube, for which metal costs $15 per square meter. The mass of air that can be stored in the tank at 530 kPa in excess of that at 400 kPa is $1.5V$ kg, where V is the volume in cubic meters.

(a) Write the expression for the total first cost of the compressor and tank in terms of S and V.

(b) Develop the constraint equation in terms of S and V to meet the operating conditions.

Ans.: (b) $S\left(1 + \dfrac{1.5V}{137S + 0.214V}\right) = 5.$

7.8. A simplified version of a combined gas- and steam-turbine plant[1] for a liquefied-petroleum-gas facility at Bushton, Kansas is shown in Fig. 7-9; the plant must meet the following minimum requirements:

	kW
Power to propane compressor	3800
Low-pressure steam equivalent for process use	6500
High-pressure steam equivalent for process use	8800

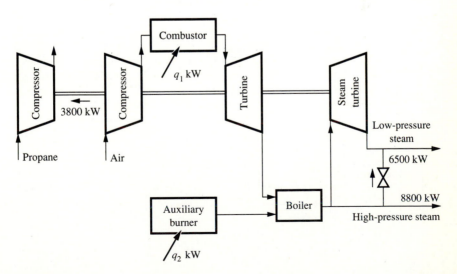

FIGURE 7-9
Schematic diagram of gas- and steam-turbine plant in Prob. 7.8.

In the gas-turbine plant, 20 percent of the heating value of natural gas is converted into mechanical power and the remaining 80 percent passes to the exhaust gas. As the exhaust gas flows through the boiler, 60 percent of its heat is converted into steam. The boiler is also equipped with an auxiliary burner, which permits 80 percent of the heating value of the natural gas to be converted into steam. High-pressure steam flows to process use and/or to the steam turbine, where 15 percent of the thermal energy is converted into mechanical power.

The heat input rates from the natural gas turbine and the boiler are designated q_1 and q_2 kW, respectively.

The plant is to be operated in such a way that it consumes a minimum total quantity of natural gas. In terms of q_1 and q_2 (a) write the objective function and (b) develop the constraint equations. Acknowledge by inequalities in the constraint equations the possibility of dumping power or steam.

Ans.: (b) $q_1 + 1.176q_2 \geq 28,090$ kW

$$q_1 + 0.441q_2 \geq 18,820 \text{ kW}$$
$$q_1 + 1.665q_2 \geq 31,880 \text{ kW}$$

7.9. Nagib[2] discusses improvement in the operating efficiency of a gas-turbine cycle by precooling the air entering the compressor. The precooler is an absorption-refrigeration unit, which is supplied with steam generated from the heat of the exhaust gas, as shown in Fig. 7-10. Achievement of this improved efficiency entails additional investment cost for the absorption unit and boiler. Furthermore, there is an optimum combination of absorption unit, boiler, and regenerator.

Performance data

Compressor power, kW $= 34(t_1 + 273)$.

Absorption unit delivers 0.6 kJ of cooling per kilojoule of heat supplied by the steam.

FIGURE 7-10
Precooler of gas-turbine inlet air in Prob. 7.9.

The flow rate of gas through the cycle is 45 kg/s, and negligible addition of mass at the combustor is assumed.

U value of the boiler based on steam-side area, 0.25 kW/(m² · K).

U value of regenerator, 0.082 kW/(m² · K).

c_p of gas throughout the cycle, 1.0 kJ/(kg · K).

Assume no heat transfer to ambient from any component.

Additional temperatures are indicated on Fig. 7-10.

Cost data

Present worth of power generated during the life of the plant, $225P$ dollars

Present worth of fuel cost for the plant life, $45Q$ dollars

First cost of absorption unit, $80A$ dollars

First cost of boiler, $1,000B$ dollars

First cost of regenerator, $90C$ dollars

where Q = heat rate at combustor, kW

P = power generated, kW

A = absorption unit size, kW of cooling

B = boiler size, m² of steam-side area

C = regenerator size, m²

(a) Develop the objective function for the total present worth of the profit of the system in terms of A, B, C, Q, and P.

(b) Develop expressions for the temperatures t_1, t_2, t_3, and t_4 in terms of the variables of part (a).

(c) Develop the constraint equations in terms of the variables of part (a).

Ans.: (b): $t_1 = 32 - 0.0222A$ $t_2 = 262.5 - 0.039A$

$$t_3 = 262.5 - 0.039A + \frac{217.5 + 0.039A}{548.8/C + 1}$$

$$t_4 = 480 - \frac{217.5 + 0.039A}{548.8/C + 1}$$

(c): $P - 4930 - 0.7548A = 0$

$$A - (27)\left(375 - \frac{217.5 + 0.039A}{548.8/C + 1}\right)(1 - e^{-0.005556B}) = 0$$

$$Q - (45)\left(557.5 + 0.039A - \frac{217.5 + 0.039A}{548.8/C + 1}\right) = 0$$

7.10. A desalination plant shown schematically in Fig. 7-11, receives sea water having a saline concentration by mass of 3% of the mixture. This sea water is mixed with some recycle solution and enters the plant which is to have a capacity of 300 kg/s. The cost of pumping and treating the raw sea water is

FIGURE 7-11
Desalination plant in Prob. 7.10.

$0.08 per Mg of mixture, and the cost of heat to operate the plant is $0.10 per Mg of feed. As the salt concentration entering the desalination plant increases, the effectiveness of the plant decreases, following the equation $y = 3.4\sqrt{x}$, so should the entering concentration reach 11.56% the plant would produce no fresh water. Develop the objective function and constraint to describe the minimum-cost operation.

   Ans.: constraint is

$$Q = \frac{(300)(11.56)(w - 300)}{11.56(w - 300) - 3w}$$

REFERENCES

1. B. G. Wobker and C. E. Knight, "Mechanical Drive Combined-Cycle Gas and Steam Turbines for Northern Gas Products," *ASME Pap.* 67-GT-39, 1967.
2. M. M. Nagib, "Analysis of a Combined Gas Turbine and Absorption–Refrigeration Cycle," *ASME Pap.* 70-PWR-18, 1970.

ADDITIONAL READINGS

Beveridge, G. S. G., and R. S. Schechter: *Optimization: Theory and Practice*, McGraw-Hill, New York, 1970.
Denn, M. M.: *Optimization by Variational Methods*, McGraw-Hill, New York, 1969.
Fox, R. L.: *Optimization Methods for Engineering Design*, Addison-Wesley, Reading, Mass., 1971.
Ray, W. H., and J. Szekely: *Process Optimization*, Wiley, New York, 1973.
Rosenbrock, H. H., and C. Storey: *Computational Techniques for Chemical Engineers*, Pergamon, New York, 1966.
Wilde, D. J., and C. S. Beightler: *Foundations of Optimization*, Prentice-Hall, Englewood Cliffs, N.J., 1967.

CHAPTER
8

LAGRANGE
MULTIPLIERS

8.1 CALCULUS METHODS OF OPTIMIZATION

Classical methods of optimization are based on calculus and specifically determine the optimum value of a function as indicated by the nature of the derivatives. In order to optimize using calculus, the function must be differentiable and any constraints must be equality constraints. That there is a need for any method other than calculus, such as linear and nonlinear programming, may arise from the appearance of inequality constraints. In addition, the fact that the function is not continuous but exists only at specific values of the parameters rules out calculus procedures and favors such techniques as search methods. On the other hand, since some of the operations in the calculus methods appear in slightly revised forms in the other optimization methods, calculus methods are important both for the cases that they can solve in their own right and to illuminate some of the procedures in noncalculus methods.

This chapter presents the Lagrange multiplier equations, proceeds to the interpretation and mechanics of optimization using these equations, and then begins the presentation of the background and visualization of constrained optimization. Since unconstrained optimization is only a special case of constrained optimization, the method of Lagrange multipliers is applicable to all situations explored in this chapter. This chapter also explains how an optimum condition can be tested to establish whether the condition is a maximum or a minimum and, finally, introduces the concept of the sensitivity coefficient.

8.2 THE LAGRANGE MULTIPLIER EQUATIONS

The mathematical statement of the constrained optimization problem first given in Eqs. (7.1) to (7.3) is repeated here:

Optimize
$$y = y(x_1, x_2, \ldots, x_n) \tag{8.1}$$

subject to
$$\phi_1(x_1, x_2, \ldots, x_n) = 0 \tag{8.2}$$
$$\cdots\cdots\cdots\cdots\cdots$$
$$\phi_m(x_1, x_2, \ldots, x_n) = 0 \tag{8.3}$$

The method of Lagrange multipliers states that the optimum occurs at values of the x's that satisfy the equations

$$\nabla y_1 \nabla \phi_1 - \lambda_m \nabla \phi_m = 0 \tag{8.4}$$
$$\phi_1(x_1, \ldots, x_n) = 0 \tag{8.5}$$
$$\cdots\cdots\cdots\cdots\cdots\cdots\cdots\cdots$$
$$\phi_m(x_1, \ldots, x_n) = 0 \tag{8.6}$$

The remainder of the chapter explains the meaning of the symbols and operations designated in the equations, the mechanics of solving the equations, and applications, examples, and geometric visualization of the Lagrange multiplier equations.

8.3 THE GRADIENT VECTOR

The inverted delta designates the *gradient vector*. A *scalar* is a quantity with a magnitude but no direction, while a *vector* has both magnitude and direction. By definition, the gradient of a scalar is

$$\nabla y = \frac{\partial y}{\partial x_1} \mathbf{i}_1 + \frac{\partial y}{\partial x_2} \mathbf{i}_2 + \cdots + \frac{\partial y}{\partial x_n} \mathbf{i}_n \tag{8.7}$$

where $\mathbf{i}_1, \mathbf{i}_2, \ldots, \mathbf{i}_n$ are unit vectors, which means that they have direction and their magnitudes are unity.

Suppose, for example, that a solid rectangular block has a temperature distribution that can be expressed in terms of the coordinates x_1, x_2, and x_3, as shown in Fig. 8-1. If the temperature t is the following function of x_1, x_2, and x_3,

$$t = 2x_1 + x_1 x_2 + x_2 x_3^2$$

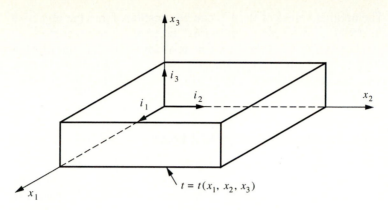

FIGURE 8-1
Solid object in which a scalar, the temperature, is expressed as a function of x_1, x_2, and x_3.

then the gradient of t, ∇t, following the definition of Eq. (8.7), is

$$\nabla t = (2 + x_2)\mathbf{i}_1 + (x_1 + x_3^2)\mathbf{i}_2 + (2x_2 x_3)\mathbf{i}_3$$

where $\mathbf{i}_1$, $\mathbf{i}_2$, and $\mathbf{i}_3$ are the unit vectors in the x_1, x_2, and x_3 directions, respectively. The gradient operation is one that converts a scalar quantity into a vector quantity.

8.4 FURTHER EXPLANATION OF LAGRANGE MULTIPLIER EQUATIONS

Also appearing in Eq. (8.4) is a series of lambdas, $\lambda_1, \ldots, \lambda_m$. These terms are constants and are the *Lagrange multipliers*, whose values are not known until the set of equations is solved.

Equation (8.4) is a vector equation, which means that for the terms on the left side of the equation to equal zero the coefficients of each unit vector must sum to zero. The vector equation (8.4) is thus a shorthand form for n scalar equations,

$$\mathbf{i}_1: \qquad \frac{\partial y}{\partial x_1} - \lambda_1 \frac{\partial \phi_1}{\partial x_1} - \lambda_m \frac{\partial \phi_m}{\partial x_1} = 0 \qquad (8.8)$$

$$\cdots\cdots\cdots\cdots\cdots\cdots\cdots\cdots\cdots\cdots$$

$$\mathbf{i}_n: \qquad \frac{\partial y}{\partial x_n} - \lambda_1 \frac{\partial \phi_1}{\partial x_n} - \lambda_m \frac{\partial \phi_m}{\partial x_n} = 0 \qquad (8.9)$$

These n scalar equations, along with the m constraint equations (8.5) to (8.6), form the set of $n + m$ simultaneous equations needed to solve for the same number of unknowns: $x_1^*, x_2^*, \ldots, x_n^*$ and $\lambda_1, \lambda_2, \ldots, \lambda_m$. The asterisk on the x's is often used to indicate the values of the variables at the optimum. The optimal values of the x's can be substituted into the objective function, Eq. (8.1), to determine the optimal value of y, designated y^*.

The number of equality constraints m is always less than the number of variables n. In the limiting case where $m = n$, the constraints (if they are independent equations) fix the values of the x's, and no optimization is possible.

8.5   UNCONSTRAINED OPTIMIZATION

The Lagrange multiplier equations can be used to attack constrained optimization problems, but the equations apply equally well to unconstrained optimization. The unconstrained optimization is a special (and simpler) case of constrained optimization. The objective function y is a function of variables $x_1, \ldots, x_n$

$$y = y(x_1, \ldots, x_n) \tag{8.10}$$

When the Lagrange multiplier equations are applied to Eq. (8.10), since there are no ϕ's, the condition for optimum is

$$\nabla y = 0 \tag{8.11}$$

or

$$\frac{\partial y}{\partial x_1} = 0, \frac{\partial y}{\partial x_2} = 0, \ldots, \frac{\partial y}{\partial x_n} = 0 \tag{8.12}$$

The state point where the derivatives are zero is called a *critical point*, and it may be a maximum or minimum (one of which we are seeking), or it may be a saddle point or a ridge or valley. Further mathematical analysis may be necessary to determine the type of critical point, although in physical situations the nature of the point will often be obvious. We shall assume in the remainder of this chapter that Eq. (8.12) describe a maximum or minimum.

A function y of two variables x_1 and x_2 can be represented graphically as in Fig. 8-2. The minimum exists where

$$\frac{\partial y}{\partial x_1} = 0 = \frac{\partial y}{\partial x_2}$$

Example 8.1. Determine the minimum value of y where

$$y = \frac{x_1}{x_2} + \frac{1}{x_1 x_3} + \frac{1}{2}x_2^2 + \frac{x_3}{16}$$

Solution. The Lagrange multiplier equations for this unconstrained optimization are

$$\nabla y = 0 \quad \text{or} \quad \frac{\partial y}{\partial x_1} = \frac{\partial y}{\partial x_2} = \frac{\partial y}{\partial x_3} = 0$$

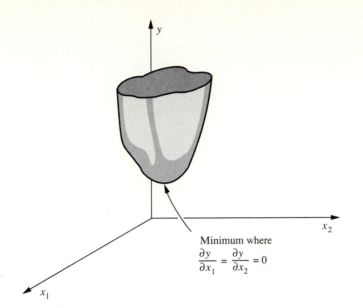

FIGURE 8-2
Unconstrained optimum occurs where partial derivatives equal zero.

$$\frac{\partial y}{\partial x_1} = \frac{1}{x_2} - \frac{1}{x_1^2 x_3} = 0$$

$$\frac{\partial y}{\partial x_2} = -\frac{x_1}{x_2^2} + x_2 = 0$$

$$\frac{\partial y}{\partial x_3} = -\frac{1}{x_1 x_3^2} + \frac{1}{16} = 0$$

Solving simultaneously gives $x_1^* = 0.305$, $x_2^* = 0.673$, and $x_3^* = 7.243$. Substituting these values into the objective function yields $y^* = 1.585$.

Sometimes it is convenient to convert a constrained problem into an unconstrained optimization by solving for a variable in a constraint equation and substituting that expression wherever else in the remaining constraints or objective function the variable appears. The number of constraints is progressively reduced until only the unconstrained objective function remains.

Example 8.2. A total length of 100 m of tubes must be installed in a shell-and-tube heat exchanger (Fig. 8-3), in order to provide the necessary heat-transfer area. The total cost of the installation in dollars includes

1. The cost of the tubes, which is constant at $900
2. The cost of the shell $= 1100D^{2.5}L$

FIGURE 8-3
Heat exchanger in Example 8.2.

3. The cost of the floor space occupied by the heat exchanger $= 320DL$, where L is the length of the heat exchanger and D is the diameter of the shell, both in meters

The spacing of the tubes is such that 200 tubes will fit in a cross-sectional area of 1 m^2 in the shell.

Determine the diameter and length of the heat exchanger for minimum first cost.

Solution. The objective function includes the three costs,

$$\text{Cost} = 900 + 1100D^{2.5}L + 320DL$$

The constraint requires the heat exchanger to include 100 m of tubes

$$\left(\frac{\pi D^2}{4} \text{ m}^2\right)(L, \text{m})(200 \text{ tubes per m}^2) = 100 \text{ m}$$

or

$$50\pi D^2 L = 100$$

To convert this constrained optimization into an unconstrained optimization, solve for L in the constraint and substitute into the objective function

$$\text{Cost} = 900 + \frac{2200}{\pi}D^{0.5} + \frac{640}{\pi D}$$

The objective function is now in terms of D only, so differentiating and equating to zero yields

$$\frac{d(\text{cost})}{d(D)} = \frac{1100}{\pi D^{0.5}} - \frac{640}{\pi D^2} = 0$$

$$D^* = 0.7 \text{ m}$$

Substituting the optimal value of D back into the constraint gives

$$L^* = 1.3 \text{ m}$$

and the minimum cost is

$$\text{Cost*} = 900 + (1100)(0.7)^{2.5}(1.3) + (320)(0.7)(1.3)$$

$$= \$1777.45$$

8.6 CONSTRAINED OPTIMIZATION USING LAGRANGE MULTIPLIER EQUATIONS

While it is sometimes possible to convert a constrained optimization into an unconstrained one, there are severe limitations to this option. For example, it may not be possible to solve the constraint equations explicitly for variables that are to be substituted into the objective function. The more general and powerful technique is to use the classic Lagrange multiplier equations (8.4) to (8.6).

Example 8.3. Solve Example 8.2 using the Lagrange multiplier equations.

Solution. The statement of the problem is:

Minimize $\qquad\qquad y = 900 + 1100D^{2.5}L + 320DL \qquad\qquad$ (8.13)

subject to $\qquad\qquad\qquad 50\pi D^2 L = 100 \qquad\qquad\qquad$ (8.14)

The two gradient vectors are

$$\nabla y = [(2.5)(1100)D^{1.5}L + 320L]\mathbf{i}_1 + (1100D^{2.5} + 320D)\mathbf{i}_2$$

and $\qquad\qquad\qquad \nabla \phi = 100\pi DL\mathbf{i}_1 + 50\pi D^2\mathbf{i}_2$

In component form the two scalar equations are

$\mathbf{i}_1: \qquad\qquad 2750D^{1.5}L + 320L - \lambda 100\pi DL = 0$

$\mathbf{i}_2: \qquad\qquad 1100D^{2.5} + 320D - \lambda 50\pi D^2 = 0$

which along with the constraint

$$50\pi D^2 L = 100$$

provide three simultaneous equations to solve for D^*, L^*, and λ. From the

i₁ equation

$$\lambda = \frac{2750D^{1.5} + 320}{100\pi D}$$

Substituted into the **i₂** equation, this yields

$$D^* = 0.7 \text{ m}$$

Substituting this value of D into the constraint gives

$$L^* = 1.3 \text{ m}$$

and $$\lambda = \frac{(2750)(0.7)^{1.5} + 320}{100\pi(0.7)} = 8.78$$

The minimum cost is

$$y^* = 900 + (1100)(0.7)^{2.5}(1.3) + (320)(0.7)(1.3) = \$1777.45$$

The objective function, Eq. (8.13), is sketched in Fig. 8-4a. So long as either L or D is zero, the cost is \$900. As L and D increase, the cost rises, providing a warped surface that rises as it moves away from the vertical axis. The problem takes on meaning only after introduction of the constraint in Fig. 8-4b, which is a curve in the LD plane. Only the vertical projections from the constraint up to the objective function surface are permitted. The minimum value of the objective function along the projection is the constrained minimum.

Since Example 8.3 had only one constraint equation, only one λ appeared in the Lagrange multiplier equations. There will be the same

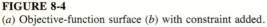

(a)   (b)

FIGURE 8-4
(a) Objective-function surface (b) with constraint added.

number of λ's as constraints. The execution of the solution requires the solution of a set of equations which are likely to be nonlinear. In Example 8.3 it was possible to solve them by substitution. When this cannot be done, the tools are now available from Chapter 6 to solve the set of nonlinear simultaneous equations by using such techniques as the Newton-Raphson method.

8.7 GRADIENT VECTOR IS NORMAL
TO THE CONTOUR SURFACE

The purpose of this section and the next is to provide a visualization of the Lagrange multiplier equations. The presentation will not be a proof[†] but a geometric display. The first step in providing the visualization is to show that the gradient vector is normal to the contour line or surface at the point where the gradient is being evaluated. If y is a function of x_1 and x_2, a contour line on the x_1 vs. x_2 graph is a line of constant y, as shown in Fig. 8-5. From calculus we recall that

$$dy = \frac{\partial y}{\partial x_1} dx_1 + \frac{\partial y}{\partial x_2} dx_2$$

[†] Further substantiation of the method is provided in Chapter 15, and a proof can be found in the book by Wilde and Beightler given in the Additional Readings at the end of the chapter.

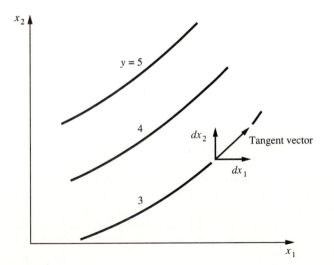

FIGURE 8-5
Contour lines (curves of constant y) when $y = f(x_1, x_2)$.

Therefore along the line of constant y, say the $y = 3$ line,

$$dy = 0 = \frac{\partial y}{\partial x_1} dx_1 + \frac{\partial y}{\partial x_2} dx_2$$

or

$$dx_1 = -dx_2 \frac{\partial y / \partial x_2}{\partial y / \partial x_1} \tag{8.15}$$

Any arbitrary vector in Fig. 8-5 is

$$dx_1 \mathbf{i}_1 + dx_2 \mathbf{i}_2$$

and any arbitrary unit vector is

$$\frac{dx_1 \mathbf{i}_1 + dx_2 \mathbf{i}_2}{\sqrt{(dx_1)^2 + (dx_2)^2}} \tag{8.16}$$

The special unit vector $\mathbf{T}$, the one that is tangent to the $y = $ constant line, has dx_1 and dx_2 related according to Eq. (8.15); substituting dx_1 from Eq. (8.15) into Eq. (8.16) yields

$$\mathbf{T} = \frac{-\dfrac{\partial y / \partial x_2}{\partial y / \partial x_1} \mathbf{i}_1 + \mathbf{i}_2}{\sqrt{\left(\dfrac{\partial y / \partial x_2}{\partial y / \partial x_1}\right)^2 + 1}} = \frac{-\dfrac{\partial y}{\partial x_2} \mathbf{i}_1 + \dfrac{\partial y}{\partial x_1} \mathbf{i}_2}{\sqrt{\left(\dfrac{\partial y}{\partial x_2}\right)^2 + \left(\dfrac{\partial y}{\partial x_1}\right)^2}} \tag{8.17}$$

Equation (8.17) is the unit vector that is tangent to the $y = $ constant line.

Returning to the gradient vector and dividing by its magnitude to obtain the unit gradient vector $\mathbf{G}$, we have

$$\mathbf{G} = \frac{\nabla y}{|\nabla y|} = \frac{\partial y / \partial x_1 \mathbf{i}_1 + \partial y / \partial x_2 \mathbf{i}_2}{\sqrt{(\partial y / \partial x_1)^2 + (\partial y / \partial x_2)^2}} \tag{8.18}$$

The relationship between the vectors represented by Eqs. (8.17) and (8.18) is that one is perpendicular to the other. If, as in Fig. 8-6, the components of vector $\mathbf{Z}$ are b and c, the perpendicular vector $\mathbf{Z}_1$ has components c and $-b$. Thus, the components are interchanged and the sign of one is reversed.

The important conclusion reached at this point is that since the gradient vector is perpendicular to the tangent vector, *the gradient vector is normal to the line or surface of constant y.*

A similar conclusion is applicable in three and more dimensions. In three dimensions, for example, where $y = y(x_1, x_2, x_3)$, the curves of constant value of y become surfaces, as shown in Fig. 8-7. In this case the

FIGURE 8-6
Components of perpendicular vectors.

FIGURE 8-7
A gradient vector in three dimensions.

gradient vector ∇y, which is

$$\nabla y = \frac{\partial y}{\partial x_1}\mathbf{i}_1 + \frac{\partial y}{\partial x_2}\mathbf{i}_2 + \frac{\partial y}{\partial x_3}\mathbf{i}_3 \qquad (8.19)$$

is normal to the $y = $ constant surface that passes through that point.

The magnitude of the gradient vector indicates the rate of change of the dependent variable with respect to the independent variables. Thus, if the surfaces of constant y in Fig. 8-7 are spaced wide apart, the absolute magnitude of the gradient is small. For the time being, however, we are interested only in the fact that the gradient vector is in a *direction* normal to the constant curve or surface.

8.8   VISUALIZATION OF THE LAGRANGE MULTIPLIER METHOD IN TWO DIMENSIONS

An optimization of a problem with two variables of optimization can be displayed graphically to help clarify the operation of the Lagrange multiplier equations. Suppose that the function y is to be minimized, where

$$y = 2x_1 + 3x_2 \tag{8.20}$$

subject to the constraint

$$x_1 x_2^2 = 48 \tag{8.21}$$

The Lagrange multiplier equations are

$$(2 - \lambda x_2^2)\mathbf{i}_1 + (3 - 2\lambda x_1 x_2)\mathbf{i}_2 = 0 \qquad \text{and} \qquad x_1 x_2^2 = 48$$

which when solved yield the solution $x_1^* = 3$, $x_2^* = 4$ and $y^* = 18$. Some lines of constant y and the constraint equation are shown in Fig. 8-8. With such a graph available it is possible almost instinctively to locate

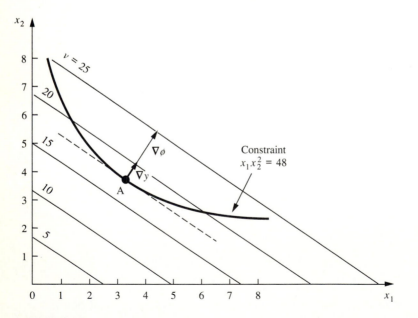

FIGURE 8-8
Optimum occurs where the constraint and the lines of constant y have a common normal.

the (x_1, x_2) position that provides the constrained optimum, point A in Fig. 8-8. The process of visually arriving at this solution might be one of following along the constraint line until the constraint and the line of constant y are "parallel." Stated more precisely, the tangent vectors to the curves have the same direction.

A more convenient means of requiring that the tangent vectors have the same direction is to require that the normal vectors to the curves have the same direction. Since the gradient vector is normal to a contour line, the mathematical statement of this requirement is

$$\nabla y - \lambda \nabla \phi = 0 \qquad (8.22)$$

The introduction of a constant, λ, is necessary because the magnitudes of ∇y and $\nabla \phi$ may be different and the vectors may even be pointing in opposite directions. The only requirement is that the two gradient vectors be collinear. The constraint equation is satisfied because the only points being considered are those along the constraint equation.

8.9 TEST FOR MAXIMUM OR MINIMUM

In mathematics books that treat optimization, test procedures for determining the nature of the critical point receive a major emphasis. These test procedures decide whether the point is a maximum, minimum, saddle point, ridge, or valley. These procedures are generally not so important in physical problems, which are our concern, because the engineer usually has an insight into whether the result is a maximum or minimum. It also seems that the occurrence of saddle points, ridges, and valleys is rare in physical problems. In addition, the tests based on pure calculus usually become prohibitive when dealing with more than about three variables anyway, and we are especially interested in systems with large numbers of components. If a test is made, it usually consists of testing points in the neighborhood of the optimum.

On the other hand, a brief discussion of the classical tests for maximum and minimum provide further insight into the nature of the optimization process. The discussion will be limited to the optimization of one and two variables of an unconstrained function.

Consider first the case of one independent variable $y = y(x)$ for which the minimum is sought. Suppose that the point $x = a_1$ is the position at which the minimum is expected to occur. To test whether the value of y at this point $y(a_1)$ is truly a minimum, move slightly in both possible directions from $x = a_1$ to see if a lower value of y can be found. If a lower value is available, $y(a_1)$ is not the minimal value. For the mathematical check, expand y in a Taylor series (see Sec. 6.9) about point $x = a_1$

$$y(x) = y(a_1) + \frac{dy}{dx}(x - a_1) + \frac{1}{2}\frac{d^2y}{dx^2}(x - a_1)^2 + \cdots \qquad (8.23)$$

First examine moves so small that the $(x - a_1)^2$ terms and higher-order terms can be ignored; if $dy/dx > 0$, then $y(x) > y(a_1)$ for a value of $x > a_1$ and $y(a_1)$ is still the acknowledged minimum. When x moves to a value less than a_1, however, $y(x) < y(a_1)$ and $y(a_1)$ will not be a minimum. In a similar manner, it can be shown that when $dy/dx < 0$, a lower value than $y(a_1)$ can be found for $y(x)$. The only solution to the dilemma is for dy/dx to equal zero, which is the classical requirement for the optimum.

Including the influence of the next term of Eq. (8.23), $(\frac{1}{2})(d^2y/dx^2)(x - a_1)^2$, we observe that a move of x in either direction from a_1 results in a positive value of $(x - a_1)^2$, so that the sign of the second derivative decides whether the optimum is a maximum or minimum.

Minimum: $\dfrac{dy}{dx} = 0$ and $\dfrac{d^2y}{dx^2} > 0$

Maximum: $\dfrac{dy}{dx} = 0$ and $\dfrac{d^2y}{dx^2} < 0$

The foregoing line of reasoning will now be extended to a function of two variables $y(x_1, x_2)$. Suppose that the expected minimum occurs at the point (a_1, a_2) and that to verify this position the point is shifted infinitesimally away from (a_1, a_2) in all possible directions. The Taylor expansion for a function of two variables is

$$y(x_1, x_2) = y(a_1, a_2) + y_1'(x_1 - a_1) + y_2'(x_2 - a_2) + \left(\tfrac{1}{2}\right) y_{11}''(x_1 - a_1)^2$$

$$+ y_{12}''(x_1 - a_1)(x_2 - a_2) + \left(\tfrac{1}{2}\right) y_{22}''(x_2 - a_2)^2 + \cdots \qquad (8.24)$$

where the prime on the y refers to a partial differentiation with respect to the subscript.

When x_1 and x_2 move slightly off (a_1, a_2), both of the first derivatives y_1' and y_2' must be zero in order to avoid some position where $y(x_1, x_2) < y(a_1, a_2)$.

The second-order terms decide whether the optimum is a maximum or minimum. If the combination is always positive regardless of the signs of $x_1 - a_1$ and $x_2 - a_2$, the optimum is a minimum. If the combination is always negative, the optimum is a maximum.

The test for maximum and minimum is as follows. The second derivatives are structured in matrix form and the value of the determinant is called D, where

$$D = \begin{vmatrix} y_{11}'' & y_{12}'' \\ y_{21}'' & y_{22}'' \end{vmatrix}$$

Then if $D > 0$ and $y_{11}'' > 0$, the optimum is a minimum. If $D > 0$ and $y_{11}'' < 0$, the optimum is a maximum.

Example 8.4. Determine the optimal values of x_1 and x_2 for the function

$$y = \frac{x_1^2}{4} + \frac{2}{x_1 x_2} + 4x_2$$

and test whether the point is a maximum or minimum.

Solution. The first derivatives are

$$\frac{\partial y}{\partial x_1} = \frac{x_1}{2} - \frac{2}{x_1^2 x_2} \quad \text{and} \quad \frac{\partial y}{\partial x_2} = -\frac{2}{x_1 x_2^2} + 4$$

Equating these derivatives to zero and solving simultaneously yields

$$x_1^* = 2 \quad \text{and} \quad x_2^* = \frac{1}{2}$$

The second derivatives evaluated at x_1^* and x_2^* are

$$\frac{\partial^2 y}{\partial x_1^2} = \frac{3}{2} \quad \frac{\partial^2 y}{\partial x_2^2} = 16 \quad \frac{\partial^2 y}{\partial x_1 \partial x_2} = 2$$

The determinant

$$\begin{vmatrix} \frac{3}{2} & 2 \\ 2 & 16 \end{vmatrix} > 0 \quad \text{and} \quad y_{11}'' > 0$$

so this optimal point is a minimum.

8.10 SENSITIVITY COEFFICIENTS

There is often an additional valuable step beyond determination of the optimal value of the objective function and the state point at which the optimum occurs. After the optimum is found, subject to one or more constraints, the question that logically arises is: What would be the effect on the optimal value of slightly relaxing the constraint? In a physical situation this question occurs, for example, in analyzing how much the capacity of a system could be increased by enlarging one of the components whose performance characteristic is one of the constraint equations.

In Example 8.3, where the cost of the heat exchanger

$$\text{Cost} = 900 + 1100D^{2.5}L + 320DL$$

was minimized subject to the constraint

$$50\pi D^2 L = 100$$

the question might be phrased: What would be the increase in minimum cost, cost*, if 101 linear meters of tubes were required rather than the original 100? To analyze this particular example, replace the specific value of 100 by a general symbol H and perform the optimization by the method of Lagrange multipliers. The result would be

$$D^* = 0.7 \text{ m} \qquad L^* = 0.013H$$

$$\text{Cost}^* = 900 + (1100)(0.7)^{2.5}(0.013H) + (320)(0.7)(0.13H)$$

$$= 900 + 8.78H$$

We are interested in the variation of cost* with H or, more specifically, a term called the *sensitivity coefficient* (SC), which is

$$SC = \frac{\partial(\text{cost}^*)}{\partial H}$$

In this example, SC $= 8.78$; thus at the optimal proportions an extra meter of tube for the heat exchanger would cost $8.78 more than the original cost. Referring back to the solution of Example 8.3, we note the remarkable fact that the sensitivity coefficient is precisely equal to the Lagrange multiplier λ. This equality of the SC to λ is true not only for this particular example but in general. Also, if there is more than one constraint, the various sensitivity coefficients are equal to the corresponding Lagrange multipliers, $SC_1 = \lambda_1, \ldots, SC_m = \lambda_m$.

The optimization process by Lagrange multipliers therefore offers an additional piece of useful information for possible adjustment of the physical system following preliminary optimization.

8.11   INEQUALITY CONSTRAINTS

The method of Lagrange multipliers appliers only to the situations where the constraints are equalities, and the method cannot be used directly with inequality constraints. This limitation should not completely rule out the employment of the method when inequalities arise, because a combination of intuition and several passes at the problem with the method may yield a solution.

As an example, suppose that the capacity of a system is said to be equal to or greater than 150 kW. It would be a rare case when a lower-cost system would provide 160 kW compared with the one providing 150 kW. This constraint would almost certainly be used as an equality constraint of 150 kW. As a further example, suppose that the temperature at some point in the process must be equal to or less than 320°C. In the first attempt at the problem, ignore the temperature constraint and after the optimal conditions are determined check to see if the temperature in question is above 320°C. If it is not, the constraint is not effective. If the temperature is above 320°C, rework the problem with the equality constraint of 320°C.

PROBLEMS

8.1. For the function

$$y = \frac{1}{4x_1} + 4x_1^2 x_2 + \frac{1}{x_2}$$

(a) Determine the positive optimum values of x_1 and x_2.
(b) Use the techniques of Sec. 8.9 to test the solution to determine whether the optimum is a maximum or minimum.

Ans.: (a) $x_1^* = \frac{1}{4}, x_2^* = 2$.

8.2. The dimensions of a rectangular duct are to be chosen so that the duct of maximum cross-sectional area can be placed, as shown in Fig. 8-9, in the opening in a bar joist.

(a) Set up the objective function and constraint in terms of h and w.
(b) Using the method of Lagrange multipliers for constrained optimization, determine the optimal values of h and w.

Ans.: $h^* = 0.3$ m.

8.3. Two parallel pump-pipe assemblies shown in Fig. 8-10 deliver water from a common source to a common destination. The total volume flow rate required at the destination is 0.01 m³/s. The drops in pressure in the two lines are

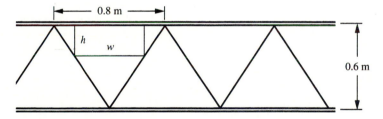

FIGURE 8-9
Duct in bar joist in Prob. 8.2.

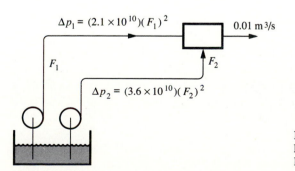

FIGURE 8-10
Parallel pump-pipe assemblies in Prob. 8.3.

functions of the square of the flow rates,

$$\Delta p_1, \text{Pa} = 2.1 \times 10^{10} F_1^2 \quad \text{and} \quad \Delta p_2, \text{Pa} = 3.6 \times 10^{10} F_2^2$$

where F_1 and F_2 are the respective flow rates in cubic meters per second. The two pumps have the same efficiency, and the two motors that drive the pumps also have the same efficiency.

(a) It is desired to minimize the total power equipment. Set up the objective function and constraints in terms of F_1 and F_2.

(b) Solve for the optimal values of the flow rates that result in minimum total water power using the method of Lagrange multipliers.

   Ans.: $F_1 = 0.00567$.

8.4. A steel framework, as in Fig. 8-11, is to be constructed at a minimum cost. The cost in dollars of all the horizontal members in one orientation is $200x_1$ and in the other horizontal orientation $300x_2$. The cost in dollars of all vertical members is $500x_3$. The frame must enclose a total volume of 900 m³.

(a) Set up the objective function for total cost and the constraint(s) in terms of x_1, x_2, and x_3.

(b) Using the method of Lagrange multipliers for constrained optimization, determine the optimal values of the dimensions and the minimum cost.

   Ans.: Minimum cost = $9000.

8.5. A flow rate of 15 m³/s of gas at a temperature of 50°C and a pressure of 175 kPa is to be compressed to a final pressure of 17,500 kPa. The choice of compressor type is influenced by the fact that centrifugal compressors can handle high-volume flow rates but develop only low pressure ratios per stage. The reciprocating compressor, on the other hand, is suited to low-volume

FIGURE 8-11
Steel framework in Prob. 8.4.

FIGURE 8-12
Staged compression in Prob. 8.5.

flow rates but can develop high pressure ratios. To combine the advantages of each, the compression will be carried out by a centrifugal compressor in series with a reciprocating compressor, as shown in Fig. 8-12.

The intercooler returns the temperature of the gas to 50°C. Assume that the gas obeys perfect-gas laws. The equations for the costs of the compressors are

$$C_c = 70Q_0 + 1600 \, \frac{p_1}{p_0} \qquad C_r = 200Q_1 + 800 \, \frac{p_2}{p_1}$$

where C_c = first cost of centrifugal compressor, dollars
C_r = first cost of reciprocating compressor, dollars
Q = volume flow rate, m³/s

(*a*) Set up the objective function for the total first cost and the constraint equation in terms of the pressure ratios.
(*b*) Using the method of Lagrange multipliers for constrained optimization, solve for the optimal pressure ratios and minimum total cost.
   Ans.: Minimum cost = $24,100.

8.6. The packing rings in a distillation tower are in the shape of a torus (doughnut) as shown in Fig. 8-13. The outside diameter of the ring is to be 20 mm, and the values of D_1 and D_2 are sought that yield maximum surface area where the area $= \pi^2 D_1 D_2$. Structure the problem as a constrained optimization and solve by the method of Lagrange multipliers.
   Ans.: $D_1^* = 10$ mm.

8.7. In a cascade refrigeration system shown in Fig. 8-14 that is used for low-temperature applications, the condenser of the low-stage unit is the evaporator of the high-stage unit. The area of this interstage condenser-evaporator is to be chosen such that minimum lifetime costs result. The costs associated with the decision of the size of the condenser-evaporator and the temperatures are:

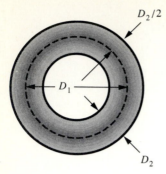

$D_2/2$

D_1

D_2

FIGURE 8-13
Doughnut-shaped ring with maximum surface area in Prob. 8.6.

cost of the condenser-evaporator $= 75\,A$

present worth of the lifetime energy
costs of the low-stage unit $= -37,000 + 150\,T_c$

present worth of the lifetime costs of
energy for the high-stage unit $= 189,000 - 0.3\,T_e^2$

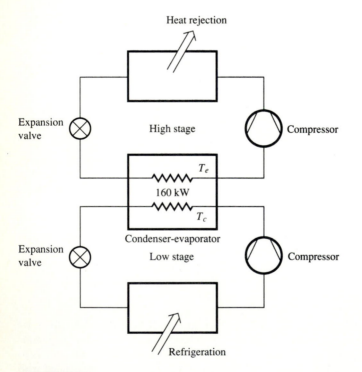

Heat rejection

Expansion
valve

High stage

Compressor

T_e

160 kW

T_c

Condenser-evaporator

Expansion
valve

Low stage

Compressor

Refrigeration

FIGURE 8-14
Cascade refrigeration system in Prob. 8.7.

where A = heat-transfer area, m^2
      T_c = condensing temperature, K
      T_e = evaporating temperature, K

The U value of the condenser-evaporator is 0.8 kW/(m$^2 \cdot$ K), and the rate of heat that must be transferred is 160 kW.

Using the method of Lagrange multipliers, determine the values of A, T_c, and T_e that result in minimum total present worth of costs.

Ans.: A^* = 20m^2.

8.8. A solar collector and storage tank, shown in Fig. 8-15a, is to be optimized to achieve minimum first cost. During the day the temperature of water in the storage vessel is elevated from 25°C (the minimum useful temperature) to t_{max}, as shown in Fig. 8-15b. The collector receives 260 W/m^2 of solar energy, but there is heat loss from the collector to ambient air by convection. The convection coefficient is 2 W/(m$^2 \cdot$ K), and the average temperature difference during the 10-hour day is $(25 + t_{max})/2$ minus the ambient temperature of 10°C.

The energy above the minimum useful temperature of 25°C that is to be stored in the vessel during the day is 200,000 kJ. The density of water is 1000 kg/m^3, and its specific heat is 4.19 kJ/(kg $\cdot$ K). The cost of the solar collector in dollars is 20A, where A is the area in square meters, and the cost of the storage vessel in dollars is 101.5V, where V is the volume in cubic meters.

(a) Using A and V as the variables, set up the objective function and constraint to optimize the first cost.

(b) Develop the Lagrange multiplier equations and verify that they are satisfied by V = 1.2 m^3 and A = 29.2 m^2.

Ans.: (a) y = 20A + 101.5V , subject to $A(230 - 47.7/V)$ = 5555.

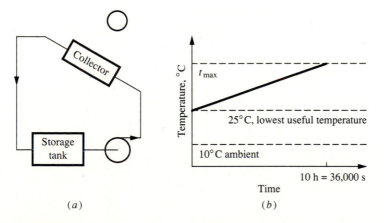

(a)   (b)

FIGURE 8-15
(a) Collector and storage tank (b) temperature variation in Prob. 8.8.

FIGURE 8-16
Air duct in Prob. 8.9.

8.9. Determine the diameters of the circular air duct in the duct system shown schematically in Fig. 8-16 so that the drop in static pressure between points A and B will be a minimum.

Further information

> Quantity of sheet metal available for the system, 60 m².
> Pressure drop in a section of straight duct, $f(L/D)(V^2/2)\rho$.
> Use a constant friction factor $f = 0.02$.
> Air density ρ, 1.2 kg/m³.
> Neglect the influence of changes in velocity pressure.
> Neglect the pressure drop in the straight-through section past an outlet.

(a) Set up the objective function and constraint(s) in terms of D_1, D_2, and D_3.
(b) Using the method of Lagrange multipliers for constrained optimization, determine the optimal values of the dimensions and the minimum cost.

   Ans.: Minimum cost = $9000.

8.10. Load dispatchers for electric utilities attempt to operate the most efficient combination of generating units that satisfies the total load requirement. If the input-output curves for two units are as shown in Fig. 8-17, where O_1 and O_2 are the outputs of units 1 and 2, respectively, and I_1 and I_2 are the inputs, respectively, and where the total load is T, use the method of Lagrange multipliers to prove that the optimum loading O_1^* and O_2^* occur when the slopes of the curves are equal.   •

8.11. A cylindrical oil-storage tank is to be constructed for which the following costs apply:

	Cost per square meter
Metal for sides	$30.00
Combined costs of concrete base and metal bottom	37.50
Top	7.50

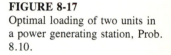

FIGURE 8-17
Optimal loading of two units in a power generating station, Prob. 8.10.

The tank is to be constructed with dimensions such that the cost is minimum for whatever capacity is selected.

(*a*) One possible approach to selecting the capacity is to build the tank large enough for an additional cubic meter of capacity to cost $8. (Note that this does not mean $8 per cubic meter average for the entire tank.) What is the optimal diameter and optimal height of the tank?

Ans.: 15 m, 11.25 m.

(*b*) Instead of the approach used in part (*a*), the tank is to be of such a size that the cost will be $9 per cubic meter average for the entire storage capacity of the tank. Set up the Lagrange multiplier equations and verify that they are satisfied by an optimal diameter of 20 m and an optimal height of 15 m.

8.12. A rectangular duct mounted beneath the beams in a building, as shown in Fig. 8-18, is to have a cross-sectional area of 0.8 m^2. The cost of the duct for the required length is $150 per meter of perimeter. The building must be heightened by the amount of the height of the duct, and that cost is $0.80 per millimeter. After the duct has been sized to provide minimum total cost, the possibility of enlarging the duct is explored. What is the additional cost per square meter of cross-sectional area of a very small increase in the area?

Ans.: $642 per square meter.

8.13. An electric-power generating and distribution system consists of two generating plants and three loads, as shown in Fig. 8-19. The loads are as follows:

FIGURE 8-18
Duct mounted under beams in Prob. 8.12.

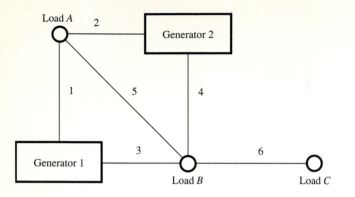

FIGURE 8-19
Generating and distribution system in Prob. 8.13.

load A = 40 MW, load B = 60 MW, and load C = 30 MW. The losses in the lines are given in Table 8.1, where the loss in line i is a function of the power p_i in megawatts carried by the line. To be precise, the line loss should be specified as a function of the power at a certain point in the line, e.g., the entrance or exit, but since the loss will be small relative to the power carried, use p_i at the point in the line most convenient for calculation.

As a first approximation in the load balances, assume that p_5 leaving load A equals p_5 entering load B, and recalculate, if necessary, after the first complete solution.

Assuming that the two generating plants are equally efficient, use the method of Lagrange multipliers to compute the optimum amount of power to be carried by each of the lines for the most efficient operation.

Ans.: 24.3, 20.3, 40, 46.4, 4.6, 31 MW.

8.14. The power-distribution system shown in Fig. 8-20 has a source voltage of 220 V at point 1 and must supply power to positions 3 and 4 at 210 and 215 V, respectively, with a current of 200 and 300 A, respectively. The electrical resistance $R\Omega$ is a function of the area and length of the conductor: $R, \Omega = 17.2 \times 10^{-9} L/A$, where L is the length of conductor in meters and A is the area of conductor in square meters.

TABLE 8.1
Line losses in Prob. 8.13

Line	Loss, MW
1	$0.0010p_1^2$
2	$0.0012p_2^2$
3	$0.0007p_3^2$
4	$0.0006p_4^2$
5	$0.0008p_5^2$
6	$0.0011p_6^2$

FIGURE 8-20
Power-distribution network in Prob. 8.14.

(*a*) Set up the objective function for the total volume of conductor and the constraint(s) in terms of A_{1-2}, A_{2-4}, and A_{2-3}.

(*b*) Verify, using all the Lagrange multiplier equations, that $A_{1-2}^* = 0.00273$ m^2.

ADDITIONAL READINGS

Bowman, F., and F. A. Gerard: *Higher Calculus*, Cambridge University Press, London, 1967.

Brand, L.: *Advanced Calculus*, Wiley, New York, 1955.

Kaplan, W.: *Advanced Calculus*, Addison-Wesley, Reading, Mass., 1952.

Taylor, A. E.: *Advanced Calculus*, Ginn, Boston, Mass., 1955.

Wilde, D. J., and C. S. Beightler: *Foundations of Optimization*, Prentice-Hall, Englewood Cliffs, N.J., 1967.

CHAPTER
9

SEARCH
METHODS

9.1 OVERVIEW OF SEARCH METHODS

The name *search method* should not suggest a helter-skelter wandering through the variables of optimization until a favorable value of the objective function, subject to the constraints, has been found. Instead, search methods generally fall into two categories, *elimination* and *hill-climbing techniques*. In both there is a progressive improvement throughout the course of the search.

In one respect search methods are a paradox. They are often the ultimate approach if other methods of optimization (such as Lagrange multipliers, dynamic programming, etc.) fail. From another standpoint search methods are unsatisfying because there is no one systematic procedure that is followed and the literature is laden with different variations of search techniques. When a problem is found in which the standard techniques experience difficulty, another variation is developed. In the short treatment of this chapter we shall adhere to several approaches that are quite classical, have broad applicability, and flow naturally from the calculus principles presented in Chapter 8. Chapter 17 puts the concepts of search methods as applied to constrained problems into a sharper focus through the use of penalty techniques and the popular reduced-gradient method.

The outline of the coverage in this chapter is as follows:

1. Single variable
   a. Exhaustive
   b. Efficient
      i. Dichotomous
      ii. Fibonacci
2. Multivariable, unconstrained
   a. Lattice
   b. Univariate
   c. Steepest ascent
3. Multivariable, constrained
   a. Penalty functions
   b. Search along a constraint

In the calculus method of optimization presented in Chapter 8, calculating the numerical value of the objective function was virtually the last step in the process. The major effort in the optimization was determining the values of the independent variables that provide the optimum. In optimization by means of search methods, an opposite sequence is followed in that values of the objective function are determined and conclusions are drawn from the values of the function at various combinations of independent variables.

9.2   INTERVAL OF UNCERTAINTY

An accepted feature of search methods is that the precise point at which the optimum occurs will never be known, and the best that can be achieved is to specify the *interval of uncertainty*. This is the range of the independent variable(s) in which the optimum is known to exist. An interval of uncertainty prevails because the search method computes the value of the function only at discrete values of the independent variables.

9.3   EXHAUSTIVE SEARCH

Of the various search methods used in single-variable problems, the exhaustive search is the least imaginative but most widely used, and justifiably so. The method consists of calculating the value of the objective function at values of x that are spaced uniformly throughout the interval of interest. The interval of interest I_0 (Fig. 9-1) is divided here into eight equal intervals. Assume that the values of y are calculated at the seven positions shown. In this example the maximum lies between x_A and x_B, so the final interval of uncertainty I is

$$I = \frac{2I_0}{8} = \frac{I_0}{4}$$

FIGURE 9-1
Exhaustive search.

If two observations are made, the final interval of uncertainty is $2I_0/3$; if three observations are made, the final interval of uncertainty is $2I_0/4$, and in general

$$\text{Final interval of uncertainty} = I = \frac{2I_0}{n+1} \tag{9.1}$$

9.4 UNIMODAL FUNCTIONS

The dichotomous and Fibonacci methods, introduced next, are applicable to *unimodal* functions. A unimodal function is one having only one peak (or valley) in the interval of interest. Figure 9-2 shows several unimodal functions. The dichotomous and Fibonacci methods can successfully handle not only the smooth curve like Fig. 9-2*a* but also the nondifferentiable function like Fig. 9-2*b* or even a discontinuous function like Fig. 9-2*c*.

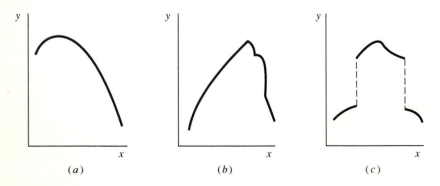

FIGURE 9-2
Unimodal functions.

In optimizing nonunimodal functions where there are several peaks (or valleys), the function can be subdivided into several parts and each part processed separately as a unimodal function.

9.5 ELIMINATING A SECTION BASED ON TWO TESTS

Knowledge of the value of the objective function at two different positions in the interval of interest is sufficient to eliminate a portion of the region of a unimodal function. Suppose that a maximum value of y is sought in the function that exists in Fig. 9-3a. The magnitude of y is known at two values of x, designated x_A and x_B. From this information it is possible to eliminate the region to the left of x_A. The region to the right of x_A must still be retained. It cannot be determined whether the maximum lies between x_A and x_B or to the right of x_B because the maximum could reside in either interval, as shown in Fig. 9-3b.

9.6 DICHOTOMOUS SEARCH

The concept of the *dichotomous search* follows closely the discussion of placing two test points. It asks the following: Where should the two points be placed in the interval of uncertainty in order to eliminate the largest possible region? A little reflection will show that placing the points as near the center as possible while maintaining distinguishability of the y values will result in elimination of almost half the original interval of uncertainty. Figure 9-4 shows the two points placed symmetric to the center with a spacing of ϵ between. With the values of y as shown, the region to the left of x_A can be eliminated, so the remaining interval of uncertainty is $(I_0 + \epsilon)/2$.

FIGURE 9-3
Two test points on a unimodal function.

FIGURE 9-4
Dichotomous search.

The next pair of observation is made in a similar manner in the remaining interval of uncertainty, resulting in the further reduction of the interval by nearly one-half. In general, the remaining interval of uncertainty I is

$$I = \frac{I_0}{2^{n/2}} + \epsilon\left(1 - \frac{1}{2^{n/2}}\right) \tag{9.2}$$

where n is the number of tests (2, 4, 6, etc.).

9.7 FIBONACCI SEARCH

The most efficient of the single variable search techniques is the Fibonacci method. This method was first presented by Kiefer,[1] who applied the Fibonacci number series, which was named after the thirteenth-century mathematician. The rule for determining a Fibonacci number F is

$$F_0 = 1$$

$$F_1 = 1$$

$$F_i = F_{i-2} + F_{i-1} \qquad \text{for } i \geq 2$$

Thus, after the first two Fibonacci numbers are available, each number thereafter is found by summing the two preceding numbers. The Fibonacci series starting with the index zero is therefore 1, 1, 2, 3, 5, 8, 13, 21, 34, 55, $F_{10} = 89$, etc.

The steps in executing a Fibonacci search are as follows:

1. Decide how many observations will be made and call this number n.
2. Place the first observation in I_0 so that the distance from one end is $I_0(F_{n-1}/F_n)$.
3. Place the next observation in the interval of uncertainty at a position

that is symmetric to the existing observation. According to the relative values of these observations, eliminate either the region to the right of the right point or to the left of the left point. Continue placing a point and eliminating a region until one point remains to be placed. At this stage there will be one observation directly in the center of the interval of uncertainty.

4. Place the last observation as close as possible to this center point and eliminate half the interval.

Example 9.1. Perform a Fibonacci search to find the maximum of the function $y = -(x)^2 + 4x + 2$ in the interval $0 \leq x \leq 5$.

Solution

1. Arbitrarily choose $n = 4$.
2. Place the first observation a distance $I_0(F_3/F_4)$ from the left end, as in Fig. 9-5. This distance is $\frac{3}{5} I_0$ or $(\frac{3}{5})(5)$. The current interval of uncertainty is 0 to 5 with an observation at 3.
3. The next observation symmetric in the interval of uncertainty to 3 locates this observation at $x = 2$. By making use of the relative values of y at $x = 2$ and $x = 3$, the section $3 < x \leq 5$ can be eliminated. The interval of uncertainty is now $0 \leq x \leq 3$, with the observation at $x = 2$ available. Placing the third point symmetric to the $x = 2$ observation locates it at $x = 1$. The relative values of y at $x = 1$ and $x = 2$ permit elimination of $0 \leq x < 1$.

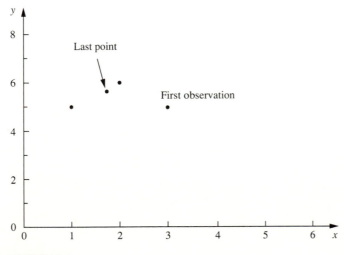

FIGURE 9-5
Fibonacci search in Example 9.1.

4. The final point is positioned as close as possible to the $x = 2$ point, which is now in the center of the existing interval of uncertainty. If this final point is placed at $x = 2 - \epsilon$, the final interval of uncertainty is $2 - \epsilon \leq x \leq 3$.

The final interval of uncertainty in Example 9.1 was $I_0/5 + \epsilon$. In general, if n observations are made, the final interval of uncertainty is $I_0/F_n + \epsilon$.

9.8 COMPARATIVE EFFECTIVENESS OF SEARCH METHODS

A measure of the efficiency of a search method is the reduction ratio (RR), proposed by Wilde,[2] which is defined as the ratio of the original interval of uncertainty to the interval after n trials, I_n:

$$RR = \frac{I_0}{I_n} \tag{9.3}$$

If the penalty on the reduction ratio imposed by ϵ in the dichotomous and Fibonacci methods is neglected, the reduction ratios are

$$RR = \begin{cases} \dfrac{n + 1}{2} & \text{exhaustive} & (9.4) \\[2ex] 2^{n/2} & \text{dichotomous} & (9.5) \\[2ex] F_n & \text{Fibonacci} & (9.6) \end{cases}$$

The RR of the Fibonacci method, for example, is 377 after 13 trials. The comparison is shown graphically in Fig. 9-6.

FIGURE 9-6
Comparison of reduction ratios of search methods.

9.9 ASSESSMENT OF SINGLE-VARIABLE SEARCHES

The efficiency of such methods as the dichotomous and Fibonacci searches compared with the exhaustive search is impressive. Realistically, however, the single-variable optimizations are not the ones where high efficiencies are needed. In most single-variable optimizations in thermal systems, if the search is performed on a computer, the exhaustive search requires only a few more dollars of computer time and is easier to program. Only if the calculations required for each point are extremely lengthy is there any significant advantage in one of the efficient search methods, and such extensive calculations for single-variable optimizations are rare.

Another potential use of efficient single-variable searches is as a step in a multivariable search. An additional restriction arises, however, and that is the risk of the function being nonunimodal. Later in this chapter a single-variable search will be shown (Fig. 9-11, for example) making a cut across some contours in a multivariable problem. If those contours are kidney-shaped, the single-variable cut may describe a nonunimodal function to which the dichotomous and Fibonacci methods do not apply.

We move now to multivariable problems, where exhaustive searches may be prohibitively expensive and efficient techniques are demanded.

9.10 MULTIVARIABLE, UNCONSTRAINED OPTIMIZATION

High efficiency in multivariable searches may be crucial. The most significant optimizations in thermal systems involve many components and thus many variables. Furthermore, the complexity of the equations of a many-component system makes the algebra extremely tedious if a calculus method is used. This complexity breeds errors in formulation. In many multivariable situations, the number of calculations using an exhaustive search introduces considerations of computer time. Suppose, for example, that the optimum temperatures are sought for a seven-stage heat-exchanger chain. Assume that 10 different outlet temperatures for each heat exchanger will be investigated in all combinations with the outlet temperatures of the other heat exchangers. The total number of combinations explored will be 10^7, which, if the calculation is at all complex makes computer time a definite concern.

The single-variable dichotomous and Fibonacci methods are elimination methods, because after each new point in the Fibonacci search (or pair of points in the dichotomous search) a portion of the interval of uncertainty is eliminated. The three methods for optimization of an unconstrained multivariable problem, to be explained next, are *hill-climbing techniques*, because the calculation is always moving in a way that improves the objective function. The three methods are (1) lattice search, (2) univariate search, and (3) steepest ascent.

9.11   LATTICE SEARCH

The procedure in the lattice search is to start at one point in the region of interest and check a number of points in a grid surrounding the central point. The surrounding point having the largest value (if a maximum is being sought) is chosen as the central point for the next search. If no surrounding point provides a greater value of the function than the central point, the central point is the maximum. A frequent practice is first to use a coarse grid and after the maximum has been found for that grid subdivide the grid into smaller elements for a further search, starting from the maximum of the coarse grid.

As an example of the progression to the maximum of a function of two variables, a grid is superimposed over the contour lines of a function in Fig. 9-7. The starting point can be selected near the center of the region unless there is some advance knowledge of where the maximum exists. Call the first point 1 and evaluate the function at points 1 to 9. In this case, the maximum value of the function occurs at point 5, so point 5 becomes the central point for the next search.

Our special objective in applying the lattice search is to improve the efficiency compared with an exhaustive search. Since efficiency is a goal, it is attractive to consider saving information that has already been calculated and can be used again. For example, in Fig. 9-7 after moving from point 1 to point 5, the values of the function at points 1, 4, 5, and 6 are needed in deciding which way to move from point 5. Writing a computer program to accomplish this task is tricky, however, particularly if

FIGURE 9-7
A lattice search.

the number of independent variables exceeds the number of dimensions possible for subscripted variables on the computer being used. It is therefore likely that the functions will be evaluated at all nine points, in the case of two variables, for each central-point location.

No definite statement can be made about the comparative efficiency of the lattice search and the exhaustive search because the nature of the function dictates the number of trials required in the lattice search. In general, however, the ratio of the lattice trials to exhaustive trials decreases as the number of variables increases and the grid becomes finer.

9.12 UNIVARIATE SEARCH

In the univariate search, the function is optimized with respect to one variable at a time. The starting procedure is to substitute trial values of all but one independent variable in the function and optimize the resulting function in terms of the one remaining variable. That optimal value is then substituted into the function and the function optimized with respect to another variable. The function is optimized with respect to each variable in sequence with the optimal value of a variable substituted into the function for the optimization of the succeeding variables. The process is shown graphically in Fig. 9-8 for a function of two variables. Along the line of constant x_1, which is the initial choice, the value of x_2 giving the optimal value of y is determined. This position is designated as point 1. With the value of x_2 at point 1 substituted into the function, the function is optimized with respect to x_1, which gives point 2. The process continues until the successive change of the dependent or independent variables is less than a specified tolerance.

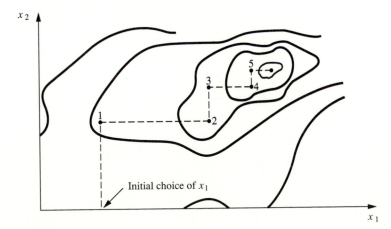

FIGURE 9-8
Univariate search.

The method chosen for performing the single-variable optimization may be to use calculus, where the task becomes one of solving one equation (usually a nonlinear one) for one variable. It is also possible to use a single-variable search, exhaustive or efficient, e.g., Fibonacci or dichotomous. It is this use of the efficient single-variable search that is probably the most significant application of this type of search method in thermal-system optimization.

The univariate search can fail when a ridge occurs in the objective function. In the function in Fig. 9-9, for example, if a trial value of x_1 is selected as shown, the optimal value of x_2 lies on the ridge. When this optimal value of x_2 is substituted, the attempt to optimize with respect to x_1 does not dislodge the point from the ridge even though the optimum has not been reached.

From the purely mathematical point of view, getting hung up on the ridge would appear to be a serious deficiency of the univariate search. In physical systems, however, the occurrence of ridges is rare because nature avoids discontinuities of both functions and derivatives of functions. Caution is needed, however, because even though a true ridge does not exist, problems may arise if the contours are very steep. If the interval chosen for the univariate search is too large, the process may stop at a nonoptimal point, such as point A in Fig. 9-10.

Another note of caution is that even though ridges may not occur in the physical system, the equations used to represent the physical system may accidentally contain ridges.

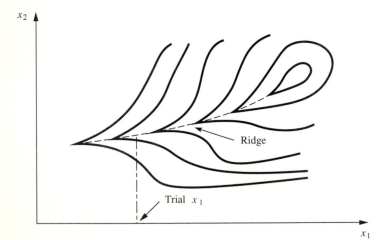

FIGURE 9-9
Failure of the invariate search at a ridge.

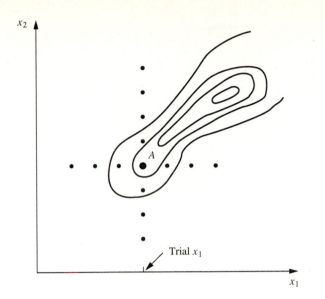

FIGURE 9-10
Erroneous conclusion because the interval of search is too large.

9.13 STEEPEST-ASCENT METHOD

As the name implies, this multivariable search method moves the state point in such a direction that the objective function changes at the greatest favorable rate. As shown in Sec. 8.7, the gradient vector is normal to the contour line or surface and therefore indicates the direction of maximum rate of change. In the function of two variables whose contour lines are shown in Fig. 9-11, the gradient vector ∇y at point A is normal to the contour line at that point and indicates the direction in which y increases at the greatest rate with respect to distance on the $x_1 x_2$ plane. The equation for ∇y is

$$\nabla y = \frac{\partial y}{\partial x_1} i_1 + \frac{\partial y}{\partial x_2} i_2 \qquad (9.7)$$

where i_1 and i_2 are unit vectors in the x_1 and x_2 directions, respectively.

The essential steps in executing the steepest ascent method are as follows:

1. Select a trial point.
2. Evaluate the gradient at the current point and the relationship of the changes of the x variables.
3. Decide in which direction to move along the gradient.
4. Decide how far to move and then move that distance.

FIGURE 9-11
Steepest-ascent method.

5. Test to determine whether the optimum has been achieved. If so, terminate, otherwise return to step 2.

The first three steps are standard, but there are many variations of steps 4 and 5. The individual steps will now be discussed.

Step 1. The trial point should be chosen as near to the optimum as possible, but usually such insight is not available and the point is selected arbitrarily.

Step 2. The partial derivatives can be extracted mathematically, or it may be more convenient to compute them numerically by resorting to the equation for the partial derivative,

$$\frac{\partial y}{\partial x_i} = \frac{y(x_1, \ldots, x_i + \Delta, \ldots, x_n) - y(x_1, \ldots, x_i, \ldots, x_n)}{\Delta}$$

$$(9.8)$$

where Δ is a very small value. In order to move in the direction of the gradient, the relationship of the changes in the x's, called Δx's, is

$$\frac{\Delta x_1}{\partial y / \partial x_1} = \frac{\Delta x_2}{\partial y / \partial x_2} = \ldots = \frac{\Delta x_n}{\partial y / \partial x_n} \qquad (9.9)$$

Step 3. It is understood that Δx_i indicates the change in x_i, but a decision must be made whether to increase x_i by Δx_i or to decrease x_i. That decision is controlled by whether a maximization or minimization is being performed. If a maximization is in progress and $\partial y / \partial x_i$ is negative, Δx_i must be negative in order for y to increase as a result of the move.

Step 4. Only two of the numerous methods for deciding how far to move in the direction of steepest ascent will be presented. The first method is to select arbitrarily a step size for one variable, say Δx_i, and compute Δx_2 to Δx_n from Eq. (9.9). This method usually works well until one or more of the partial derivatives approaches zero.

The second method of deciding how far to move is to proceed until reaching an optimal value in the direction of the gradient. Then another gradient vector is calculated, and once again the position moves until an optimal value in that direction is achieved. From point B in Fig. 9-11 the point moves to C and thence to D. The process of determining the optimal value in the gradient direction can be converted into a single-variable search by using Eq. (9.9) to eliminate all x's except one in the objective function. The Fibonacci method can then be used as an efficient single-variable search.

Example 9.2. An insulated steel tank storing ammonia, as shown in Fig. 9-12, is equipped with a recondensation system which can control the pressure and thus the temperature of the ammonia.[3] Two basic decisions to make in the design of the tank are the shell thickness and insulation thickness.

If the tank operates with a temperature near ambient, the pressure in the tank will be high and a heavy expensive vessel will be required. On the other hand, to maintain a low pressure in the tank requires more operation of the recondensation system because there will be more heat transferred from the environment unless the insulation is increased, which also adds cost.

Determine the optimum operating temperature and insulation thickness if the following costs and other data apply:

Vessel cost, $1000 + 2.2(p - 100)^{1.2}$ dollars for $p \geq 200$ kPa
Insulation cost for the 60 m^2 of heat-transfer area, $21x^{0.9}$ dollars

FIGURE 9-12
Ammonia storage tank in Example 9.2.

Recondensation cost, 2.5 cents per kilogram of ammonia
Lifetime hours of operation, 50,000 h
Ambient temperature, 25°C
Average latent heat of vaporization of ammonia, 1200 kJ/kg
Conductivity of insulation k, 0.04 W/(m · K)
Pressure-temperature relation for ammonia

$$\ln p = -\frac{2800}{t + 273} + 16.33$$

Solution. The total lifetime cost is the sum of three individual costs, the vessel, the insulation, and the lifetime cost of recondensation. All these costs will be expressed in terms of the operating temperature t°C and the insulation thickness x mm.

The insulation cost is

$$IC = 21x^{0.9}$$

The saturation pressure is a function of temperature

$$p = e^{-2800/(t+273) + 16.33}$$

and so the vessel cost VC is

$$VC = 1000 + 2.2(e^{-2800/(t+273) + 16.33} - 100)^{1.2}$$

Recondensation cost RC is

$$RC = (w \text{ kg/s})(0.25 \text{ \$/kg})(3600 \text{ s/h})(50,000 \text{ h}) \tag{9.10}$$

where w is the evaporation and recondensation rate in kilograms per second. But also

$$w = \frac{q \text{ kW}}{1200 \text{ kJ/kg}} \tag{9.11}$$

where q is the rate of heat transfer from the environment to the ammonia. Assuming that only the insulation provides any significant resistance to heat transfer, we have

$$q, \text{kW} = \frac{25 - t}{(x \text{ mm})/1000} [0.00004 \text{ kW/(m · K)} (60 \text{ m}^2) \tag{9.12}$$

Combining Eqs. (9.10) to (9.12) results in the expression for the recondensation cost

$$RC = \frac{9000(25 - t)}{x}$$

The total cost is the sum of the individual costs,

$$\text{Total cost } C = IC + VC + RC$$

The search method chosen to perform this optimization will be the steepest descent. The position will be moved along the gradient direction until a minimum is reached; then a new gradient will be established. The partial derivatives of the total cost C with respect to x and t are

$$\frac{\partial C}{\partial x} = (0.9)(21)x^{-0.1} - \frac{9000(25 - t)}{x^2}$$

$$\frac{\partial C}{\partial t} = (2.2)(1.2)(e^A - 100)^{0.2}e^A \frac{2800}{(t + 273)^2} - \frac{9000}{x}$$

where

$$A = -\frac{2800}{t + 273} + 16.33$$

For the first point, arbitrarily select $x = 100$ mm and $t = 5°C$. At this position $C = \$7237.08$, $\partial C/\partial x = -6.075$, and $\partial C/\partial t = 77.347$. The derivative with respect to x is negative and with respect to t is positive; therefore to decrease C the value of x must be increased and t must be decreased. Furthermore, to move along the direction of the gradient, the changes in x and t, designated Δx and Δt, should bear the relation

$$\frac{\Delta x}{-6.075} = \frac{\Delta t}{77.347}$$

The minimum value of C achieved by moving in the direction of this gradient is $\$6675.58$, where $x = 101.23$ and $t = -10.66$. New partial derivatives are computed at this point and the position changed according to the new gradient. Table 9.1 presents a summary of the calculations.

The minimum cost is $\$5986.04$ when the insulation thickness is 196.8 mm and the operating temperature is $-23.3°C$. Both partial derivatives are nearly zero at this point. The steepest-descent calculation required an appreciable number of steps before finally homing in on the optimum. The reason is that the route passes through a curved valley, and the minimum point along the gradient moved from one side of the valley to the other, as indicated by the alternation in sign of $\partial C/\partial t$.

TABLE 9.1
Steepest-descent search in Example 9.2

Iteration	x, mm	t, °C	C	$\partial C/\partial x$	$\partial C/\partial t$
0	100.00	5.00	$7237.08	-6.075	77.347
1	101.23	-10.66	6675.58	-19.409	-1.455
2	142.83	-7.54	6334.98	-2.845	37.505
3	143.62	-17.78	6152.85	-7.165	-0.634
4	168.92	-15.54	6067.68	-1.471	16.145
5	169.40	-20.80	6026.09	-3.051	-0.294
6	182.87	-19.50	6005.85	-0.749	7.477
. . .	. . . . .	. . . . .	. . . . .	. . . . .	. . . . .
32	196.56	-23.24	5986.07	-0.090	0.189
33	196.62	-23.36	5986.05	-0.112	-0.123
34	196.69	-23.28	5986.05	-0.086	0.111
35	196.77	-23.38	5986.04	-0.100	-0.143
36	196.83	-23.30	5986.04	-0.076	0.088

9.14   SCALES OF THE INDEPENDENT VARIABLES

The name steepest ascent implies the best possible direction in which to move. The meaning of this statement is that for a given distance $\sqrt{\Delta x_1^2 + \Delta x_2^2 + \ldots}$ the objective function will experience a maximum change. As Fig. 9-13a shows, however, it may be desirable for there to be large changes in x_1 compared with those of x_2. Wilde[2] extends the conclusions of Buehler, Shah, and Kampthorne[4] to recommend that the scales be chosen so that the contours are as spherical as possible in order to accelerate the convergence. In Fig. 9-13b, for example, the original equation would be revised with a new variable x_2' replacing x_2 so that $x_2' = 4x_2$ and the contours would thus cover the same range as x_1.

9.15   CONSTRAINED OPTIMIZATION

Constrained optimizations are probably the most frequent and most important ones encountered in the design of thermal systems. Numerous methods and variations of search methods applicable to constrained optimizations are presented in the literature, most of them prompted by attempts to accelerate the search. In some cases new methods were developed to prevent the search from failing on a certain function or problem. It is this multitude of techniques that blunts the satisfaction of using search methods, because there is not just one fundamental concept, as there is with Lagrange multipliers.

Two methods of constrained searches will be presented in the next several sections, (1) conversion to unconstrained by use of penalty functions and (2) searching along the constraint. Only equality constraints (and

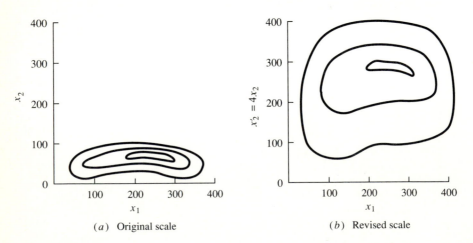

(*a*)   Original scale   (*b*)   Revised scale

FIGURE 9-13
Effect of scale of independent variables.

not inequality constraints) will be considered, although the user of a search could adapt the equality-constraint technique by making an inequality constraint active or inactive depending upon whether the constraint is violated or not.

9.16   PENALTY FUNCTIONS

If a function is to be maximized

$$y = y(x_1, x_2, \ldots, x_n) \rightarrow \text{maximum}$$

subject to the constraints

$$\phi_1(x_1, x_2, \ldots, x_n) = 0$$
$$\cdots \cdots \cdots \cdots \cdots \cdots \cdots \cdots$$
$$\phi_m(x_1, x_2, \ldots, x_n) = 0$$

a new unconstrained function can be constructed

$$Y = y - P_1(\phi_1)^2 - P_2(\phi_2)^2 - \cdots - P_m(\phi_m)^2$$

If the function is to be minimized, the $P_i(\phi_i)^2$ terms would be added to the original objective function to construct the new unconstrained function. The underlying principle of the technique is valid, but care must be exercised in the execution to maintain proper relative influence of the function being optimized to that of the constraints. The choice of the P terms provides the relative weighting of the two influences, and if P is too high, the search will satisfy the constraint but move very slowly in optimizing the function. If P is too small, the search may terminate without satisfying the constraints adequately. One suggestion is to start with small values of the P's and gradually increase the values as the magnitudes of the ϕ's become small.

9.17   OPTIMIZATION BY SEARCHING ALONG A CONSTRAINT–HEMSTITCHING

The next search technique for constrained optimization to be explained is the "search along the constraint(s)" or "hemstitching" method.[5,6] The technique consists of starting at a trial point and first driving directly toward the constraint(s). Once on the constraint(s), the process is one of optimizing along the constraint(s). For nonlinear constraints, a tangential move starting on a constraint moves slightly off the constraint, so after each tangential move it is necessary to drive back onto the constraint. This search method is one of many that are available but is effective in most problems and offers the further satisfaction of building logically on the principles of Lagrange multipliers. Hemstitching introduces the flavor of the generalized-reduced gradient method that will be explored in Section 17.4.

Three different cases will be explored in the next several sections: (1) two-variable problem with one constraint, (2) three-variable problem with one constraint, and (3) three-variable problem with two constraints. All of the constraints will be equalities. These three cases lead us through the several fundamental operations that appears in larger problems as well, namely, (1) driving toward constraint(s), (2) moving in a favorable direction with respect to the objective function under complete restriction of the constraint(s) and (3) moving in a favorable direction with respect to the objective function with two or more degrees of freedom.

9.18   DRIVING TOWARD THE CONSTRAINT(S)

The number of constraints m will always be less than the number of variables n, so if m variables are held constant at the current location, the remaining $n - m$ variable(s) can be solved to bring the point back onto the constraint(s).

> **Example 9.3.** A constraint in an optimization problem is $x_1^2 x_2 = 8$. Locate a point on the constraint using a trial value of (a) $x_1 = 2$, and (b) $x_2 = 1$.
>
> **Solution.** (a) $x_2 = 8/x_1^2 = 0.5$, and (b) $x_1 = (8/x_2)^{0.5} = 2.83$. The purpose of the trivial calculation in Example 9.3 is to draw attention to the process of returning to the constraint following a tangential move, as illustrated in Fig. 9-14. The choice of whether to hold x_1 or x_2 constant when returning to the constraint is arbitrary, but in certain problems the choice may influence the rate of convergence to the optimum.

9.19   HEMSTITCHING SEARCH WHEN n − m = 1

One constraint in a two-variable problem or two constraints in a three-variable problem establish the vector for a tangential move along the constraint(s). Only the direction of the move along the vector is open to question, and this direction is chosen to achieve a favorable change in the objective function. In the two-variable problem, where the constraint $\phi(x_1, x_2) = 0$, the tangent attempts to maintain $\phi = 0$ during the move, so

$$\Delta\phi \approx \frac{\partial\phi}{\partial x_1}\Delta x_1 + \frac{\partial\phi}{\partial x_2}\Delta x_2 = 0 \tag{9.13}$$

so the relationship of Δx_1 and Δx_2 in the tangential move must satisfy

$$\frac{\Delta x_1}{\Delta x_2} = -\frac{\partial\phi/\partial x_2}{\partial\phi/\partial x_1} \tag{9.14}$$

Thus, to determine whether to increase or decrease x_2, return to the objective function $y(x_1, x_2)$, where

$$\Delta y \approx \frac{\partial y}{\partial x_1}\Delta x_1 + \frac{\partial y}{\partial x_2}\Delta x_2 \tag{9.15}$$

FIGURE 9-14
Hemstitching search in a two-variable problem with (a) return to the constraint with fixed x_1 (b) return to the constraint with fixed x_2.

which in combination with Eq. (9.14) gives

$$\Delta y = \left(\frac{\partial y}{\partial x_2} - \frac{\partial y}{\partial x_1} \frac{\partial \phi/\partial x_2}{\partial \phi/\partial x_1} \right) \Delta x_2 = G \Delta x_2 \qquad (9.16)$$

In a minimization, for example, when $G > 0$, Δx_2 should be negative, and thus x_2 should be decreased.

Example 9.4. The objective function associated with the constraint of Example 9.3, $x_1^2 x_2 - 8 = 0$, is

$$y = 3x_1^2 + x_2^2$$

Minimize this function starting with a trial value of $x_2 = 1.6$ choosing a step size $|\Delta x_2| = 0.05$, and returning to the constraint by holding x_2 constant.

Solution. With the trial value of $x_2 = 1.6$, the constraint requires $x_1 = 2.236$, so the first point (2.236, 1.6) gives $y = 17.56$ and $\phi = 0$. Table 9.2 shows the progression of moves, starting tangent to the constraint followed by a move back to the constraint at a constant value of x_2. The initial value of G is -6.175, but steadily decreases until it passes through zero and attains a positive value of 0.0631 on the 17th cycle. The zero-value of G is critical, because this position satisfies the Lagrange multiplier equations that establish the optimum. This fact can be shown by arbitrarily defining a term λ as

$$\lambda = \frac{\partial y/\partial x_1}{\partial \phi/\partial x_1}$$

TABLE 9.2
Hemstitching search in Example 9.4

Cycle	Before move	x_1	x_2	y	ϕ	G
1	tangent	2.236	1.60	17.560	0	-6.175
	return	2.271	1.65	18.194	0.509	
2	tangent	2.202	1.65	17.268	0	-5.515
	return	2.235	1.70	17.880	0.494	
3	tangent	2.169	1.70	17.008	0	-4.904
	return	2.201	1.75	17.598	0.479	
. . . .						. . . .
16	tangent	1.886	2.25	15.729	0	-0.241
	return	1.907	2.30	16.195	0.361	
17	tangent	1.865	2.30	15.725	0	0.0631
	return	1.845	2.25	15.272	-0.343	
18	tangent	1.886	2.25	15.729	0	-0.241
	return	1.907	2.30	16.195	0.361	

When $G = 0$, λ also equals $(\partial y/\partial x_2)/(\partial \phi/\partial x_2)$. From the 17th cycle in Table 9.2 an oscillation about the optimum begins, and it would be here that the step size should be reduced.

The three-variable problem where $n - m = 1$ is

$$y(x_1, x_2, x_3) \rightarrow \text{optimize} \tag{9.17}$$

subject to

$$\phi_1(x_1, x_2, x_3) = 0 \tag{9.18}$$

and

$$\phi_2(x_1, x_2, x_3) = 0 \tag{9.19}$$

When on the constraint, a move that starts tangent to both constraints requires that

$$\frac{\partial \phi_1}{\partial x_1} \Delta x_1 + \frac{\partial \phi_1}{\partial x_2} \Delta x_2 + \frac{\partial \phi_1}{\partial x_3} \Delta x_3 = 0 \tag{9.20}$$

$$\frac{\partial \phi_2}{\partial x_1} \Delta x_1 + \frac{\partial \phi_2}{\partial x_2} \Delta x_2 + \frac{\partial \phi_2}{\partial x_3} \Delta x_3 = 0 \tag{9.21}$$

For a chosen step size of one of the variables, x_3, for example, Δx_1 and Δx_2 must satisfy the matrix equation,

$$\begin{bmatrix} \dfrac{\partial \phi_1}{\partial x_1} & \dfrac{\partial \phi_1}{\partial x_2} \\ \dfrac{\partial \phi_2}{\partial x_1} & \dfrac{\partial \phi_2}{\partial x_2} \end{bmatrix} \begin{bmatrix} \Delta x_1 \\ \Delta x_2 \end{bmatrix} = - \begin{bmatrix} \dfrac{\partial \phi_1}{\partial x_3} \\ \dfrac{\partial \phi_2}{\partial x_3} \end{bmatrix} \Delta x_3 \tag{9.22}$$

The direction of the tangential move along the vector to achieve a favorable

Δy in the equation

$$\Delta y = \frac{\partial y}{\partial x_1}\Delta x_1 + \frac{\partial y}{\partial x_2}\Delta x_2 + \frac{\partial y}{\partial x_3}\Delta x_3$$

determines whether Δx_3 should be positive or negative.

Following the move tangent to the two constraints, the position is brought back onto the constraints by simultaneous solution of Eqs. (9.18) and (9.19) holding one of the variables constant.

9.20 MOVING TANGENT TO A CONSTRAINT IN THREE DIMENSIONS

A special situation occurs in a three-variable optimization having one constraint, when the direction of the move must be decided such that the objective function changes in the most favorable manner. The move could be visualized as in Fig. 9-15, where a pointer (the vector) is pinned onto the surface (the constraint) and is free to turn in a plane. The direction to be taken by the pointer is such that the maximum change of y occurs, where

$$\Delta y = \frac{\partial y}{\partial x_1}\Delta x_1 + \frac{\partial y}{\partial x_2}\Delta x_2 + \frac{\partial y}{\partial x_3}\Delta x_3 \qquad (9.23)$$

FIGURE 9-15
Vector giving maximum rate of change of y, but also tangent to a constraint.

but even more specifically the maximum *rate of change* of y is sought. This means that the maximum Δy is to be determined subject to the constraint of a given length of the move

$$\Delta x_1^2 + \Delta x_2^2 + \Delta x_3^2 = r^2 \tag{9.24}$$

where r^2 is an arbitrary length that will ultimately cancel out, since we are interested only in the relative magnitudes of Δx_1, Δx_2, and Δx_3. The other requirement is that the vector has an initial direction tangent to the constraint, so that

$$\frac{\partial \phi}{\partial x_1}\Delta x_1 + \frac{\partial \phi}{\partial x_2}\Delta x_2 + \frac{\partial \phi}{\partial x_3}\Delta x_3 = 0 \tag{9.25}$$

We are confronted with an optimization problem to find the relative values of the Δx's that maximize Δy as expressed in Eq. (9.23), subject to the two constraints, Eqs. (9.24) and (9.25).

Choosing the method of Lagrange multipliers to perform this optimization yields three equations:

$$\frac{\partial y}{\partial x_1} - \lambda_1(2)\Delta x_1 - \lambda_2 \frac{\partial \phi}{\partial x_1} = 0 \tag{9.26}$$

$$\frac{\partial y}{\partial x_2} - \lambda_1(2)\Delta x_2 - \lambda_2 \frac{\partial \phi}{\partial x_2} = 0 \tag{9.27}$$

$$\frac{\partial y}{\partial x_3} - \lambda_1(2)\Delta x_3 - \lambda_2 \frac{\partial \phi}{\partial x_3} = 0 \tag{9.28}$$

Multiply Eq. (9.26) by $\partial \phi / \partial x_1$, Eq. (9.27) by $\partial \phi / \partial x_2$, and Eq. (9.28) by $\partial \phi / \partial x_3$, then total the first terms, the second terms, and the third terms of the three equations,

$$\left(\frac{\partial y}{\partial x_1}\frac{\partial \phi}{\partial x_1} + \frac{\partial y}{\partial x_2}\frac{\partial \phi}{\partial x_2} + \frac{\partial y}{\partial x_3}\frac{\partial \phi}{\partial x_3} \right) - 2\lambda_1 \left(\frac{\partial \phi}{\partial x_1}\Delta x_1 + \frac{\partial \phi}{\partial x_2}\Delta x_2 \right.$$
$$\left. + \frac{\partial \phi}{\partial x_3}\Delta x_3 \right) - \lambda_2 \left[\left(\frac{\partial \phi}{\partial x_1}\right)^2 + \left(\frac{\partial \phi}{\partial x_2}\right)^2 + \left(\frac{\partial \phi}{\partial x_3}\right)^2 \right] = 0 \tag{9.29}$$

From Eq. (9.25), the second group in the summation is zero. For convenience, let

$$A = \left(\frac{\partial y}{\partial x_1}\frac{\partial \phi}{\partial x_1} + \frac{\partial y}{\partial x_2}\frac{\partial \phi}{\partial x_2} + \frac{\partial y}{\partial x_3}\frac{\partial \phi}{\partial x_3} \right)$$

and

$$B = \left(\frac{\partial \phi}{\partial x_1}\right)^2 + \left(\frac{\partial \phi}{\partial x_2}\right)^2 + \left(\frac{\partial \phi}{\partial x_3}\right)^2$$

Then, from Eq. (9.29),

$$\lambda_2 = \frac{A}{B} \tag{9.30}$$

Substitute Eq. (9.30) into each of the equations Eq. (9.26) to Eq. (9.28). This gives the following expression for the relative values of Δx_1, Δx_2, and Δx_3.

$$\frac{1}{2\lambda_1} = \frac{\Delta x_1}{\partial y/\partial x_1 - (A/B)\partial \phi/\partial x_1} = \frac{\Delta x_2}{\partial y/\partial x_2 - (A/B)\partial \phi/\partial x_2}$$

$$= \frac{\Delta x_3}{\partial y/\partial x_3 - (A/B)\partial \phi/\partial x_3} \tag{9.31}$$

When the step size of one of the variables in the move has been selected, the magnitudes of the remaining Δx's can be computed from Eq. (9.31).

9.21 SUMMARY

This chapter explored single-variable searches as well as both unconstrained and constrained multivariable searches. The types of problems for which search methods are most likely to be called into service are the difficult ones, which are probably the multivariable constrained problems. Of the many methods available for multivariable constrained optimizations, the search along the constraint(s) was chosen because of its wide applicability (although it is not necessarily the most efficient) and because it follows logically from calculus methods.

The actual execution of the calculations in complicated problems would probably be carried out on a computer, and the availability of the interactive mode is particularly convenient. The possibility the searcher has of changing such quantities as the step size or starting with a new trial point facilitates convergence to the optimum in a rapid manner.

The techniques presented in this chapter are tools that can solve some realistic engineering problems. The chapter also leads into extensions, such as those that appear in Chapter 17, where penalty methods are explored in greater depth, and the reduced-gradient technique is shown to be a generalization of some of the specific situations examined in this chapter.

PROBLEMS

9.1. The function

$$y = \frac{(\ln x)\sin(x^2/25)}{x}$$

where $x^2/25$ is in radians, is unimodal in the range $1.5 \le x \le 10$.

(a) If a Fibonacci search is employed to determine the maximum, how many points will be needed for the final interval of uncertainty of x to be 0.3 or less?

(b) Using the number of points determined in part (a), conduct the Fibonacci search and determine the interval in which the maximum occurs.

Ans.: (b) 5.75 to 6.00.

9.2. One of the strategies in some search methods is to first use a coarse subdivision to determine the approximate region of the optimum and then a fine subdivision for a second search. For a single-variable search 16 points total are to be applied. Compare the ratio of the initial to final interval of uncertainty if (a) all 16 points are used in one Fibonacci search and (b) 8 points in a Fibonacci search are used to determine an interval of uncertainty of reduced size on which another 8-point Fibonacci search is applied.

9.3. An economic analysis of a proposed facility is being conducted to select an operating life such that the maximum uniform annual income is achieved. A short life results in high annual amortization costs, but the maintenance costs become excessive for a long life. The annual income after deducting all operating expenses, except maintenance costs, is $180,000. The first cost of the facility is $500,000 borrowed at 10 percent interest compounded annually.

The maintenance costs are zero at the end of the first year, $10,000 at the end of the second, $20,000 at the end of the third, etc. To express these maintenance charges on an annual basis the gradient present-worth factor of Sec. 3.8 can be multiplied by the capital-recovery factor, which for the 10 percent interest is presented in Table 9.3.

Use a Fibonacci search for integer years between 0 and 21 to find the life of the facility which results in the maximum annual profit. Omit the last calculation of the Fibonacci process since we are interested only in integer-year results.

Ans.: 12 years, $62,760 annual income.

9.4. The exhaust-gas temperature leaving a continuously operating furnace is 260°C, and a proposal is being considered to install a heat exchanger in the exhaust-gas stream to generate low-pressure steam at 105°C. The question to be investigated is whether it is economical to install such a heat exchanger, and, if so, to find its optimum size. The following data apply:

TABLE 9.3
Factors for conversion of gradient series to an annual cost

Year	Factor	Year	Factor	Year	Factor
1	0.000	8	3.008	15	5.275
2	0.476	9	3.376	16	5.552
3	0.937	10	3.730	17	5.801
4	1.379	11	4.060	18	6.058
5	1.810	12	4.384	19	6.295
6	2.224	13	4.696	20	6.500
7	2.622	14	5.002	21	6.703

Flow of exhaust gas, 7.5 kg/s.

Specific heat of exhaust gas, 1.05 kJ/(kg·K).

Value of the heat in the form of steam, $1.50 per gigajoule.

U value of heat exchanger based on gas-side area, 23 W/(m²·K).

Cost of the heat exchanger including installation based on gas-side area, $90 per square meter.

Interest rate, 8 percent.

Life of installation, 5 years.

Saturated liquid water enters heat exchanger at 105°C and leaves as saturated vapor.

(a) Develop the equation for the savings as a function of the area, expressed as a uniform annual amount.

(b) What is the maximum permitted area if the exit-gas temperature is to be above 120°C in order to prevent condensation of water vapor from the exhaust gas?

(c) Use a seven-point Fibonacci search and set up a table to simplify calculation of the optimum heat-transfer area.

Ans.: Optimum area between 686 and 724 m².

9.5. Perform a univariate search to find the minimum value of the function

$$y = x_1 + \frac{16}{x_1 x_2} + \frac{x_2}{2}$$

using only integer values of x_1 and x_2 and starting with $x_2 = 3$.

Ans.: $y^* = 6$.

9.6. The minimum value of the function

$$y = \frac{72x_1}{x_2} + \frac{360}{x_1 x_3} + x_1 x_2 + 2x_3$$

is to be sought using the method of steepest descent. If the starting point is $x_1 = 5$, $x_2 = 6$, $x_3 = 8$ ($y = 115$ at this point) and x_1 is to be changed by 1.0, what is the location of the next point?

Ans.: New $y = 98.1$.

9.7. The function

$$y = 2x_1 x_2 + \frac{3}{x_1 x_2 x_3} + x_1 x_3^2 + 3x_2$$

is to be minimized by the steepest-descent search method. From the starting point the direction of steepest descent will be determined, and the search is to move in that direction until a minimum is reached, whereupon a new location is to be ascertained. If the arbitrarily chosen first point is (1,1,1) what is the second point?

Ans.: y at second point $= 8.548$.

9.8. A pipe carrying high-temperature water is to be insulated and then mounted in a restricted space, as shown in Fig. 9-16. The choices of the pipe diameter D m and the insulation thickness x m are to be such that the OD of the insu-

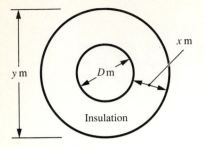

FIGURE 9-16
Insulated pipe in Prob. 9.8.

lation y is a minimum, but the total annual operating cost of the installation is limited to $40,000. This annual operating cost has two components, the water pumping cost and the cost of the heat loss:

$$\text{Pumping cost} = \frac{8}{D^5} \text{dollars}$$

$$\text{Heat cost} = \frac{1500}{x} \text{dollars}$$

(*a*) Write the objective function and the constraint.
(*b*) The values of $D = 0.2$ and $x = 0.1$ satisfy the constraint. Starting at this point move in a favorable direction tangent to the constraint by an increment of $\Delta D = 0.005$; then return in the most direct manner to the constraint. What are the new values of x and D following this hemstitch move?
  Ans.: $y = 0.3664$ m, $\phi = 19.86$.

9.9. A refrigeration plant chills water and delivers it to a heat exchanger some distance away, as shown in Fig. 9-17. The supply water temperature is t_1°C, and the return water temperature is t_2. The flow rate is w kg/s, and the pipe diameter is D m. The cooling duty at the heat exchanger is 1200 kW, and the arithmetic mean temperature, $(t_1 + t_2)/2$, must be 12°C in order to transfer the 1200 kW from the air. The pump develops a pressure rise of 100,000 Pa that may be assumed independent of flow rate. The equation for pressure

FIGURE 9-17
Refrigeration plant, heat exchanger, and interconnecting piping.

drop in a circular pipe is

$$\Delta p, \text{Pa} = f \frac{L}{D} \frac{V^2}{2} \rho$$

where $f = 0.02$
$\quad\quad L = 2100$ m
$\quad\quad \rho = 1000$ kg/m^3

The specific heat of water is 4.19 kJ/(kg · K).
   The three major costs associated with the choice of t_1, w, and D are

Cost of the pipe, 150,000D dollars
Present worth of lifetime pumping costs, 500w dollars
Present worth of lifetime chilling costs, $60,000 - 4000t_1$ dollars

(*a*) Set up the equation for the cost as the objective function, the pressure drop as constraint 1, and the heat-transfer requirements as constraint 2.
(*b*) The point $w = 26$ kg/s, $D = 0.187$ m, and $t_1 = 6.5°$C essentially satisfies both constraints. What is the new point after making a move of $w = 1$ kg/s tangent to both constraints in a direction that reduces the cost?
   Ans.: (*a*) $(2.938 \times 10^6)D^5 - w^2 = 0$ and $w(24 - 2t_1) - 286.4 = 0$.
   (*b*) $D = 0.184$, $w = 25$, $t_1 = 6.3$.

9.10. The optimization of the dimensions of the steel frame in Prob. 8.4 is to be conducted by a search method as a three-variable problem with one constraint. If the current position is $x_1 = 16$, $x_2 = 12.5$, and $x_3 = 4.5$ m, what are the new values of the x's following a step of $\Delta x_1 = 0.2$ m, if the move is tangent to the constraint and in the direction of the most favorable rate of reduction of cost?
   Ans.: cost reduced $66 in the step.

REFERENCES

1. J. Kiefer, "Sequential Minimax Search for a Maximum," *Proc. Am. Math. Soc.*, vol. 4, p. 502, 1953.
2. D. J. Wilde, *Optimum Seeking Methods*, Prentice-Hall, Englewood Cliffs, N.J., 1964.
3. N. McCloskey, "Storage Facilities Associated with an Ammonia Pipeline," *ASME Pap.* 69-Pet-21, 1969.
4. R. J. Buehler, B. V. Shah, and C. Kempthorne: *Some Properties of Steepest Ascent and Related Procedures for Finding Optimum Conditions*, Iowa State University Statistical Laboratory, pp. 8–10, April 1961.
5. G. S. G. Beveridge and R. S. Schechter: *Optimization: Theory and Practice*, McGraw-Hill, New York, 1970.
6. S. M. Roberts and H. I. Lyvers, "The Gradient Method in Process Control," *Ind. Eng. Chem.*, vol. 53, pp. 877–882, 1961.

CHAPTER
10

DYNAMIC
PROGRAMMING

10.1 UNIQUENESS OF DYNAMIC-PROGRAMMING PROBLEMS

Dynamic programming is a method of optimization that is applicable either to staged processes or to continuous functions that can be approximated by staged processes. The word "dynamic" has no connection with the frequent use of the word in engineering technology, where dynamic implies changes with respect to time.

As a method of optimization, dynamic programming is not usually interchangeable with such other forms of optimization as Lagrange multipliers and linear and nonlinear programming. Instead, it is related to the calculus of variations, whose result is an optimal *function* rather than an optimal *state point*. An optimization problem that can be subjected to dynamic programming or the calculus of variations is usually different from those suitable for treatment by Lagrange multipliers and linear and nonlinear programming. The calculus of variations is used, for example, to determine the trajectory (thus, a function in spatial coordinates) that results in minimum fuel cost of a spacecraft. Dynamic programming can attack this same problem by dividing the total path into a number of segments and considering the continuous function as a series of steps or stages. In such an application, the finite-step approach of dynamic programming is an approximation of the calculus-of-variations method. In many engineering situations, on the other hand, the problem consists of analysis of discrete stages, such as a series of compressors, heat exchangers, or reactors. These cases fit dynamic programming exactly, and here the calculus of variations would only approximate the result.

10.2 SYMBOLIC DESCRIPTION OF DYNAMIC PROGRAMMING

Figure 10-1 shows a symbolic description of the dynamic-programming problem. The decision variables are to be chosen so that for a specified input to stage n and a specified output from stage 1 the summation of returns $\sum_{i=1}^{n} r_i$ is optimum (either maximum or minimum, depending upon the problem).

The description of the calculus of variations in Section 10.1 suggests that a function, for example $y(x)$, is sought. In Fig. 10-1 that goal is to find the optimal state variables S for the various stages, where the stage corresponds to the x variable. The decision variables control the change in S through the stage and also determine the return from a stage. The calculus of variations seeks a function that optimizes an *integral*, while dynamic programming seeks to optimize a summation, denoted here as $\sum_{i=1}^{n} r_i$. In the calculus of variations the terminal points of the function y are specified. In dynamic programming, also, S_n and S_i' are specified.

Often some insight is needed to recognize that a physical problem fits into the mold of dynamic programming. Some hint is provided when the problem involves sequences of stages, such as a chain of heat exchangers, reactors, compressors, etc.

So far only the nature of the problem has been described. The next section shows how dynamic programming solves the problem.

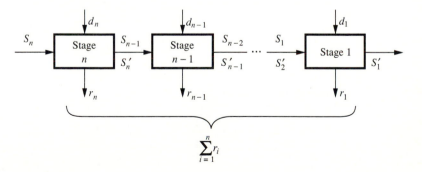

FIGURE 10-1
Pictorial representation of problem that can be solved by dynamic programming; S = state of the input to a stage, S' = state of the output from a stage, d = decision variable, and r = return from a stage.

10.3   CHARACTERISTICS OF THE DYNAMIC-PROGRAMMING SOLUTION

The trademark of dynamic programming in arriving at an overall optimal plan is to establish optimal plans for subsections of the problem. In succeeding evaluations the optimal plans for the subsections are used, and all nonoptimal plans are ignored. The mechanics can be illustrated by a problem of determining the optimal route between two points as in Example 10.1.

> **Example 10.1.** A minimum-cost pipeline is to be constructed between points A and E, passing successively through one node of each, B, C, and D, as shown in Fig. 10-2. The costs from A to B and from D to E are shown in Fig. 10-2, and the costs between B and C and between C and D are given in Table 10.1.
>
> **Solution.** The strategy in dynamic programming is to begin with one of the terminal stages (either A-B or D-E in this problem) and then progressively

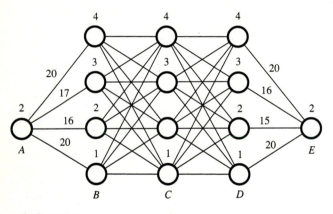

FIGURE 10-2
Dynamic programming used to minimize the cost between points A and E.

TABLE 10.1
Costs from B to C and C to D in Fig. 10-2

	To			
From	1	2	3	4
1	12	15	21	28
2	15	16	17	24
3	21	17	16	15
4	28	24	15	12

analyze cumulative sets of stages. In this problem it makes no difference whether we start with A-B or with D-E. In some problems the form of the existing data will make starting at one end or the other more convenient.

In this problem we shall start at the right end and work toward the left. Thus the first table will be for stage D-E, the next table for the cumulative stages C-D and D-E, and so on until the last table applies to the entire system, A-E.

Table 10.2 simply shows the costs from the various D positions to the end, which is at E. Table 10.2 assumes the form that will be used in subsequent tables; i.e., it designates the optimum manner in which to pass from $D1$, $D2$, $D3$, and $D4$ to the next stage, which here is E.

Table 10.2 is trivial because there is only one way to pass from $D1$ to the terminus, and that route must be optimum. Future tables will rule out nonoptimal paths, but no such selection can be made in Table 10.2 because it is invalid to say, for example, that the route from $D2$ is less costly than from $D1$, which would permit ruling out $D1$. The consequences of the choice of the routes from $D2$ over that from $D1$ may impose overriding penalties later, so all must be kept in consideration.

It is in Table 10.3 from C to E that dynamic programming begins to show its benefit. From position $C1$, for example, there are four paths by which to reach E, through $D1$, $D2$, $D3$, and $D4$. The total costs are shown in the appropriate column in Table 10.3, and the least costly route is the one passing through $D2$, resulting in a cost of 30. Within that block there is a common basis of comparison: all the paths start from $C1$. Table 10.3 also denotes the optimal choices when starting at the other C positions. In the path from $C3$ there is a tie (cost of 32), so either choice can be made. Now that the preferred route from $C1$ to E has been decided, all the nonoptimal routes are henceforth ignored.

Advancing (backward) to include stage B-C in the assembly, Table 10.4 shows the accumulation from B to E. The costs are shown for the possible paths starting at all of the B points. From $B1$ it is possible to pass through $C1$ and then to the end, through $C2$ and then to the end, etc. The total costs from $B1$ through $C1$ to the end, for example, are computed by summing the cost of 12 from $B1$ to $C1$ (from Table 10.1) with the cost from $C1$ to the end. The cost of $C1$ to the end is 30, which is available from Table 10.3 as the minimum cost of the four possibilities in the $C1$ block. Of the various paths from $B1$ to the end, one is optimum, with the cost of 42. Table 10.4 shows

TABLE 10.2
Example 10.1, D to E

From	Through	Cost	Optimum
$D1$	–	20	x
$D2$	–	15	x
$D3$	–	16	x
$D4$	–	20	x

TABLE 10.3
Example 10.1, C to E

			Cost		
From	Through	C to D	D to E	Total	Optimum
$C4$	$D4$	12	20	32	
	$D3$	15	16	31	x
	$D2$	24	15	39	
	$D1$	28	20	48	
$C3$	$D4$	15	20	35	
	$D3$	16	16	32	x
	$D2$	17	15	32	x
	$D1$	21	20	41	
$C2$	$D4$	24	20	44	
	$D3$	17	16	33	
	$D2$	16	15	31	x
	$D1$	15	20	35	
$C1$	$D4$	28	20	48	
	$D3$	21	16	37	
	$D2$	15	15	30	x
	$D1$	12	20	32	

TABLE 10.4
Example 10.1, B to E

			Cost		
From	Through	B to C	C to E	Total	Optimum
$B4$	$C4$	12	31	43	x
	$C3$	15	32	47	
	$C2$	24	31	55	
	$C1$	28	30	58	
$B3$	$C4$	15	31	46	x
	$C3$	16	32	48	
	$C2$	17	31	48	
	$C1$	21	30	51	
$B2$	$C4$	24	31	55	
	$C3$	17	32	49	
	$C2$	16	31	47	
	$C1$	15	30	45	x
$B1$	$C4$	28	31	59	
	$C3$	21	32	53	
	$C2$	15	31	46	
	$C1$	12	30	42	x

TABLE 10.5
Example 10.1, A to E

| To E from | Through | Cost | | | Optimum |
		A to B	B to E	Total	
$A2$	$B4$	20	43	63	
	$B3$	17	46	63	
	$B2$	16	45	61	x
	$B1$	20	42	62	

four state variables, $B1$, $B2$, $B3$, and $B4$, and the optimum cost from each of these states to the end.

The final step in the solution is provided by picking up the first state (A to B) in the accumulation, and the result is Table 10.5. There is only one state variable, the position $A2$. From $A2$ the options are to pass through $B1$, $B2$, $B3$, or $B4$. The cost from $A2$ to B is found in Fig. 10-2 and the cost from B to $E2$ is the optimal one from the appropriate B position from Table 10.4.

The minimum cost from $A2$ to $E2$, thus the minimum total cost, is 61. To identify which route results in this minimum cost, it is only necessary to trace back through the tables and note which choices were optimum: $A2$, $B2$, $C1$, $D2$, $E2$.

The key feature of dynamic programming is that after an optimal policy has been determined from an intermediate state to the final state, future calculations passing through that state use only the optimal policy.

10.4 EFFICIENCY OF DYNAMIC PROGRAMMING

The combination of tables necessary to solve Example 10.1 may seem to constitute a lengthy and tedious solution. Comparatively speaking, however, dynamic programming is efficient, particularly in large problems. If we call each line in Tables 10.2 to 10.5 a *calculation*, a total of 40 calculations was required. If an exhaustive examination of all possible routes between A and B had been made, the total number would have been the product of the number of possibilities. From $A2$ to B there are four possibilities, from each of the B points there are four possibilities of passing to C, and similarly from C to D. From D to $E2$ there is just one possibility. The number of possible routes if all are considered is therefore $(4)(4)(4)(1) = 64$.

The saving of effort would be more impressive if the problem had included another stage consisting of four positions. The number of calculations by dynamic programming would have been the current number of 40 plus an additional 16, for a total of 56. Examining all possible routes would require $(64)(4) = 256$ calculations.

FIGURE 10-3
Chain of ultrafilters to separate protein from lactose in whey.

Example 10.2. Whey is a by-product of cheese manufacture and contains, among other constituents, protein and lactose. The two substances are valuable when separated, protein-rich whey for yogurt and lactose for ethyl alcohol. One method of separation is ultrafiltration, where the separation occurs on the basis of molecular size and shape. A series of ultrafilters is employed (four in the chain shown in Fig. 10-3) to separate the protein and lactose progressively. Klinkowski[1] shows the operating cost of a stage to be a function of the inlet and outlet protein concentrations, as shown in Table 10.6. Use dynamic programming to solve for the concentrations leaving each stage that results in the minimum total cost.

Solution. The dynamic-programming calculations start, arbitrarily, with stage IV in Table 10.7 and proceed back through the stages until Table 10.10, which includes all the stages. The minimum operating cost is $34.42, which is achieved by operating the filtration plant with concentrations 0.6, 0.9, 1.8, 3.6 to 6.0.

TABLE 10.6
Operating cost of one stage in a protein-lactose separator in Example 10.2, dollars

Entering protein concentration, %	Leaving protein concentration, %									
	0.9	1.2	1.8	2.4	3.0	3.6	4.2	4.8	5.4	6.0
0.6	5.53	10.77	20.24	28.38	35.20	40.70	44.88	47.74	49.28	49.50
0.9		3.73	10.77	17.23	23.10	28.38	33.07	37.18	40.70	43.63
1.2			5.54	10.78	15.67	20.24	24.47	28.38	31.95	35.20
1.8				3.74	7.33	10.79	14.00	17.23	20.24	23.10
2.4					2.82	5.55	8.21	10.80	13.27	15.67
3.0						2.26	4.47	6.63	8.73	10.81
3.6							1.89	3.75	5.56	7.33
4.2								1.62	3.21	4.78
4.8									1.42	2.82
5.4										1.26
6.0										

TABLE 10.7
Example 10.2, stage IV

Concentration entering Stage IV	Through	Cost
1.8	–	$23.10
2.4	–	15.67
3.0	–	10.81
3.6	–	7.33
4.2	–	4.78
4.8	–	2.82
5.4	–	1.26

TABLE 10.8
Example 10.2, stages III and IV

Concentration entering III	Through	Cost
1.2	1.8	5.54 + 23.10 = $28.64
	2.4	10.78 + 15.67 = 26.45*
	3.0	15.67 + 10.81 = 26.48
	3.6	20.24 + 7.33 = 27.57
	4.2	24.47 + 4.78 = 29.25
	4.8	28.38 + 2.82 = 31.20
	5.4	31.95 + 1.26 = 33.21
1.8	2.4	3.74 + 15.67 = 19.41
	3.0	7.33 + 10.81 = 18.14
	3.6	10.79 + 7.33 = 18.12*
	4.2	14.00 + 4.78 = 18.78
	4.8	17.23 + 2.82 = 20.05
	5.4	20.24 + 1.26 = 21.50
2.4	3.0	2.82 + 10.81 = 13.63
	3.6	5.55 + 7.33 = 12.88*
	4.2	8.21 + 4.78 = 12.99
	4.8	10.80 + 2.82 = 13.62
	5.4	13.27 + 1.26 = 14.53
3.0	3.6	2.26 + 7.33 = 9.59
	4.2	4.47 + 4.78 = 9.25*
	4.8	6.63 + 2.82 = 9.45
	5.4	8.73 + 1.26 = 9.99
3.6	4.2	1.89 + 4.78 = 6.67
	4.8	3.75 + 2.82 = 6.57*
	5.4	5.56 + 1.26 = 6.82
4.2	4.8	1.62 + 2.82 = 4.44*
	5.4	3.21 + 1.26 = 4.47
4.8	5.4	1.42 + 1.26 = 2.68*

TABLE 10.9
Example 10.2, stages II, III, and IV

Concentration entering II	Through	Cost
0.9	1.2	3.73 + 26.45 = $30.18
	1.8	10.77 + 18.12 = 28.89*
	2.4	17.23 + 12.88 = 30.11
	3.0	23.10 + 9.25 = 32.35
	3.6	28.38 + 6.57 = 34.95
	4.2	33.07 + 4.44 = 37.51
	4.8	37.18 + 2.68 = 39.86
1.2	2.4	5.54 + 18.12 = 23.66*
	3.0	10.78 + 12.88 = 23.66*
	3.6	15.67 + 9.25 = 24.92
	4.2	20.24 + 6.57 = 26.81
	4.8	24.47 + 4.44 = 28.91
	5.4	28.38 + 2.68 = 31.06
1.8	2.4	3.74 + 12.88 = 16.62
	3.0	7.33 + 9.25 = 16.58*
	3.6	10.79 + 6.57 = 17.36
	4.2	14.00 + 4.44 = 18.44
	4.8	17.23 + 2.68 = 19.91
2.4	3.0	2.82 + 9.25 = 12.07*
	3.6	5.55 + 6.57 = 12.12
	4.2	8.21 + 4.44 = 12.65
	4.8	10.80 + 2.68 = 13.48
3.0	3.6	2.26 + 6.57 = 8.83*
	4.2	4.47 + 4.44 = 8.91
	4.8	6.63 + 2.68 = 9.31
3.6	4.2	1.89 + 4.44 = 6.33*
	4.8	3.75 + 2.68 = 6.43
4.2	4.8	1.62 + 2.68 = 4.30*

TABLE 10.10
Example 10.2, stages I to IV

Concentration entering I	Through	Cost
0.6	0.9	5.53 + 28.89 = $34.42*
	1.2	10.77 + 23.66 = 34.43
	1.8	20.24 + 16.58 = 36.82
	2.4	28.38 + 12.07 = 40.45
	3.0	35.20 + 8.83 = 44.03
	3.6	40.70 + 6.33 = 47.03
	4.2	44.88 + 4.30 = 49.18

10.5   THE PATTERN OF THE DYNAMIC-PROGRAMMING SOLUTION

Examples 10.1 and 10.2 begin to display a pattern of the nature of the problem and the characteristics of the dynamic-programming solution. The solutions can be shown graphically as in Fig. 10-4a and b. Example 10.1 is a geometric problem, and the ordinate is a position y. The ordinate in Example 10.2 is a physical variable, the protein concentration. The abscissas are the stages; A-B is the first stage in Example 10.1, and the first stage in Example 10.2 lies between positions 1 and 2.

Another characteristic of the problems is that the terminal points are specified ($A2$ and $E2$ in Fig. 10-4a and 0.6 and 6 percent in Fig. 10-4b). An optimal path is sought that minimizes a summation which in both problems is a cost.

The first table in the dynamic-programming solution is a necessary routine, but it is in the second table that dynamic programming becomes effective. In Example 10.2 from a given state variable entering stage III, e.g., a concentration of 3.6 percent, the various possible routes to 6 percent at position 5 are investigated. Once the optimum route from 3.6 percent at position 3 to 6 percent at position 5 has been identified, the nonoptimal routes are discarded. Any future path that passes through position 3 at 3.6 percent uses only the optimal route.

The choice of the increments of protein concentration in Example 10.2 (0.6, 0.9, 1.2, 1.8, etc.) was somewhat arbitrary and introduced an approximation. More precision would have resulted if 0.1 percent increments had been chosen and the performance data that yielded Table 10.6

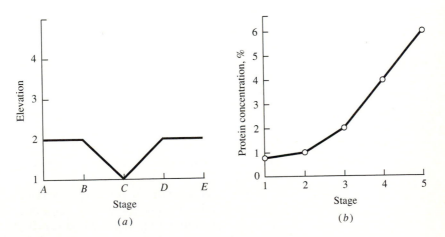

FIGURE 10-4
Graphic display of dynamic-programming (a) Example 10.1 (b) Example 10.2.

TABLE 10.11
First table of Example 10.2 proceeding forward, stage I

Leaving stage I	Entering stage I	Cost, $
0.9	0.6	5.53
1.2	0.6	10.77
1.8	0.6	20.24
2.4	0.6	28.38
etc.		

TABLE 10.12
Second table of Example 10.2 proceeding forward, stages I and II

Leaving stage II	Through ___, at position 2	Cost, $		
1.2	0.9	3.73 +	5.53 =	9.26
1.8	0.9	5.53 +	10.77 =	16.30*
	1.2	10.77 +	5.54 =	16.31
2.4	0.9	5.53 +	17.23 =	22.76
	1.2	10.77 +	10.78 =	21.55*
	1.8	20.24 +	3.74 =	23.98
etc.				

used to compute a more detailed table. The coarser grid was chosen simply to lessen the number of calculations. After having established the approximate optimal path a recalculation could have been made using a finer grid but considering paths only in the neighborhood of the preliminary optimum.

It has already been pointed out that the form of the available data may in some problems make starting at the front preferable and in other cases starting at the end and working backward. In Examples 10.1 and 10.2 it was immaterial which direction was chosen. It should be realized, however, that the form of the tables will differ depending upon whether the progression is forward or backward, because the state variable will be different. If the progression in Example 10.2 were forward, the first table would have the form shown in Table 10.11 and the second table would be as shown in Table 10.12.

The state variable in Table 10.12 is the protein percentage leaving stage II, and once it has been determined that the optimal way to achieve a protein concentration of 2.4, for example, through the first two stages (0.6 to 1.2 in the first stage and 1.2 to 2.4 in the second stage, as indicated by Table 10.12), the nonoptimal routes are neglected.

10.6 APPARENTLY CONSTRAINED PROBLEMS

An important class of problems in dynamic programming is that of constrained optimization, where a function $y(x)$ is sought that minimizes a summation $\Sigma g(y,x)$ but in addition some other summation is specified $\Sigma h(y,x) = H$, where the functions g, h, and the numerical term H are known. This class of constrained problems will be treated in Chapter 18. Another class of problems at first glance may seem to be constrained, but they can be converted into a form identical to that used in Examples 10.1 and 10.2. This class may be called *apparently constrained* and is illustrated by Examples 10.3 and 10.4.

Example 10.3. An evaporator which boils liquid inside tubes consists of four banks of tubes. Each bank consists of a number of tubes in parallel, and the banks are connected in series, as shown in Fig. 10-5. A mixture of liquid and vapor enters the first bank with a fraction of vapor $x = 0.2$, and the fluid leaves the evaporator as saturated vapor, $x = 1.0$. The flow rate is 0.5 kg/s, and each tube is capable of vaporizing 0.01 kg/s and thus of increasing x by 0.02.

Forty tubes are to be arranged in the banks so that the minimum total pressure drop prevails in the evaporator. The pressure drop in a bank is approximately proportional to the square of the velocity, and a satisfactory expression for the pressure drop Δp is

$$\Delta p, \text{kPa} = 720\left(\frac{x_i}{n}\right)^2 \tag{10.1}$$

FIGURE 10-5
Evaporator in Example 10.3.

where x_i = vapor fraction entering bank
n = number of tubes in bank

Use dynamic programming to determine the distribution of the 40 tubes so that the total pressure drop in the evaporator is minimum.

Solution. Selection of the number of tubes in a stage (bank) as the state variable is unproductive because the terminal points in graphs comparable to Fig. 10-4 are not fixed. Furthermore, the choice of number of tubes as the state variable does not account for the constraint that precisely 40 tubes are available.

The difficulties are overcome by choosing as the state variable cumulative tubes committed, which results in coordinates as shown in Fig. 10-6. After stage 0 (before stage I) no tubes have been committed, and following stage IV 40 tubes have been committed.

Table 10.13 shows pressure drops for the first bank or stage for several different choices of tubes. Table 10.14 uses as the state variable the total number of tubes committed in the first two stages and permits an optimal selection. For example, if 13 tubes are used in the first two stages, the optimal distribution is to allot 5 tubes in bank I and 8 tubes in bank II to achieve a total

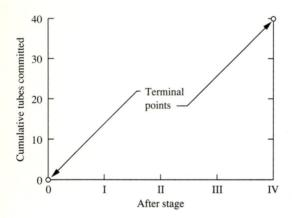

FIGURE 10-6
State variable of cumulative number of tubes committed in Example 10.3.

TABLE 10.13
Example 10.3, stage I

Total tubes committed	Tubes in stage I	Total Δp, kPa
2	2	7.20
3	3	3.20
4	4	1.80
5	5	1.15
6	6	0.80

TABLE 10.14
Example 10.3, stages I and II

Total tubes	Tubes in II	Total Δp, kPa
11	5	$0.80 + 2.95 = 3.75$
	6	$1.15 + 1.80 = 2.95^*$
	7	$1.80 + 1.15 = 2.95^*$
	8	$3.20 + 0.76 = 3.96$
	9	$7.20 + 0.51 = 7.70$
12	6	$0.80 + 2.05 = 2.85$
	7	$1.15 + 1.32 = 2.47^*$
	8	$1.80 + 0.88 = 2.68$
	9	$3.20 + 0.60 = 3.68$
13	7	$0.80 + 1.50 = 2.30$
	8	$1.15 + 1.01 = 2.16^*$
	9	$1.80 + 0.73 = 2.53$
	10	$3.20 + 0.49 = 3.69$
14	7	$0.59 + 1.70 = 2.29$
	8	$0.80 + 1.15 = 1.95^*$
	9	$1.15 + 0.80 = 1.95^*$
	10	$1.80 + 0.56 = 2.36$
15	8	$0.59 + 1.30 = 1.89$
	9	$1.15 + 0.80 = 1.71^*$
	10	$1.15 + 0.65 = 1.80$

Δp in banks I and II of 2.16 kPa. The pressure drop in stage II is computed from Eq. (10.1) using $x_i = 0.20 + (0.02)$(number of tubes in stage I).

Table 10.15 determines the optimum number of tubes in bank III for various total numbers of tubes in the first three banks. Table 10.16 is the final table and indicates that 17 is the optimum number of tubes in bank IV.

The optimal distribution of tubes is 5, 7, 11, 17, resulting in a total pressure drop of 4.71 kPa.

A further illustration of the solution by dynamic programming of an apparently constrained problem is shown in Example 10.4 in the optimization of feedwater heating. Heating the boiler feedwater with extraction steam, as shown in Fig. 10-7, improves the efficiency of a steam-power cycle and is a common practice in large central power stations. Some plants use more than half a dozen heaters, which draw off extraction steam at as many different pressures. That feedwater heaters improve the efficiency of a cycle can be shown by a calculation of a specific case, but a qualitative explanation may provide a better sense of this improvement. First, recall

TABLE 10.15
Example 10.3, stages I and III

Total tubes	Tubes in III	Total Δp, kPa
22	9	2.16 + 1.88 = 4.04
	10	2.47 + 1.39 = 3.86*
	11	2.95 + 1.05 = 4.00
23	9	1.95 + 2.05 = 4.00
	10	2.16 + 1.52 = 3.68
	11	2.47 + 1.15 = 3.62*
	12	2.95 + 0.88 = 3.83
24	10	1.95 + 1.66 = 3.61
	11	2.16 + 1.26 = 3.42*
	12	2.47 + 0.97 = 3.44
	13	2.95 + 0.75 = 3.70
25	10	1.71 + 1.80 = 3.51
	11	1.95 + 1.37 = 3.32
	12	2.16 + 1.06 = 3.22*
	13	2.47 + 0.82 = 3.29
26	11	1.71 + 1.49 = 3.20
	12	1.95 + 1.15 = 3.00*
	13	2.16 + 0.90 = 3.06

TABLE 10.16
Example 10.3, stages I to IV

Total tubes	Tubes in IV	Total Δp, kPa
40	13	2.93 + 2.33 = 5.26
	14	3.00 + 1.90 = 4.90
	15	3.22 + 1.57 = 4.79
	16	3.42 + 1.30 = 4.72
	17	3.62 + 1.09 = 4.71*
	18	3.86 + 0.91 = 4.77

that in the steam-power cycle approximately 3 J of heat is supplied at the boiler for every joule of work at the turbine shaft. The difference of 2 J is the amount rejected at the condenser, which usually represents a loss. The proposal to try to use some of the heat rejected at the condenser for boiler-water heating is doomed because if we tried, for example, to heat the feedwater with exhaust steam from the turbine, there would be no temperature difference between the exhaust system and the feedwater to provide the driving force for heat transfer.

FIGURE 10-7
Selection of optimum areas of feedwater heaters in Example 10.4.

Extraction steam, however, has a higher temperature than exhaust steam and can be used for the heating. Concentrating on 1 kg of extraction steam leaving the boiler, we find that it performs some work in the turbine before extraction and then uses the remainder of its energy above saturated liquid at the condensing temperature to heat the feedwater. In effect, then, all the heat supplied to that kilogram of steam in the boiler is eventually converted into work. The practice of feedwater heating by extraction steam raises the effectiveness of the cycle compared to rejecting 2J of boiler heat per joule of work.

It is further to be expected that the high-pressure steam is more valuable than the low-pressure steam because the steam extracted at high pressure would have been able to deliver additional work at the turbine shaft.

Example 10.4. An economic analysis has determined that a total of 1000 m² of heat-transfer area should be used in the four feedwater heaters shown in Fig. 10-7. This 1000 m² can be distributed in the four heaters in 100-m² increments. The overall heat-transfer coefficient of all heaters is 2800 W/(m² · K). The cost of heat at the boiler is 60 cents per gigajoule, and the worth of the extraction steam determined by thermodynamic calculations is listed in Table 10.17. The flow rate of feedwater is 100 kg/s.

Use dynamic programming to determine the optimum distribution of the area.

Solution. Before beginning the solution it may be instructive to try to predict the nature of the optimal solution. It is desirable to use the lowest-cost steam

TABLE 10.17
Extraction-steam data

Extraction point and heater number	Saturation temperature, °C	Worth of extraction steam, cents per gigajoule
1	90	23
2	120	28
3	160	38
4	215	47

possible, which would suggest a large area in stage I, but each additional unit of area in that stage is less effective than the previous unit area because the temperature difference between the steam and feedwater is less. Thus there must be a compromise between trying to use the low-cost steam and maintaining a high temperature difference.

This problem is apparently constrained because the total area is specified, but it can be converted into the unconstrained form by using as the state variable the *total area committed*. In this problem it is advantageous to start at the front (with respect to the feedwater flow), because the inlet temperature of the feedwater is known there (32°C). Table 10.18 shows the outlet temperatures from stage I for various areas in that stage. As is typical of dynamic programming, this first table is routine. The temperatures are computed by use of Eq. (5.10) for a condenser, and the saving is the value of the heat saved at the boiler less the cost of extraction steam used.

Table 10.19 uses as the state variable the total area committed in the first two stages. If, for example, 1000 m² is available for the first two stages, the optimum distribution is to allot 400 m² in the first stage and 600 m² in the second, resulting in a saving of $1.285 per second.

Table 10.20 shows various area distributions in the first three stages.

TABLE 10.18
Example 10.4, stage I

Total area committed, m²	Area for stage I	t_1	Saving per second
0	0	32.00	$.000
100	100	60.27	.438
200	200	74.76	.663
300	300	82.19	.778
400	400	86.00	.837
500	500	87.95	.867
600	600	88.95	.883
700	700	89.46	.891
800	800	89.72	.895
900	900	89.86	.897
1000	1000	89.93	.898

TABLE 10.19
Example 10.4, stage I and II

Total area	Area in II	t_2	Savings per second
0	0	32.00	$ .000
100	0	60.27	.438
	100	74.89	.575*
200	0	74.76	.663
	100	89.38	.829
	200	96.88	.870*
. . . .
1000	0	89.93	.898
	100	104.55	1.094
	200	112.04	1.194
	300	115.89	1.245
	400	117.86	1.270
	500	118.87	1.282
	600	119.38	1.285*
	700	119.65	1.280
	800	119.78	1.267
	900	119.85	1.237
	1000	119.89	1.178

TABLE 10.20
Example 10.4, stage I to III

Total area	Area in III	t_3	Savings per second
0	0	32.00	$ .000
100	0	74.89	.575*
	100	94.39	.575*
200	0	96.88	.870
	100	116.37	.957*
	200	126.37	.870
. . . .
1000	0	119.38	1.285
	100	138.88	1.462
	200	149.14	1.548
	300	154.26	1.590
	400	157.02	1.598*
	500	158.37	1.597
	600	159.13	1.566
	700	159.48	1.537
	800	159.70	1.449
	900	159.79	1.358
	1000	159.84	1.178

TABLE 10.21
Example 10.4, stages I to IV

Total area	Area stage IV	t_4	Savings per second
1000	0	157.02	$1.598
	100	185.18	1.738
	200	198.92	1.801
	300	206.72	1.804*
	400	210.68	1.790
	500	212.54	1.773
	600	213.70	1.682
	700	214.19	1.625
	800	214.53	1.492
	900	214.66	1.336
	1000	214.77	.996

Finally, Table 10.21 where the full area of 1000 m^2 is committed, indicates that the optimum distribution of area is 100, 300, 300, 300 for a total saving of $1.804 per second.

10.7   SUMMARY

When optimizing a system that consists of a chain of events or components where the output condition from one unit forms the input to the next, dynamic programming should be explored. In large problems the amount of calculation may be extensive, even though it represents only a fraction of the effort of conducting an exhaustive exploration. The systematic nature of dynamic programming lends itself to development of a computer program to perform the calculations. The major challenge usually appears in setting up the tables, and especially in identifying the state variable.

Chapter 18 will revisit dynamic programming and focus on its relationship to calculus of variations. The insight provided by calculus of variations suggests procedures by which dynamic programming is able to solve constrained optimization problems.

PROBLEMS

10.1. The total pressure drop from point 1 to point 5 in the multi-branch duct system shown in Fig. 10-8 is to be 500 Pa. Table 10.22 presents the costs for various duct sizes in each of the sections as a function of the pressure drop in the section. Use dynamic programming to determine the pressure drop in each section that results in the minimum total cost of the system.
Ans.: $599.

10.2. A truck climbs a hill that consists of three sections. The fuel consumption in each section is a function of the time required for the distance in the

FIGURE 10-8
Duct system in Prob. 10.1.

TABLE 10.22
Pressure drop and costs of sections of duct in Prob. 10.1

Section	Pressure drop, Pa	Cost, $
	100	222
1–2	150	205
	200	193
	100	180
2–3	150	166
	200	157
	100	135
3–4	150	125
	200	117
	100	93
4–5	150	86
	200	81

section to be covered, as shown in Table 10.23. A total of 25 s is available for the climb of the hill. Use dynamic programming to determine the time allocation to each section that results in minimum total fuel consumption.

Ans.: 119 g.

10.3. Use dynamic programming to determine the flight plan for a commercial airliner flying between two cities 1200 km apart so that the minimum amount of fuel is consumed during the flight. Specifically, the altitudes at locations B through F in Fig. 10-9 during the course of the flight are to be specified.

Table 10.24 shows the fuel consumption for 200-km ground distances as a function of the climb or descent during that distance. Determine the flight plan and the minimum fuel cost.

Ans.: 2770 kg.

10.4 A minimum-cost pipeline is to be constructed between positions A and G in Fig. 10-10 and can pass through any of six locations in the successive stages, B, C, D, E, and F.

TABLE 10.23
**Fuel consumption in various sections of
hill climb in Prob. 10.2**

Section	Time, s	Fuel consumption, g
	7	40
A–B	8	34
	9	29
	10	25
	7	61
B–C	8	52
	9	45
	10	38
	7	49
C–D	8	41
	9	35
	10	30

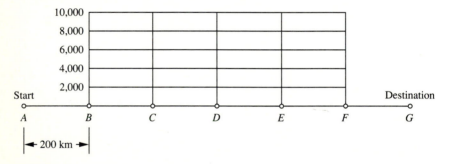

FIGURE 10-9
Altitudes and distances in Prob. 10.3.

TABLE 10.24
Fuel consumption per 200 km of ground travel, kg

From altitude, m	To altitude, m					
	0	2000	4000	6000	8000	10,000
0	–	1500	1800	2070	2300	2500
2,000	300	600	1000	1500	1950	2300
4,000	120	180	300	840	1160	1730
6,000	0	60	120	200	600	1200
8,000	0	0	30	80	180	600
10,000	0	0	0	0	60	90

FIGURE 10-10
Pipeline route in Prob. 10.4.

(a) If all possible combinations of routes are investigated, how many different paths must be examined?
(b) If dynamic programming is used, how many calculations must be made if one calculation consists of one line in a table?
   Ans.: 7776 and 156.

10.5. The maintenance schedule for a plant is to be planned so that a maximum total profit will be achieved during a 4-year span that is part of the life of the plant. The income level of the plant during a given year is a function of the condition of the plant carried over from the previous year and the amount of maintenance expenditures at the beginning of the year. Table 10.25 shows the necessary maintenance expenditures that result in a certain income level during the year for various income levels carried over from the previous year. The income level at the beginning of year 1 just before the

TABLE 10.25
Maintenance expenditures made at beginning of year, thousands of dollars

Income level carried over from previous year	Income level during year					
	$30	$32	$34	$36	$38	$40
$30	$2	$4	$7	$11	$16	$23
32	2	3	5	9	13	18
34	1	2	4	7	10	14
36	0	1	2	5	8	10
38	X	0	1	2	6	9
40	X	X	0	1	4	8

FIGURE 10-11
Chain of heat exchangers in Prob. 10.6.

maintenance expenses are made is $36,000, and the income level specified during and at the end of year 4 is to be $34,000. The profit for any one year will be the income during the year less the expenditure made for maintenance at the beginning of the year. Use dynamic programming to determine the plan for maintenance expenditures that results in maximum profit for the 4 years.
Ans.: Maximum profit = $130,000.

10.6. Four heat exchangers (or fewer) in series, as shown in Fig. 10-11, are each served by steam at a different temperature and heat water from 50 to 300°C. The sums of the first costs of the heat exchangers and present worths of the lifetime steam costs are shown in Table 10.26. Use dynamic programming to determine the outlet temperature from each heat exchanger that results in the minimum total present worth of costs.
Ans.: $254,300.

TABLE 10.26
Present worth of heat exchanger and lifetime costs of steam, thousands of dollars

Heat exchanger	Inlet temp., °C	Outlet temperature, °C					
		50	100	150	200	250	300
1	50	0	$20.8	$58.0	X	X	X
2	50	0	23.1	62.6	$132.3	X	X
	100		0	36.1	93.6	X	X
	150			0	62.8	X	X
3	50	0	24.8	79.9	129.9	$177.0	$308.3
	100		0	41.1	94.3	141.7	266.7
	150			0	51.2	103.0	223.6
	200				0	57.6	176.4
4	50						372.4
	100						309.3
	150						243.7
	200						173.7
	250						94.4
	300						0

10.7. A cooling pond serving a power plant is equipped with circulating pumps and sprays to enhance the rate of heat rejection from the pond. Furthermore, the pumps are to be operated so that the heat rejection is accomplished with minimal pumping energy. On one particular day the pond temperature is 28.5°C at 1800 hours, and the pond temperature is to be reduced to 21.5°C by 0600 the next morning. The rate of decrease of temperature of the pond water is a function of the temperature difference between the pond water and the ambient wet-bulb temperature as well as the intensiveness of pumping. The pumping energies during a 3-h period are shown in Table 10.27.

  The ambient wet-bulb temperatures at the beginning of each 3-h period are 1800, 25.0°C; 2100, 23.0°C; 2400, 21.5°C; and 0300, 20.0°C. Use dynamic programming to determine the pond-water temperature at each 3-h interval that results in minimum pumping energy.

  Ans.: Minimum energy = 139.

10.8. A rocket starting from rest carries an initial fuel charge of 10,000 kg, which is to be burned in 120 s at such a rate that the maximum velocity of the rocket is to be achieved at the end of the burning time. The 120-s burning time is divided into four 30-s intervals. Table 10.28 presents the increase in velocity in each 30-s interval as a function of the mass of fuel in the rocket at the start of the interval and the mass of fuel burned during the interval. All 10,000 kg is to be expended in 120 s, and at least 1000 kg is to be burned each time interval.

  Use dynamic programming to determine the fuel-burning plan that results in the highest velocity of the rocket in 120 s.

  Ans.: 1533 m/s.

10.9. Hydrazine, N_2H_4, is a possible fuel for an emergency-use gas turbine because in the presence of a catalyst it decomposes in an exothermic reaction

$$2N_2H_4 \rightarrow 2NH_3 + N_2 + H_2$$

TABLE 10.27
Pumping energy

Temperature difference at start of 3-h period, $t_w - t_{wb}$†	Drop in t_w in 3-h period, °C					
	0	**0.5**	**1.0**	**1.5**	**2.0**	**2.5**
1.5	0	2	16	54	X	X
2.0		1	12	45	X	X
2.5		0	8	36	96	X
3.0			4	27	80	166
3.5			0	18	64	139
4.0				9	48	111
4.5				0	32	83
5.0					16	55
5.5					0	27
6.0						0

†t_w = temperature of water in pond, °C

t_{wb} = wet-bulb temperature of ambient air, °C

TABLE 10.28
Increase in velocity in 30-s interval, m/s

Initial mass of fuel, kg	Fuel burned in 30 s, kg						
	1000	2000	3000	4000	5000	6000	7000
1,000	226						
2,000	200	424					
3,000	180	377	599				
4,000	165	341	536	756			
5,000	152	312	486	679	895		
6,000	143	288	445	617	808	1024	
7,000	136	269	412	567	737	926	1140
8,000	131	253	385	525	679	847	1034
9,000	127	241	362	491	630	781	947
10,000	124	231	343	462	589	726	875

The reactor is to consist of four stages, as shown in Fig. 10-12. A particular reactor is limited to a total of 14 kg of catalyst, which is available in 1-kg packages. Each stage can accommodate 0 to 5 kg of catalyst. The fraction of hydrazine undecomposed in a stage is given by

$$\frac{y_0}{y_i} = 0.5 + 0.5e^{-my_i}$$

where y_i = fraction of undecomposed hydrazine entering stage
y_0 = fraction of undecomposed hydrazine leaving stage
m = mass of catalyst used in stage

Use dynamic programming to determine the optimum distribution of the 14 kg of catalyst. To shorten the calculation effort, construct the tables only in the neighborhood of the answer given below.
        Ans.: 2, 3, 4, 5.

10.10. Some impurities in industrial waste water can be removed[2] by passing the water through a series of adsorbers containing activated carbon. In a certain installation the adsorbers are arranged in four stages with waste water having

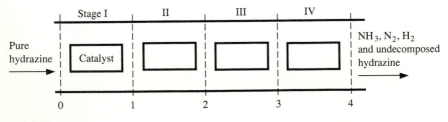

FIGURE 10-12
Four-stage hydrazine reactor.

an initial contamination of 2000 ppm entering the first stage. A total of 140 kg of activated carbon is available to be distributed in the four stages in quantities of 10, 20, 30, 40, or 50 kg in each stage. The performance of each stage is expressed by

$$x_0 = x_i\left(1 - \frac{m}{60} + \frac{m^2 x_i}{20,000,000}\right)$$

where x_0 = contamination at outlet of stage, ppm
x_i = contamination at inlet of stage, ppm
m = mass of activated carbon in the stage, kg

Use dynamic programming to establish the distribution that results in the minimum contamination of the waste water leaving stage 4. To shorten the calculation effort, construct the tables only in the neighborhood of the answer given below.

Ans.: 10, 30, 50, 50.

REFERENCES

1. P. R. Klinkowski, "Ultrafiltration: An Emerging Unit Operation," *Chem. Eng.*, vol. 85, no. 11, pp. 164–173, May 6, 1978.
2. J. L. Rizzo and A. R. Shepherd, "Treating Industrial Wastewater with Activated Carbon," *Chem. Eng.*, vol. 84, no. 1, pp. 95–100, Jan. 3, 1977.

ADDITIONAL READINGS

Bellman, R. E.: *Dynamic Programming*, Princeton University Press, Princeton, N.J., 1957.
Bellman, R. E., and S. Dreyfus: *Applied Dynamic Programming*, Princeton University Press, Princeton, N.J., 1962.
Denn, M. M.: *Optimization by Variational Methods*, McGraw-Hill, New York, 1969.
Gluss, B.: *An Elementary Introduction to Dynamic Programming; a State Equation Approach*, Allyn and Bacon, Boston, 1972.
Hastings, N. A. J.: *Dynamic Programming with Management Applications*, Crane, New York, 1973.
Nemhauser, G. L.: *Introduction to Dynamic Programming*, Wiley, New York, 1960.
Roberts, S.: *Dynamic Programming in Chemical Engineering and Process Control*, Academic, New York, 1964.

CHAPTER
11

GEOMETRIC PROGRAMMING

11.1 INTRODUCTION

Geometric programming is one of the newest methods of optimization. Clarence Zener first recognized the significance of the geometric and arithmetic mean in cases of unconstrained optimization. Since then others have extended the methods to accommodate more general optimization problems. The form of problem statement that is particularly adaptable to treatment by geometric programming is a sum of polynomials for both the objective function and the constraint equations. These polynomials can be made up of combinations of variables of either positive or negative noninteger or integer exponents. After seeing the usefulness of polynomial representations in Chapter 4, we recognize that the ability to optimize such functions is clearly of engineering importance.

A feature of geometric programming is that the first stage of the solution is to find the optimum value of the function, rather than first to determine the values of the independent variables that give the optimum. This knowledge of the optimum value may be all that is of interest, and the calculation of the values of the variables can be omitted.

This chapter first presents the form of a geometric programming problem and then defines *degree of difficulty*, because problems with zero degree of difficulty are ideally suited to geometric programming. Furthermore, optimization problems with degree of difficulty greater than zero are probably best solved by some other method of optimization. Next the mechanics of the method will be illustrated with several examples. Gradually this chapter weaves in some explanation and proof of geometric programming so that the user will have confidence in a technique that at first may seem to be a dark art.

Unconstrained optimizations are the first ones attacked, and later the study moves to constrained optimization with equality constraints. Geometric programming is capable of optimizing objective functions subject to inequality constraints, but this application is beyond the scope of this introductory chapter.

In addition to providing optimal values of the objective function and independent variables, geometric programming supplies additional insight into the solution. For example, the solution by geometric programming also shows how the total cost is divided among the various contributors.

11.2 FORM OF THE OBJECTIVE FUNCTION AND CONSTRAINTS

Geometric programming is adaptable to problems where the objective function and constraints are sums of polynomials of the variables. The variables can be taken to integer or noninteger positive or negative exponents. The following examples of unconstrained objective functions can be solved by geometric programming:

$$\text{Minimize} \qquad y = 5x + \frac{10}{\sqrt{x}} \qquad (11.1)$$

$$\text{Maximize} \qquad y = 6 + 3x - x^{1.8} \qquad (11.2)$$

$$\text{Minimize} \qquad y = 2x_1 x_2^2 + \frac{5}{\sqrt{x_1 x_2}} + 10x_1^{0.8} - 4x_2 \qquad (11.3)$$

An example of a constrained optimization adaptable to geometric programming is

$$\text{Minimize} \qquad y = 5x_1\sqrt{x_2} + 2x_1^2 + x_2^{3/2} \qquad (11.4)$$

$$\text{subject to} \qquad x_1 x_2 = 50$$

11.3 DEGREE OF DIFFICULTY

Duffin, Peterson, and Zener[1] define degree of difficulty when applied to geometric programming problems as $T - (N + 1)$, where T is the number of terms in the objective function plus those in the constraints and N is the number of variables. The degree of difficulty of the objective function in Eq. (11.1) is $2 - (1 + 1) = 0$. The degree of difficulty of the objective function in Eq. (11.2) is also zero, since only $3x = x^{1.8}$ will be optimized. The degree of difficulty of Eq. (11.3) is unity. In the constrained optimization of the objective function in Eq. (11.4), the number of terms is $3 + 1 = 4$ (three in the objective function and one in the constraint), and the number of variables is 2, so the degree of difficulty is 1.

When the degree of difficulty is zero, geometric programming is often the simplest method available for solution. For degrees of difficulty greater than zero, geometric programming will work,[2] but the method involves the solution of nonlinear equations, which will probably be more time-consuming than if some other method, such as Lagrange multipliers, is used. Henceforth in this chapter, only problems of zero degree of difficulty will be considered.

11.4   MECHANICS OF SOLUTION FOR ONE INDEPENDENT VARIABLE, UNCONSTRAINED

Before providing the support for the method, the mechanics will be presented through several examples. The optimal value $y*$ will be sought for the function

$$y = c_1 x^{a_1} + c_2 x^{a_2} \tag{11.5}$$

The individual terms will be designated by the symbol u; thus

$$u_1 = c_1 x^{a_1} \qquad u_2 = c_2 x^{a_2} \qquad y = u_1 + u_2$$

Geometric programming asserts that the optimal value $y*$ can also be represented in product form by an expression that we shall call $g*$,

$$y* = g* = \left(\frac{c_1 x^{a_1}}{w_1}\right)^{w_1} \left(\frac{c_2 x^{a_2}}{w_2}\right)^{w_2} \tag{11.6}$$

provided that $w_1 + w_2 = 1$ \hfill (11.7)

and $a_1 w_1 + a_2 w_2 = 0$ \hfill (11.8)

A consequence of Eq. (11.8) is that the x's cancel out of Eq. (11.6), so the solution is

$$y* = g* = \left(\frac{c_1}{w_1}\right)^{w_1} \left(\frac{c_2}{w_2}\right)^{w_2} \tag{11.9}$$

where w_1 and w_2 are specified by Eqs. (11.7) and (11.8).

A further significance of w_1 and w_2 is that at the optimum

$$w_1 = \frac{u_1^*}{u_1^* + u_2^*} = \frac{u_1^*}{y*} \tag{11.10}$$

$$w_2 = \frac{u_2^*}{u_1^* + u_2^*} = \frac{u_2^*}{y*} \tag{11.11}$$

Equations (11.10) and (11.11) may be useful in solving for $x*$.

Example 11.1. Determine the optimum pipe diameter which results in minimum first plus operating cost for 100 m of pipe conveying a given water flow rate. The first cost of the installed pipe in dollars is $160D$, where D is the pipe diameter in millimeters. The lifetime pumping cost is $(32 \times 10^{12})/D^5$ dollars. It is to be expected that the pumping cost will be proportional to D^{-5} because the pumping cost for a specified flow rate Q, number of hours of operation, pump efficiency, motor efficiency, and electric rate is proportional to the pressure drop in the pipe. Further, this pressure drop Δp is

$$\Delta p, \text{Pa} = f \frac{L}{D} \frac{V^2 \rho}{2} = f \frac{L}{D} \left(\frac{Q}{\pi D^2/4} \right)^2 \frac{\rho}{2} = \frac{\text{const}}{D^5}$$

The objective function, the cost y, in terms of the variable D is then

$$y = 160D + \frac{32 \times 10^{12}}{D^5} \tag{11.12}$$

Solution 1. To provide a check on the geometric-programming method, optimize by calculus,

$$\frac{dy}{dD} = 160 - \frac{(5)(32 \times 10^{12})}{D^6} = 0$$

Then $\qquad D^* = 100\text{mm} \qquad$ and $\qquad y^* = \$19,200$

Next optimize Eq. (11.12) by geometric programming.

Solution 2

$$y^* = 160D^* + \frac{32 \times 10^{12}}{(D^*)^5} = g^*$$

$$y^* = g^* = \left(\frac{160D}{w_1} \right)^{w_1} \left(\frac{32 \times 10^{12}}{D^5 w_2} \right)^{w_2}$$

provided that w_1 and w_2 are chosen so that

$$a_1 w_1 + a_2 w_2 = w_1 - 5w_2 = 0$$

and $\qquad\qquad\qquad\qquad w_1 + w_2 = 1$

Solving gives $\quad w_1 = \dfrac{5}{6} \quad$ and $\quad w_2 = \dfrac{1}{6}$

Substituting these values for w_1 and w_2 into the expression for g^* results in the cancellation of the D's, leaving

$$y^* = g^* = \left(\frac{160}{5/6} \right)^{5/6} \left(\frac{32 \times 10^{12}}{1/6} \right)^{1/6}$$

$$y^* = g^* = \$19,200$$

The value of $D*$ can be found by applying Eq. (11.10)

$$w_1 = \frac{5}{6} = \frac{u_1^*}{u_1^* + u_2^*} = \frac{u_1^*}{y^*} = \frac{160D*}{19,200}$$

and so $D* = 100$ mm $=$ optimum diameter

The following example is another illustration of the mechanics of geometric programming applied to an unconstrained function of one independent variable. It differs from Example 11.1 in that the objective function now contains a negative term.

Example 11.2. The torque T in newton-meters developed by a certain internal combustion engine is represented by

$$T = 23.6\,\omega^{0.7} - 3.17\omega$$

where ω is the rotative speed in radians per second. Determine the maximum power of which this engine is capable and the rotative speed at which the maximum occurs.

Solution. The power P in watts is the product of the torque and the rotative speed,

$$P = T\omega = 23.6\,\omega^{1.7} - 3.17\omega^2$$

Applying geometric programming, we have

$$y* = g* = \left(\frac{23.6}{w_1}\right)^{w_1}\left(\frac{-3.17}{w_2}\right)^{w_2} \tag{11.13}$$

provided that

$$w_1 + w_2 = 1 \qquad \text{and} \qquad 1.7w_1 + 2w_2 = 0$$

Solving the two equations for w_1 and w_2 gives

$$w_1 = 6.667 \qquad \text{and} \qquad w_2 = -5.667$$

Substituting these values of w_1 and w_2 into Eq. (11.13) gives

$$g* = \left(\frac{23.6}{6.667}\right)^{6.667}\left(\frac{-3.17}{-5.667}\right)^{-5.667}$$

$$= 122,970 \text{ W} = 123.0 \text{ kW} = y*$$

To compute the value of $\omega*$, use the fact that

$$u_1^* = 23.6\,\omega^{*1.7} = y*w_1 = (122,970)(6.667)$$

so $\omega* = \left[\dfrac{(122,970)(6.667)}{23.6}\right]^{1/1.7} = 469$ rad/s $= 74.6$ r/s

An alternate means of determining ω^* would be to use u_2^*,

$$u_2^* = (-3.17)(\omega^*)^2 = 122{,}970w_2 = (122{,}970)(-5.667)$$

$$(\omega^*)^2 = 219{,}830 \qquad \omega^* = 469 \text{ rad/s}$$

The existence of the negative coefficient in the objective function caused no special difficulty.

A further example will illustrate yet another class of problems and also a situation which at first may seem to make the problem insoluble.

Example 11.3. Find y^* and x^* for the function

$$y = -4x + x^{1.6}$$

Solution

$$y^* = g^* = \left(\frac{-4}{w_1}\right)^{w_1}\left(\frac{1}{w_2}\right)^{w_2}$$

provided that $w_1 + w_2 = 1 \qquad$ and $\qquad w_1 + 1.6w_2 = 0$

The values of the w's are $w_1 = 2.667$ and $w_2 = -1.667$, and so

$$y^* = g^*\left(\frac{-4}{2.667}\right)^{2.667}\left(\frac{1}{-1.667}\right)^{-1.667} \qquad (11.14)$$

Each combination of numbers in the parentheses in Eq. (11.14) is negative, and a negative number taken to a noninteger power is a complex number. The difficulty can be resolved, however, by extracting the negative sign from both terms in parentheses,

$$y^* = (-1)^{2.667}\left(\frac{4}{2.667}\right)^{2.667}(-1)^{-1.667}\left(\frac{1}{1.667}\right)^{-1.667}$$

$$= (-1)\left(\frac{4}{2.667}\right)^{2.667}\left(\frac{1}{1.667}\right)^{-1.667} \qquad = -6.90$$

The optimal value of x can be found next

$$-4x = u_1^* = w_1 y^* = (2.667)(-6.90)$$

$$x^* = 4.6$$

The optimal value of y in this example is negative, which makes the components complex numbers in Eq. (11.14). It is possible, however, to extract the negative terms and group them as negative unity taken to the power of unity.

11.5   WHY GEOMETRIC PROGRAMMING WORKS; ONE INDEPENDENT VARIABLE

The previous section presented the mechanics of optimizing a function of one independent variable by using geometric programming but gave no substantiation for the method. This section will prove the validity of the geometric-programming method for the one-independent-variable problem. The form of the objective function is the same as that of Eq. (11.5)

$$y = c_1 x^{a_1} + c_2 x^{a_2} = u_1 + u_2 \tag{11.15}$$

Next fabricate a function g such that

$$g = \left(\frac{u_1}{w_1}\right)^{w_1}\left(\frac{u_2}{w_2}\right)^{w_2} = \left(\frac{c_1 x^{a_1}}{w_1}\right)^{w_1}\left(\frac{c_2 x^{a_2}}{w_2}\right)^{w_2} \tag{11.16}$$

where

$$w_1 + w_2 = 1 \tag{11.17}$$

A certain combination of values of w_1 and w_2 will provide a maximum value of g. To determine these values of w_1 and w_2, apply the method of Lagrange multipliers to Eq. (11.16) subject to the constraint of Eq. (11.17). The maximum values of g and $\ln g$ both occur at the same value of the w's; since it is more convenient to optimize $\ln g$:

Maximize $$\ln g = w_1(\ln u_1 - \ln w_1) + w_2(\ln u_2 - \ln w_2) \tag{11.18}$$

subject to $$w_1 + w_2 - 1 = \phi = 0 \tag{11.19}$$

Use the method of Lagrange multipliers

$$\nabla(\ln g) - \lambda \nabla \phi = 0$$

$$\phi = 0$$

which provides the three equations:

w_1: $$\ln u_1 - 1 - \ln w_1 - \lambda = 0$$

w_2: $$\ln u_2 - 1 - \ln w_2 - \lambda = 0$$

$$w_1 + w_2 - 1 = 0$$

The unknowns are w_1, w_2, and λ, and the solutions for w_1 and w_2 are

$$w_1 = \frac{u_1}{u_1 + u_2} \tag{11.20}$$

and $$w_2 = \frac{u_2}{u_1 + u_2} \tag{11.21}$$

Substituting these values of w_1 and w_2 into Eq. (11.16) gives

$$g = \left[\frac{u_1}{u_1/(u_1 + u_2)}\right]^{u_1/(u_1 + u_2)}\left[\frac{u_2}{u_2/(u_1 + u_2)}\right]^{u_2/(u_1 + u_2)}$$

Thus
$$g = u_1 + u_2$$

Let us pause at this point to assess our status. By the choice of w_1 and w_2 according to Eqs. (11.20) and (11.21) the value of g is made equal to that of $u_1 + u_2$ and therefore also to y. Any other combination of w_1 and w_2 results in a value of g that is less than $u_1 + u_2 = y$.

Since our original objective is to minimize y, the next step is to use the value of x in Eq. (11.15) that results in the minimum value of $u_1 + u_2$. The value of g at this condition g^* will therefore be the one where the w's are chosen such that g always equals y but also with the value of x chosen so that y is a minimum. This value of x^* can be found by equating the derivative of Eq. (11.15) to zero

$$a_1 c_1 x^{(a_1 - 1)} + a_2 c_2 x^{(a_2 - 1)} = 0$$

Multiplying by x gives

$$a_1 c_1 x^{a_1} + a_2 c_2 x^{a_2} = 0$$

and so
$$a_1 u_1^* + a_2 u_2^* = 0 \tag{11.22}$$

where u_1^* and u_2^* are the values of u_1 and u_2 at the minimum value of y.
From Eq. (11.22)

$$u_1^* = \frac{a_2 u_2^*}{a_1}$$

which, when substituted into Eqs. (11.20) and (11.21), yields

$$w_1 = \frac{-(a_2/a_1)u_2^*}{-(a_2/a_1)u_2^* + u_2^*} = \frac{-a_2}{a_1 - a_2} \tag{11.23}$$

and
$$w_2 = \frac{u_2^*}{-(a_2/a_1)u_2^* + u_2^*} = \frac{a_1}{a_1 - a_2}$$

When these values of w_1 and w_2 are substituted into Eq. (11.16) for the exponents, the expression for g^* results

$$g^* = \left(\frac{c_1 x^{a_1}}{w_1}\right)^{-a_2/(a_1 - a_2)} \left(\frac{c_2 x^{a_2}}{w_2}\right)^{a_1/(a_1 - a_2)}$$

Of special importance is the fact that x can be canceled out, leaving

$$g^* = \left(\frac{c_1}{w_1}\right)^{-a_2/(a_1 - a_2)} \left(\frac{c_2}{w_2}\right)^{a_1/(a_1 - a_2)}$$

or
$$g^* = \left(\frac{c_1}{w_1}\right)^{w_1} \left(\frac{c_2}{w_2}\right)^{w_2} = y^* \tag{11.24}$$

In executing geometric programming the values of w proportion themselves so that the x's cancel in the expression for g.

11.6   SOME INSIGHTS PROVIDED BY GEOMETRIC PROGRAMMING

Geometric programming not only yields optimal values of the variables and objective function but can provide some insight into the solution. Consider, for instance, Example 11.1, where the life-cycle cost of the pump-pipe system was optimized:

$$y = \text{pipe cost} + \text{present worth of lifetime energy cost}$$

$$= 160D + \frac{32 \times 10^{12}}{D^5}$$

For this objective function the optimal value of diameter $D^* = 100$ mm and $y^* = \$19,200$. The values of the weighting factors, w_1 and w_2 of $\frac{5}{6}$ and $\frac{1}{6}$, respectively, are also of interest, because at the optimum five-sixths of the total cost is devoted to the pipe and one-sixth to the energy.

Let us suppose that the cost of the energy increases. What would be the effect on the optimum? What happens to the solution, for example, if the lifetime energy cost becomes $(50 \times 10^{12})/D^5$? Of particular importance is the fact that the distribution of the total cost remains unchanged (five-sixths and one-sixth) because the exponents of D that control w_1 and w_2 remain unchanged. Both y^* and D^* increase; the optimum total cost now becomes \$20,680, and $D = 107.7$ mm. The increase in optimal diameter probably conforms to our expectation that upon an increase in the energy cost the diameter responds by increasing. But regardless of what happens to the unit cost of energy or pipe, the distribution between the first cost of the pipe and the energy remains constant.

Suppose, however, that an exponent of D changes, e.g., the cost of the pipe increases at a more rapid rate than linearly. If the cost of the pipe is proportional to $D^{1.2}$, for example, the distribution of costs between the pipe and energy changes because the new value of $w_1 = 0.806$, compared with the original value of 0.833. The optimal condition responds to the more rapid increase in pipe cost by decreasing the fraction of the total cost devoted to the pipe.

11.7   UNCONSTRAINED, MULTIVARIABLE OPTIMIZATION

The geometric-programming procedures for the one independent variable extend in a logical manner to multivariable optimizations. If applications continue to concentrate on problems of zero degree of difficulty, a two-variable problem, for example, will have an objective function containing three terms.

Example 11.4. The pump and piping of Example 11.1 are actually part of a waste-treatment complex, as shown in Fig. 11-1. The system accomplishes

FIGURE 11-1
Waste-treatment system in Example 11.1.

the treatment by a combination of dilution and chemical action so that the effluent meets code requirements. The size of the reactor can decrease as the dilution increases. The cost of the reactor is $150/Q$, where Q is the flow rate in cubic meters per second. The equation for the pumping cost with Q broken out of the combined constant is $(220 \times 10^{15}Q^2)/D^5$. Use geometric programming to optimize the total system.

Solution. The total cost is the sum of the costs of the pipe, the pumping power, and the treatment plant

$$y = 160D + \frac{220 \times 10^{15}Q^2}{D^5} + \frac{150}{Q}$$

$$y* = g* = \left(\frac{160}{w_1}\right)^{w_1} \left(\frac{220 \times 10^{15}}{w_2}\right)^{w_2} \left(\frac{150}{w_3}\right)^{w_3}$$

provided that

$$w_1 + w_2 + w_3 = 1$$

to cancel

D:

$$w_1 - 5w_2 = 0$$

Q:

$$2w_2 - w_3 = 0$$

Solving gives

$$w_1 = \frac{5}{8}, \quad w_2 = \frac{1}{8}, \quad \text{and} \quad w_3 = \frac{2}{8}$$

Then

$$\left(\frac{160}{5/8}\right)^{5/8} \left(\frac{220 \times 10^{15}}{1/8}\right)^{1/8} \left(\frac{150}{1/4}\right)^{1/4} = \$30,224$$

$$u_1^* = 160D = \frac{5}{8}(30,224) \quad \text{so} \quad D* = 118 \text{ mm}$$

$$u_3^* = \frac{150}{Q*} = \frac{2}{8}(30,224) \quad \text{so} \quad Q* = 0.0198 \text{ m}^3/\text{s}$$

The core of the execution of Example 11.4 by geometric programming was the solution of *three* simultaneous *linear* equations for the w's. Had the problem been solved by Lagrange multipliers, *two* simultaneous *nonlinear* equations (from $\nabla y = 0$) would have been solved. In general, the solution by geometric programming of a zero-degree-of-difficulty problem requires the solution of one more equation in a set of simultaneous equations than required by Lagrange multipliers, *but the equations are linear*. For this special class of problems, then, solution by geometric programming is likely to be much simpler than by calculus.

The mathematical substantiation for the procedure of solving the multivariable problem, Example 11.4, has not yet been provided. It can be developed by following the pattern for the proof of the single-variable optimization presented in Sec. 11.5. The steps are as follows:

1. Propose a g function in the form of Eq. (11.16).
2. Set the sum of the w's equal to unity [Eq. (11.17)].
3. Maximize $\ln g$ subject to

$$\sum_{i=1}^{N} w_i = 1$$

 to find that the optimal w's equal the fractions that the respective u's are of the total [Eq. (11.21)]. With those optimal values of w the function g equals the original function to be optimized.
4. Optimize the function. The differentiation with respect to the x's is now a *partial* differentiation, and the derivatives are equated to zero, following which the equations are multiplied through by the appropriate x_i.
5. The result will be a summation for each variable

$$\sum_{i=1}^{T} a_{tn} w_t = 0 \tag{11.25}$$

where t = term number in object function
   T = total number of terms in objective function
   n = variable, ranging from 1 to total number of variables N

The N equations designated by Eq. (11.25) are the conditions that result in the cancellation of all the x's in the g-function formulation.

11.8   CONSTRAINED OPTIMIZATION WITH ZERO DEGREE OF DIFFICULTY

The final type of geometric-programming problem to be explored is one with an equality constraint. Only the zero-degree-of-difficulty case will be considered, and the total number of terms T means the sum of those in the objective function and the constraint. Suppose that the objective function to be minimized is

$$y = u_1 + u_2 + u_3 \tag{11.26}$$

subject to the constraint

$$u_4 + u_5 = 1 \tag{11.27}$$

where the u's are polynomials in terms of four independent variables, x_1, x_2, x_3, and x_4. The right side of the constraint equation must be unity, which poses no problem as long as one pure numerical term appears in the equation. If that number is not unity, the entire equation can be divided by the number to convert it into unity.

The objective function can be rewritten

$$y = g = \left(\frac{u_1}{w_1}\right)^{w_1} \left(\frac{u_2}{w_2}\right)^{w_2} \left(\frac{u_3}{w_3}\right)^{w_3} \tag{11.28}$$

provided that

$$w_1 + w_2 + w_3 = 1 \tag{11.29}$$

and

$$w_i = \frac{u_i}{u_1 + u_2 + u_3} \tag{11.30}$$

The constraint equation can also be rewritten as

$$u_4 + u_5 = 1 = \left(\frac{u_4}{w_4}\right)^{w_4} \left(\frac{u_5}{w_5}\right)^{w_5} \tag{11.31}$$

provided that

$$w_4 + w_5 = 1 \tag{11.32}$$

and

$$w_4 = \frac{u_4}{1} = u_4 \quad \text{and} \quad w_5 = \frac{u_5}{1} = u_5 \tag{11.33}$$

Equation (11.31) can be raised to the Mth power, where M is an arbitrary constant, and its value remains unity,

$$1 = \left(\frac{u_4}{w_4}\right)^{M w_4} \left(\frac{u_5}{w_5}\right)^{M w_5} \tag{11.34}$$

Next multiply Eq. (11.28) by Eq. (11.34)

$$y = g = \left(\frac{u_1}{w_1}\right)^{w_1} \left(\frac{u_2}{w_2}\right)^{w_2} \left(\frac{u_3}{w_3}\right)^{w_3} \left(\frac{u_4}{w_4}\right)^{M w_4} \left(\frac{u_5}{w_5}\right)^{M w_5} \tag{11.35}$$

Momentarily set aside the representations of Eqs. (11.28) to (11.35) and return to Eqs. (11.26) and (11.27) and solve by Lagrange multipliers

$$\nabla(u_1 + u_2 + u_3) - \lambda[\nabla(u_4 + u_5)] = 0 \tag{11.36}$$

$$u_4 + u_5 = 1 \tag{11.37}$$

The vector equation (11.36) represents four scalar equations, the terms of which are the partial derivatives with respect to x_1 to x_4. By taking advantage of the fact that all the u's are polynomials each of the scalar equations of Eq. (11.36) can be multiplied by the variable with respect to which it has just been differentiated. The result is

$$a_{11}u_1^* + a_{21}u_2^* + a_{31}u_3^* - \lambda a_{41}u_4^* - \lambda a_{51}u_5^* = 0$$
$$\cdots\cdots\cdots\cdots\cdots\cdots\cdots\cdots\cdots\cdots\cdots\cdots\cdots$$
$$a_{14}u_1^* + a_{24}u_2^* + a_{34}u_3^* - \lambda a_{44}u_4^* - \lambda a_{54}u_5^* = 0 \qquad (11.38)$$

The asterisk in Eqs. (11.38) indicates that these are optimal values of u. Next the results of Eqs. (11.38) can be merged with the representation of Eqs. (11.28) to (11.35). Dividing Eqs. (11.38) by y^* permits replacement of u_1^*/y^* by w_i.

Since the constant M in Eq. (11.34) was arbitrary, let it equal $-\lambda/y^*$. The revised Eqs. (11.38) are then

$$a_{11}w_1 + a_{21}w_2 + a_{31}w_3 + Ma_{41}w_4 + Ma_{51}w_5 = 0$$
$$\cdots\cdots\cdots\cdots\cdots\cdots\cdots\cdots\cdots\cdots\cdots\cdots\cdots$$
$$a_{14}w_1 + a_{24}w_2 + a_{34}w_3 + Ma_{44}w_4 + Ma_{54}w_5 = 0 \qquad (11.39)$$

Along with the conditions

$$w_1 + w_2 + w_3 = 1 \qquad (11.29)$$

and, from Eq. (11.32),

$$Mw_4 + Mw_5 = M \qquad (11.40)$$

They provide six linear simultaneous equations. These six equations can be solved for the six unknowns w_1, w_2, w_3, Mw_4, Mw_5, and M.

It is especially significant that Eqs. (11.39) require the cancellation of all the x terms in Eq. (11.35). This fact permits a convenient evaluation of the optimum value of y

$$y^* = \left(\frac{c_1}{w_1}\right)^{w_1} \left(\frac{c_2}{w_2}\right)^{w_2} \left(\frac{c_3}{w_3}\right)^{w_3} \left(\frac{c_4}{w_4}\right)^{Mw_4} \left(\frac{c_5}{w_5}\right)^{Mw_5}$$

Example 11.5. A water pipeline extends 30 km across a desert from a desalination plant at the seacoast to a city. The pipeline, as shown schematically in Fig. 11-2, conveys $0.16 \text{ m}^3/\text{s}$ of water. The first costs of the pipeline are

$$\text{Cost of each pump} = 2500 + 0.00032\Delta p^{1.2} \text{ dollars}$$

$$\text{Cost of 30 km of pipe} = 2{,}560{,}000D^{1.5} \text{ dollars}$$

where Δp = pressure drop in each pipe section, Pa
D = diameter of pipe, m

Assume a friction factor of 0.02.

Use geometric programming on a constrained objective function to select the number of pumps and the pipe diameter that results in the minimum total first cost for the system.

FIGURE 11-2
Water pipeline in Example 11.5.

Solution. If n designates the number of pump-and-pipe sections, the total cost y is

$$y = n(2500 + 0.00032\Delta p^{1.2}) + 2,560,000D^{1.5}$$

and

$$n = \frac{30,000 \text{ m}}{L}$$

where L is the length of each pipe section in meters. The pressure drop in each pipe section is

$$\Delta p = f\frac{L}{D}\frac{V^2}{2}\rho = (0.02)\frac{L}{D}\left(\frac{0.16}{\pi D^2/4}\right)^2\frac{1}{2}(1000 \text{ kg/m}^3)$$

and so

$$\frac{\Delta p D^5}{L} = 0.4150 \tag{11.41}$$

The statement of the problem is

Minimize $$y = \frac{75,000,000}{L} + \frac{9.6\Delta p^{1.2}}{L} + 2,560,000D^{1.5} \tag{11.42}$$

subject to $$\frac{2.410\Delta p D^5}{L} = 1 \tag{11.43}$$

$$y^* = \left(\frac{75,000,000}{w_1}\right)^{w_1}\left(\frac{9.6}{w_2}\right)^{w_2}\left(\frac{2,560,000}{w_3}\right)^{w_3}\left(\frac{2.41}{w_4}\right)^{Mw_4}$$

provided that

$$
\begin{array}{llll}
L: & -w_1 - & w_2 & -Mw_4 = 0 \\
\Delta p: & & 1.2w_2 & +Mw_4 = 0 \\
D: & & & 1.5w_3 + 5Mw_4 = 0 \\
& w_1 + & w_2 + & w_3 = 1 \\
& & & 5Mw_4 = M
\end{array}
$$

so $w_1 = 0.0385$, $w_2 = 0.1923$, $w_3 = 0.7692$, $M = -0.2308$, and $Mw_4 = -0.2308$,

$$y* = \left(\frac{75,000,000}{0.0385}\right)^{0.0385}\left(\frac{9.6}{0.1923}\right)^{0.1923}\left(\frac{2,560,000}{0.7692}\right)^{0.7692}\left(\frac{2.41}{1}\right)^{-0.2308}$$

$$y* = \$410,150$$

$$u_1^* = (410,150)(0.0385) = \frac{75,000,000}{L}$$

and so
$$L* = 4750 \text{ m}$$

$$u_3^* = (410,150)(0.769) = 2,560,000D^{1.5}$$

and so
$$D* = 0.246 \text{ m}$$

$$\Delta p* = \frac{L*}{2.410D^{*5}} = 2,188,000 \text{ Pa} = 2188 \text{ kPa}$$

Since the number of pump-and-pipe sections is 30,000/4750 = 6.3, six pumps could be used, which revises L to 5000 m, and this value of L can be substituted into Eqs. (11.42) and (11.43) for a reoptimization of Δp and D.

11.9 SENSITIVITY COEFFICIENT

In Sec. 8.10 the Lagrange multiplier λ emerged as the sensitivity coefficient, which was $dy*/dA$, where A is the numerical term on the right side of the constraint equation. In the development of the procedure for constrained optimization using geometric programming the expression $-\lambda/y*$ was replaced by the term $|M|$ in arriving at Eqs. (11.39). When the optimization is complete and $y*$ and M are known, the sensitivity coefficient can be determined quickly. In Example 11.5 $|M| = -0.2308$ and $y* = 410,150$, so the sensitivity coefficient $\lambda = -My* = 94,660$. This value of the sensitivity coefficient is applicable to the constraint equation in the form of Eq. (11.43). The value of the sensitivity coefficient for the original form of the constraint in Eq. (11.41) is $94,660/0.4150 = 228,100$.

The physical interpretation of the sensitivity coefficient in Example 11.5 is that it indicates the rate of change of the total first cost with respect to the change in the constant 0.4150. If, for example, the water flow rate were changed from 0.160 to 0.161 m^3/s, the constant would change from 0.4150 to $(0.4150)(0.161/0.160)^2 = 0.4202$. The increase in cost of the system is $(228,100)(0.4202 - 0.4150) = \$1187$.

11.10 HIGHER DEGREES OF DIFFICULTY AND EXTENSIONS OF GEOMETRIC PROGRAMMING

The only optimization problems to which geometric programming was applied in this chapter were those of zero degree of difficulty. For such problems geometric programming is probably the simplest means of solution.

Geometric programming also provides some physical insight into the solution not usually offered by other methods of optimization.

Geometric programming is not limited to zero-degree-of-difficulty problems; it can accommodate higher degrees of difficulty but the need to solve a set of nonlinear simultaneous equations arises in such problems. Much of the effort by operations-research workers who are concentrating on geometric programming is devoted to developing efficient and reliable computer programs that include the solution of these nonlinear equations. These computer programs can also handle inequality constraints, as well as the equality constraints presented in this chapter.

Geometric programming is a useful tool to carry in the optimization kit for those special situations to which it is particularly adaptable.

PROBLEMS

Solve the following problems by geometric programming.

11.1. The thickness of the insulation of a hot-water tank is to be selected so that the total cost of the insulation and standby heating for the 10-year life of the facility will be minimum.

Data

> Average water temperature, 60°C
> Average ambient temperature, 24°C
> Conductivity of insulation, 0.036 W/(m · K)
> Cost of heat energy, \$4 per gigajoule
> Cost of insulation, where x = insulation thickness, mm, $0.5x^{0.8}$ dollars

per square meter

The operation is continuous. Assume that the only resistance to heat transfer is the insulation.

(a) What is the minimum total cost of insulation plus standby heat loss per square meter of heat-transfer area for 10 years, neglecting interest charges?

(b) What is the optimum insulation thickness?

Ans.: (b) 99.3 mm.

11.2. A hydraulic power system must provide 300 W of power, where the power is the product of the volume flow rate Q m³/s and the pressure buildup Δp Pa. The cost of the hydraulic pump is a function of both the flow rate and pressure buildup:

$$\text{Cost} = 1200Q^{0.4}\sqrt{10 + (\Delta p \times 10^{-4})} \text{ dollars}$$

Convert to a single-variable unconstrained problem and use geometric programming to determine the minimum cost of the pump and the optimum values of Q and Δp.

Ans.: $\Delta p^* = 400$ kPa.

11.3. Schlichting[3] presents the following equation for the velocity in a free-stream jet that issues into a large space, as illustrated in Fig. 11-3:

$$u = \frac{7.41 u_o \sqrt{A}}{x[1 + 57.5(r/x)^2]^2}$$

where u = velocity in the x direction, m/s
      r = radial distance measured from the centerline, m
      u_0 = outlet velocity, m/s
      A = outlet area, m^2
      x = distance in x direction measured from wall, m

The position of maximum u at a radial distance of 0.5 m from the centerline is to be determined in a certain jet with given values of u_0 and A. Specifically, use geometric programming to find the value of x at which u is maximum when $r = 0.5$ m. *Hint:* The problem can be structured as one of zero degree of difficulty.

     Ans.: 6.57 m.

11.4. The total annual cost of an insulated facility is the sum of the annual cost of the insulation plus the annual cost of the energy. Specifically,

$$\text{cost, \$/m}^2 = c_1 x^{0.5} + \frac{c_2}{x}$$

where x = insulation thickness, mm

If the unit energy cost increases by 10%, what must be the percent decrease in unit insulation cost so that the total annual cost remains constant?

11.5. A hot-water boiler consists of a combustion chamber and a heat exchanger arranged as shown in Fig. 11-4. Fuel at a flow rate of 0.0025 kg/s with a heating value of 42,000 kJ/kg burns in the combustion chamber. Thereafter the flue gases are cooled to 150°C in the process of heating the water. The combustion efficiency increases with an increase in the rate of airflow according to the equation

$$\text{Efficiency, \%} = 100 - \frac{0.023}{(m_a + 0.0025)^2}$$

FIGURE 11-3
Free-stream jet in Prob. 11.3.

FIGURE 11-4
Hot-water boiler in Prob. 11.5.

The specific heat of the gas mixture is 1.05 kJ/(kg · K). Determine the value of m_a that results in the maximum rate of heat transfer to the water. *Suggestion:* Use $m_a + 0.0025$ as the variable to be optimized.

  Ans.: 0.0695 kg/s.

11.6. In a three-stage compression system air enters the first stage at a pressure of 100 kPa and a temperature of 20°C. The exit pressure from the third stage is 6400 kPa. Between stages the air is passed through an intercooler that brings the temperature back to 20°C. The expression for the work of compression per unit mass in an ideal process is

$$\frac{k}{k-1}RT_1\left[1 - \left(\frac{p_2}{p_1}\right)^{(k-1)/k}\right]$$

where subscript 1 refers to the entering conditions and subscript 2 to the leaving conditions.

(a) With the intermediate pressures chosen so that the total work of compression is a minimum, what is the minimum work required to compress 1 kg of air, assuming that the compressions are reversible and adiabatic?

(b) What are the intermediate pressures that result in this minimum work of compression?

  Ans.: 429 kJ; 400 and 1600 kPa.

11.7. The heat-rejection system for a condenser of a steam power plant (Fig. 11-5) is to be designed for minimum first plus pumping cost. The heat-rejection rate from the condenser is 14 MW. The following costs in dollars must be included:

  First cost of cooling tower, $800A^{0.6}$, where $A = $ area, m^2

  Lifetime pumping cost, $0.0005w^3$, where $w = $ flow rate of water, kg/s

  Lifetime penalty in power production due to elevation of temperature of cooling water, $270t$, where $t = $ temperature of water entering the condenser, °C

  The rate of heat transfer from the cooling tower can be represented adequately by the expression q, $W = 3.7(w^{1.2})tA$.

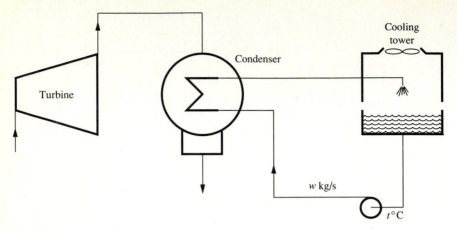

FIGURE 11-5
Heat-rejection system in Prob. 11.7.

    (*a*) Set up an unconstrained objective function in terms of the variables A and w.
    (*b*) Determine the minimum lifetime cost.
    (*c*) Calculate the optimal values of A and w.
        Ans.: (*c*) $w = 202.6$ kg/s.

11.8. The total cost of a rectangular building shell and the land it occupies is to be minimized for a building that must have a volume of 14,000 m³. The following costs per square meter apply: land, $90; roof, $14; floor, $8; and walls, $11. Set up the problem as one of constrained optimization and determine the minimum cost and the optimal dimensions of the building.
        Ans.: Height $= 71.4$ m.

11.9. Newly harvested grain often has a high moisture content and must be dried to prevent spoilage. This drying can be achieved by warning ambient air and blowing it through a bed of the grain. The seasonal operating cost in dollars per square meter of grain bed for such a dryer consists of the cost of heating the air

$$\text{Heating cost} = 0.002Q\Delta t$$

and the blower operating cost

$$\text{Blower cost} = 2.6 \times 10^{-9}Q^3$$

where $Q =$ air quantity delivered through the bed during season, m³/m²
             bed area
    $\Delta t =$ rise in temperature through heater, °C

The values of Q and Δt also influence the time required for adequate drying of the grain according to the equation

$$\text{Drying time} = \frac{80 \times 10^6}{Q^2\Delta t} \text{ days}$$

Using the geometric-programming method of constrained optimization, compute the minimum operating cost and the optimum value of Q and Δt that will achieve adequate drying in 60 days.

Ans.: $\Delta t = 2.28°C$.

REFERENCES

1. R. J. Duffin, E. L. Peterson, and C. M. Zener, *Geometric Programming*, Wiley, New York, 1967.
2. W. F. Stoecker, *Design of Thermal Systems*, 1st ed., McGraw-Hill, New York, 1971.
3. H. Schlichting, *Boundary Layer Theory*, 5th ed., McGraw-Hill, New York, 1958.

BIBLIOGRAPHY

C. Zener, *Engineering Design by Geometric Programming*, Wiley-Interscience, New York, 1971.
C. Beightler and D. T. Phillips, *Applied Geometric Programming*, Wiley, New York, 1976.

CHAPTER
12

LINEAR PROGRAMMING

12.1 THE ORIGINS OF LINEAR PROGRAMMING

Linear programming is an optimization method applicable where both the objective function and the constraints can be expressed as linear combinations of the variables. The constraint equations may be equalities or inequalities. Linear programming first appeared in Europe in the 1930s, when economists and mathematicians began working on economic models. During World War II the United States Air Force sought more effective procedures for allocating resources and turned to linear programming. In 1947 a member of the group working on the Air Force problem, George Dantzig, reported the *simplex method* for linear programming, which was a significant step in bringing linear programming into wider use.

Economists and industrial engineers have applied linear programming more in their fields of work than most other technical groups. Decisions about time allocations of machines to various products in a manufacturing plant, for example, lend themselves neatly to linear programming. In thermal systems linear programming has become an important tool in the petroleum industry and is now being applied in other thermal industries. Most large oil companies use linear-programming models to determine the quantities of the various products that will result in optimum profit for the entire operation. Within the refinery itself, linear programming helps determine where the bottlenecks to production exist and how much the total output of the plant could be increased, for instance, by enlarging a heat exchanger or by increasing a certain rate of flow.

Entire books are devoted to linear programming, but it is possible in one chapter to explain the physical situation that results in a linear-programming structure and to solve practical problems with the technique. The emphasis will be on obtaining a geometric feel for the linear-programming situation, and the simplex algorithm for solving both maximization and minimization problems in linear programming will be applied.

12.2 SOME EXAMPLES OF LINEAR PROGRAMMING

Some classic uses of linear programming are to solve (1) the blending problem, (2) machine allocation, (3) inventory and production planning, and (4) the transportation problem.

The oil-company application mentioned in Sec. 12.1 is typical of the blending application. The oil company has a choice of buying crude from several different sources with different compositions and at differing prices. It has a choice of manufacturing various quantities of aviation fuel, automobile gasoline, diesel fuel, and oil for heating. The combinations of these products are restricted by material balances based on the incoming crude and by the capacity of such components in the refinery as the cracking unit. A mix of purchased crude and manufactured products is sought that gives maximum profit.

The machine-allocation problem occurs where a manufacturing plant has a choice of making several different products, each of which requires varying machine times of different machines such as lathes, screw machines, and grinders. The machine time of some or all of these different machines is limited. The goal of the analysis is to determine the production quantities of each product that result in maximum profits.

The sales of some manufacturing firms fluctuate, often according to a seasonal pattern. The company can build up an inventory of manufactured products to carry it through the period of peak sales, but carrying an inventory costs money. Or it can pay overtime rates in order to step up its production during the period of peak sales, which also entails an additional expense. Finally, the company can simply plan on losing some sales because it does not meet the sales demand at the times that it exists, thus losing a potential profit. Linear programming can incorporate the various cost and loss factors and arrive at the most profitable production plan.

The fourth application, the transportation problem, occurs when an organization has several production plants distributed throughout a geographical area and a number of differently distributed warehouses. Each plant has a certain production capability, and each warehouse has a certain requirement. Any one warehouse may receive the production from one or more plants. The object is to determine how much of each plant's production should be shipped to each warehouse in order to minimize the total manufacturing and transportation cost.

Simple linear-programming problems can be done in a hit-or-miss fashion, but those with three or more variables require systematic procedures. Even when using methodical techniques, the magnitude of a problem that can be solved by hand is limited. Large problems (with several thousand variables) require computer programs, which are currently available as library routines.

12.3 MATHEMATICAL STATEMENT OF THE LINEAR-PROGRAMMING PROBLEM

The form of the statement is typical of the optimization problem in that it consists of an objective function and constraints. The objective function which is to be minimized (or maximized) is

$$y = c_1 x_1 + c_2 x_2 + \cdots + c_n x_n \tag{12.1}$$

and the constraints are

$$\phi_1 = a_{11} x_1 + a_{12} x_2 + \cdots + a_{1n} x_n \geq r_1$$
$$\cdots\cdots\cdots\cdots\cdots\cdots\cdots\cdots\cdots \tag{12.2}$$
$$\phi_m = a_{m1} x_1 + a_{m2} x_2 + \cdots + a_{mn} x_n \geq r_m$$

Furthermore, if the x's represent physical quantities, they are likely to be non-negative, so that $x_1, \cdots, x_n \geq 0$. The c values and a values are all constants, which make both the objective function and the constraints linear; hence the name linear programming. The values of c and a may be positive, negative, or zero. The inequalities in the constraints can be in either direction and can even be strict equalities.

At first glance, this problem might seem readily soluble by Lagrange multipliers, but we recall that the method of Lagrange multipliers is applicable where *equality* constraints exist. Furthermore, Lagrange multipliers apply where $n > m$, but in linear programming n can be greater than, equal to, or less than m. The significance of $n < m$ will be discussed in Sec. 12.14.

12.4 DEVELOPING THE MATHEMATICAL STATEMENT

The translation of the physical conditions into a mathematical statement of linear-programming form will be illustrated by an example.

Example 12.1. A simple power plant consists of an extraction turbine that drives a generator, as shown in Fig. 12-1. The turbine receives 3.2 kg/s of steam, and the plant can sell either electricity or extraction steam for

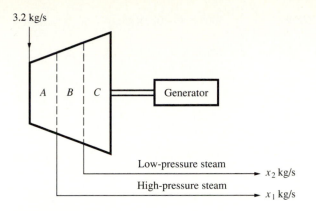

FIGURE 12-1
Power plant in Example 12.1.

processing purposes. The revenue rates are

Electricity, $0.03 per kilowatthour
Low-pressure steam, $1.10 per megagram
High-pressure steam, $1.65 per megagram

The generation rate of electric power depends upon the flow rate of steam passing through each of the sections A, B, and C; these flow rates are w_A, w_B, and w_C, respectively. The relationships are

$$P_A, \text{ kW} = 48w_A$$

$$P_B, \text{ kW} = 56w_B$$

$$P_C, \text{ kW} = 80w_C$$

where the w's are in kilograms per second. The plant can sell as much electricity as it generates, but there are other restrictions.

To prevent overheating the low-pressure section of the turbine, no less than 0.6 kg/s must always flow through section C. Furthermore, to prevent unequal loading on the shaft, the permissible combination of extraction rates is such that if $x_1 = 0$, then $x_2 \leq 1.8$ kg/s, and for each kilogram of x_1 extracted 0.25 kg less can be extracted of x_2.

The customer of the process steam is primarily interested in total energy and will purchase no more than

$$4x_1 + 3x_2 \leq 9.6$$

Develop the objective function for the total revenue from the plant and also the constraint equations.

Solution. The revenue per hour is the sum of the revenues from selling the steam and the electricity.

$$\text{Revenue} = \frac{1.65}{1000}(3600x_1) + \frac{1.10}{1000}(3600x_2) + 0.03(48w_A + 56w_B + 80w_C)$$

Since $w_A = 3.2$ kg/s and from mass balances $w_B = 3.2 - x_1$ and $w_C = 3.2 - x_1 - x_2$,

$$\text{Revenue} = 17.66 + 1.86x_1 + 1.56x_2 \tag{12.3}$$

Because the constant has no effect on the state point at which the optimum occurs, the objective function to be maximized is

$$y = 1.86x_1 + 1.56x_2 \tag{12.4}$$

The three constraints are

$$x_1 + x_2 \le 2.6 \tag{12.5}$$

$$x_1 + 4x_2 \le 7.2 \tag{12.6}$$

$$4x_1 + 3x_2 \le 9.6 \tag{12.7}$$

12.5 GEOMETRIC VISUALIZATION OF THE LINEAR-PROGRAMMING PROBLEM

Since it involves only the two variables x_1 and x_2, Example 12.1 can be illustrated geometrically as in Fig. 12-2. The constraint of Eq. (12.5), for example, states that only the region on and to the left of the line $x_1 +$

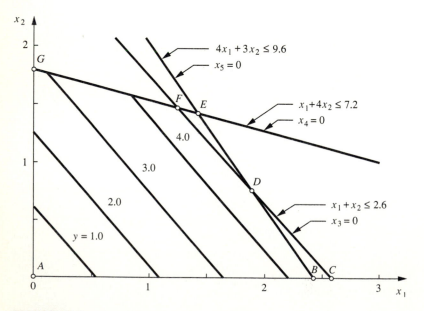

FIGURE 12-2
Constraints and lines of constant profit in Example 12.1.

$x_2 = 2.6$ is permitted. Placing the other two constraints in Fig. 12-2 further restricts the permitted region to *ABDFG*.

Next the lines of constant revenue y are plotted on Fig. 12-2. Inspection shows that the greatest profit can be achieved by moving to point D, where $x_1 = 1.8$ and $x_2 = 0.8$. An important generalization is that the *optimum solution lies at a corner*. A special case of this generalization is where the lines of constant profit is parallel to a constraint line, in which case any point on the constraint line between the corners is equally favorable.

If the objective function depends upon three variables, a three-dimensional graph is required, and then the constraint equations are represented by planes. The corner where the optimum occurs is formed by the intersection of three planes.

12.6   INTRODUCTION OF SLACK VARIABLES

The constraint equations (12.5) to (12.7) are inequalities, but they can be converted into equalities by the introduction of another variable in each equation

$$x_1 + \ \ x_2 + x_3 = 2.6 \qquad\qquad (12.8)$$

$$x_1 + 4x_2 + x_4 = 7.2 \qquad\qquad (12.9)$$

$$4x_1 + 3x_2 + x_5 = 9.6 \qquad\qquad (12.10)$$

This substitution is valid provided that $x_3 \geq 0, x_4 \geq 0$, and $x_5 \geq 0$. These new variables are called *slack variables*.

Reference to Fig. 12-2 permits a geometric interpretation of the slack variables x_3, x_4, and x_5. Any point on the graph defines specific values of x_3, x_4, and x_5. Along the $x_1 + x_2 = 2.6$ line, for example, $x_3 = 0$. The value of x_3 in the region to the right of the line is less than zero and is thus prohibited, while on the line and to the left of it $x_3 \geq 0$, and thus this region is permitted.

12.7   PREPARATION FOR THE SIMPLEX ALGORITHM

The simplex algorithm has a mathematical basis, but the mechanics will be presented before the theory. No rigorous proof will be given, but a geometric explanation will be provided to give insight into the functions the simplex algorithm is performing.

The first step in preparing for the simplex algorithm is to write the equations in table or tableau form. The reason is that equations will be written many times; instead of repeating x_1, for example, in all the equations x_1 will be used as a column heading, as in Table 12.1. The double line is interpreted as the equality sign and only the coefficients of the x terms appear in the boxes.

TABLE 12.1
Constraint equations in tableau form

x_1	x_2	x_3	x_4	x_5	
1	1	1			2.6
1	4		1		7.2
4	3			1	9.6

12.8 INCLUDING THE OBJECTIVE FUNCTION IN THE TABLEAU

The bottom line of Table 12.1 was left blank anticipating the inclusion of the objective function, Eq. (12.4). The form of the equation is revised, however, before inserting the numbers so that all the x terms are moved to the left side of the equation

$$y - 1.86x_1 - 1.56x_2 = 0 \qquad (12.11)$$

Insertion of these terms into the tableau yields the results shown in Table 12.2. The coefficients of the x terms in the bottom line for the objective function are called *difference coefficients*.

12.9 STARTING AT THE ORIGIN

The progression in linear-programming solutions is to move from one corner to the next corner, achieving an improvement in the objective function with each move. When no further improvement is possible, the optimum has been reached. The starting point is always the origin, namely, the position where the physical variables are zero. In Example 12.1 the starting point is where $x_1 = 0$ and $x_2 = 0$. One technique of indicating these values is to

TABLE 12.2
Constraint equations and objective function in tableau form

x_1	x_2	x_3	x_4	x_5	
1	1	1			2.6
1	4		1		7.2
4	3			1	9.6
-1.86	-1.56				0

TABLE 12.3
Tableau 1 of Example 12.1

$x_1 = 0$	$x_2 = 0$	x_3	x_4	x_5	
1	1	1			2.6
1	4		1		7.2
4	3			1	9.6
-1.86	-1.56				0

note them in the column heading; this converts the tableau into the form of Table 12.3, which is now the complete tableau 1.

A property of all tableaux throughout the linear-programming procedure is that the current values of all the x's and the objective function can be read immediately from the tableau. In Table 12.3, for example, x_1 and x_2 are zero; since the first line corresponds to Eq. (12.8), the value of x_3 is 2.6. Similarly, $x_4 = 7.2$ and $x_5 = 9.6$. The bottom line is the objective function in the form of Eq. (12.11), so $y = 0$. The number in the lower right corner of each tableau is the current value of the objective function.

12.10   THE SIMPLEX ALGORITHM

The simplex algorithm is a procedure whereby the successive tableaux can be developed from the first tableau. The steps are as follows:

1. Decide which of the variables that currently are zero should be programmed next. In a maximization problem the variable with the largest negative difference coefficient is chosen; in minimization the variable with the largest positive difference coefficient is chosen.
2. Determine which is the controlling constraint by selecting the constraint with the most restrictive (the smallest) quotient of the numerical term on the right side of the equality divided by the coefficient of the variable being programmed.
3. Transfer the controlling constraint to the new tableau by dividing all terms by the coefficient of the variable being programmed.
4. For all other boxes in the new tableau (including noncontrolling constraints and difference coefficients) use the following procedure:
   a. Select a box in the new tableau. Call the value in the same box of the old tableau v.
   b. Move sideways in the old tableau to the coefficient of the variable being programmed. Call this value w.

    c. In the new tableau move from the box being calculated up or down to the row which contains the previous controlling equation. Call the value in that box z.

    d. The value of the box in the new tableau is $v - wz$.

12.11   SOLUTION OF EXAMPLE 12.1

The simplex algorithm described in Sec. 12.10 will now be used to solve Example 12.1. The first tableau is complete as shown in Table 12.3 and is reproduced in Table 12.4 for transformation to the second tableau.

Step 1. The first step is to decide which of the variables currently noted in the column heading as zero (x_1 or x_2) should be programmed first. The largest negative difference is -1.86 in the x_1 column, which indicates that x_1 should be programmed (increased from its zero value). The vertical arrow indicates that x_1 is being programmed.

Step 2. The next step is to determine to what extent x_1 can be increased. The numerical terms to the right of the equality sign are divided by the coefficients of the variable being programmed in the same line, $2.6/1$, $7.2/1$, and $9.6/4$. The smallest of the resulting quotients is 2.4, as shown in the left column of Table 12.4, so the third constraint is the controlling one.

    In Fig. 12-2 the operation in moving from tableau 1 to tableau 2 is that of moving along the x_1 axis from A to B. The quotients in the left column denote values of x_1 that the respective constraints permit. The first constraint would permit x_1 to increase to 2.6, but the most restrictive is the third constraint, which permits x_1 to increase to 2.4.

    The construction of tableau 2 in Table 12.5 can now begin with the designation of the column headings. The variable x_1 was zero but is no longer, and x_2 remains zero. The variable x_5, which was nonzero, now

TABLE 12.4
Tableau 1 of Example 12.1 with indication of variable being programmed and controlling equation

	$x_1=0$ $\Downarrow$	$x_2=0$	x_3	x_4	x_5	
$2.6/1 = 2.6$	1	1	1			2.6
$7.2/1 = 7.2$	1	.4		1		7.2
$\Rightarrow 9.6/4 = 2.4$	4	.3			1	9.6
	-1.86	-1.56				0

TABLE 12.5
Tableau 2 of Example 12.1

⇓

	x_1	$x_2=0$	x_3	x_4	x_5	
⇒0.8	$1-(1)(1)=0$	$1-(1)(0.75)$ $=0.25$	$1-(1)(0)$ $=1$	$0-(1)(0)$ $=0$	$0-(1)(0.25)$ $=-0.25$	$2.6-(1)(2.4)$ $=0.20$
1.48	$1-(1)(1)=0$	$4-(1)(0.75)$ $=3.25$	$0-(1)(0)$ $=0$	$1-(1)(0)$ $=1$	$0-(1)(0.25)$ $=-0.25$	$7.2-(1)(2.4)$ $=4.8$
3.2	1	0.75	0	0	0.25	2.4
	$-1.86-$ $(-1.86)(1)$ $=0$	$-1.56-$ $(-1.86)(0.75)$ $=-0.165$	$0-$ $(-1.86)(0)$ $=0$	$0-$ $(-1.86)(0)$ $=0$	$0-$ $(-1.86)(0.25)$ $=0.465$	$0-$ $(-1.86)(2.4)$ $=4.464$

becomes zero, as indicated both by the geometry in Fig. 12-2 and by the third constraint in tableau 1. If that constraint is thought of as an equation, the only variables taking part in the action of the move are x_1 and x_5. All the other variables are either zero or have zero coefficients. The variable x_1 increases until x_5 has dropped to zero.

Step 3. The first numbers inserted in the boxes of tableau 2 come from the controlling equation of tableau 1. Dividing all the coefficients of the third constraint by 4, which is the coefficient of the variable being programmed, the numbers in the boxes of tableau 2 for the third constraint become 1, 0.75, 0, 0, 0.25, and 2.4.

Step 4. For all the other boxes in tableau 2 the $v - wz$ routine is followed. The individual calculations are shown in the boxes of tableau 2.

Tableau 2 is now complete. It contains the values of all the x's and the value of the objective function. Two of the x's, namely x_2 and x_5, are zero, as shown by the column heading. For each of the other x's a constraint equation will provide its value. In the first constraint all the variables are either zero or have a zero coefficient, except for x_3. So the equation has reduced to $x_3 = 0.20$. From the other two constraints, $x_4 = 4.8$ and $x_1 = 2.4$. The box in the bottom right corner indicates the current value of the objective function to be 4.464.

The entire simplex algorithm is repeated to transform tableau 2 into tableau 3. The largest negative difference coefficient, in fact the only negative one, is -0.165, under x_2, so x_2 is programmed next. This time the first constraint is most restrictive; it therefore becomes the controlling equation, and as x_2 increases to its limit, x_3 drops to zero, as shown by the column heading in tableau 3 (Table 12.6). The coefficients in the first constraint of tableau 2 are divided by 0.25 and the terms transferred to tableau 3.

TABLE 12.6
Tableau 3 of Example 12.1

x_1	x_2	$x_3 = 0$	x_4	$x_5 = 0$	
0	1	4	0	-1	0.8
0	0	-13	1	3	2.2
1	0	-3	0	1	1.8
0	0	0.66	0	0.3	4.596

The remaining boxes are computed according to step 4 in Sec. 12.10. The complete tableau 3 shows $x_1 = 1.8$, $x_2 = 0.8$, $x_3 = 0$, $x_4 = 2.2$, $x_5 = 0$, and $y = 4.596$.

The search for which variable to program next discloses that there are no negative difference coefficients, and this condition indicates that no further improvement is possible and the optimum has been reached. The transformation from tableau 2 to tableau 3 was a move on Fig. 12-2 from point B to point D. At the solution two of the slack variables, x_3 and x_5, are zero, but $x_4 = 2.2$. These values indicate that the second constraint has no influence on the optimum, although it might not have been possible to realize this fact in advance.

12.12 ANOTHER GEOMETRIC INTERPRETATION OF THE TABLEAU TRANSFORMATION

Figure 12-2 showed one geometric interpretation of the linear-programming problem and the progressive moves from one corner to another. Another geometric interpretation can be achieved by changing the coordinates with each transformation so that the current point is always at the origin. Since the first tableau starts at the origin of the physical variables, the coordinates x_1 and x_2 in Fig. 12-2 represent the first tableau. In moving from tableau 1 to tableau 2 the coordinate x_1 is replaced by x_5. The procedure in making this replacement is to solve for x_1 in the third constraint

$$x_1 = 2.4 - 0.75x_2 - 0.25x_5 \qquad (12.12)$$

and substitute this expression for x_1 into the first two constraints and the objective function to obtain

Constraint 1: $\qquad\qquad x_2 + 4x_3 - \qquad x_5 = 0.8 \qquad (12.16)$

Constraint 2: $\qquad - \quad 13x_3 + \quad x_4 + \quad 3x_5 = 2.2 \qquad (12.17)$

Constraint 3: $\qquad\qquad x_1 - 3x_3 + \qquad x_5 = 1.8 \qquad (12.18)$

Objective function: $\quad y + 0.66x_3 \qquad + 0.3x_5 = 4.596 \qquad (12.19)$

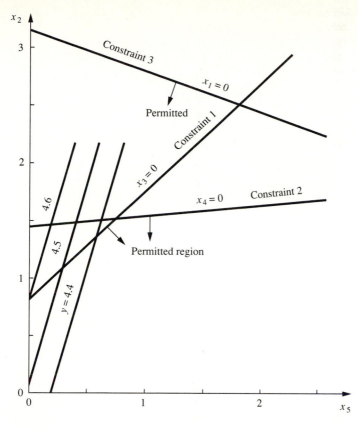

FIGURE 12-3
Tableau 2 expressed on $x_5 x_2$ coordinates.

The three constraints, Eqs. (12.12) to (12.14), are shown on the $x_5 x_2$ coordinates of Fig. 12-3, as well as a few lines of constant objective function from Eq. (12.15). An examination of Fig. 12-3 indicates that to improve (increase) the objective function x_2 should be increased until a constraint limits the advance. The restriction is imposed by constraint 1, which occurs when $x_3 = 0$, so the coordinate x_2 will now be replaced by x_3. That replacement will be performed by solving for x_2 in Eq. (12.13) and substituting into Eqs. (12.12), (12.14), and (12.15). The resulting set of equations is

Constraint 1: $\qquad x_2 + 4x_3 - \qquad x_5 = 0.8 \qquad$ (12.16)

Constraint 2: $\qquad -13x_3 + x_4 + 3x_5 = 2.2 \qquad$ (12.17)

Constraint 3: $\qquad x_1 - 3x_3 + \qquad x_5 = 1.8 \qquad$ (12.18)

Objective function: $\quad y + 0.66x_3 \qquad + 0.3x_5 = 4.596 \qquad$ (12.19)

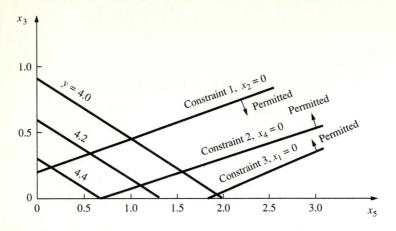

FIGURE 12-4
Tableau 3 expressed on x_5x_3 coordinates.

A graph on x_5x_3 coordinates of the constraints and constant values of the objective function is shown in Fig. 12-4. The current point is once again located at the origin, and any attempt to increase the objective function is doomed because an increase in either x_3 or x_5 decreases the value. No further improvement is possible, and the optimum has been reached. The particularly significant point about the transformations where the coordinates were progressively replaced in Figs. 12-3 and 12-4 is the equations on which the graphs are based. If Eqs. (12.12) to (12.15) are placed in a tableau form with the x's as the column headings, Table 12.7 is the result. A comparison of Table 12.7 with tableau 2 (Table 12.5) shows the two tables to be identical. Equations (12.16) to (12.19) could also be tabulated to show that these equations reproduce tableau 3. The conclusion is that the linear-programming process can be visualized as a progressive transformation of coordinates where, in each transformation, one of the coordinates is replaced by a variable indicating the limiting constraint.

TABLE 12.7
Equations (12.12) to (12.15)

x_1	x_2	x_3	x_4	x_5	
0	0.25	1		−0.25	0.2
0	3.25	0	1	−0.25	4.8
1	0.75			0.25	2.4
0	−0.165			0.465	4.464

TABLE 12.8
Compact form of Table 12.4

$\Downarrow$

		x_1	x_2	
$2.6/1 = 2.6$	x_3	1	1	2.6
$7.2/1 = 7.2$	x_4	1	0.4	7.2
$\Rightarrow 9.6/4 = 2.4$	x_5	4	0.3	9.6
		-1.86	-1.56	0

TABLE 12.9
Compact form of Table 12.6

	x_3	x_5	
x_2	4	-1	0.8
x_4	-13	3	2.2
x_1	-3	1	1.8
	0.66	0.3	4.596

12.13 COMPACT FORM OF TABLEAUX

Examination of Tables 12.4 and 12.6 shows that the coefficients in the columns headed by nonzero values of x are all zero except for a value of 1 in the position that keys to the numbers in the right-hand column. This pattern suggests that just as much information can be supplied by the tableau even if arranged in a more compact form. Tables 12.8 and 12.9 are conversions of Tables 12.4 and 12.6, respectively. In Table 12.8, for example, $x_3 = 2.6, x_4 = 7.2$, and $x_5 = 9.6$. Because x_1 is the variable to be programmed next, the right-hand column values are divided by the coefficients in the x_1 column to find the most restricitve constraint.

12.14 NUMBER OF VARIABLES AND NUMBER OF CONSTRAINTS

The relationship of the number of physical variables and the number of constraints gives an indication of the number of variables that are zero in the solution. Let the number of physical variables be denoted by n and the number of constraints (and therefore the number of slack variables) be denoted by m. There will always be n variables (physical plus slack) equal

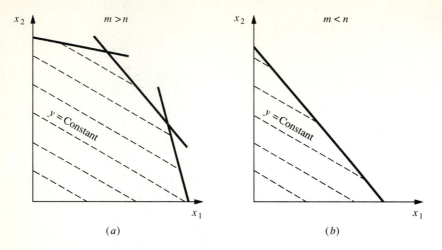

FIGURE 12-5
Relation of number of physical and slack variables.

to zero at the optimum, or for that matter at any corner. When $m > n$, as Fig. 12-5a shows, at least $m - n$ constraints play no role in the solution. In Fig. 12-5b where $m < n$, at least $n - m$ physical variables are zero.

12.15 MINIMIZATION WITH GREATER-THAN CONSTRAINTS

Solution of the maximization problem with less-than constraints consisted of moving from one corner in the feasible region to whichever adjacent corner showed the most improvement in the objective function. Since linear programming always starts at the physical origin, the origin is in the feasible region with less-than constraints. In the minimization problem with greater-than constraints, locating the first feasible point may be difficult. Admittedly, in simple problems involving a small number of variables, combinations of variables could be set to zero in the constraint equations and the other variables solved until a combination is found that violates no constraints. In large problems this method is prohibitive and a more systematic procedure must be employed. The introduction of *artificial variables* facilitates this procedure.

12.16 ARTIFICIAL VARIABLES

Suppose that an inequality constraint with a greater-than sense is to be converted into an equality but also must permit the physical variables to take on zero values. The inequality

$$3x_1 + 4x_2 \geq 12 \qquad (12.20)$$

can be converted into an equality by introducing a slack variable x_3,

$$3x_1 + 4x_2 - x_3 = 12 \qquad (12.21)$$

The slack variable takes on a negative sign, so that the constraint is satisfied when $x_3 \geq 0$.

The next requirement is the ability to set x_1 and x_2 equal to zero. Equation (12.21) will not permit zero values of x_1 and x_2, and so another variable, the artificial variable, is inserted

$$3x_1 + 4x_2 - x_3 + x_4 = 12 \qquad (12.22)$$

The geometric interpretation taken on by the slack and artificial variables with respect to the constraint is shown in Fig. 12-6. Along the constraint $3x_1 + 4x_2 = 12$, $x_3 = 0$, and $x_4 = 0$. When moving to the right of the constraint into the feasible region, x_3 takes on positive values and x_4 remains zero. When moving to the left of the constraint into the infeasible region, x_4 takes on positive values and x_3 remains zero. The result of the introduction of both the slack and artificial variable is that the position may be located anywhere on the graphs but all the variables are still abiding by the requirement that $x_i \geq 0$.

An uneasy feeling should prevail at this point because the arbitrary addition of terms to equations is not orthodox mathematics. Further treatment of the artificial variable is necessary, and this treatment will be explained as part of Example 12.2.

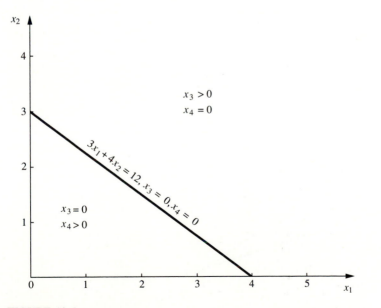

FIGURE 12-6
Slack and artificial variables.

12.17   SIMPLEX ALGORITHM APPLIED TO MINIMIZATION WITH GREATER-THAN CONSTRAINTS

Example 12.2. Determine the minimum value of y and the magnitudes of x_1 and x_2 at this minimum, where

$$y = 6x_1 + 3x_2$$

subject to the constraints

$$5x_1 + x_2 \geq 10$$
$$9x_1 + 13x_2 \geq 74$$
$$x_1 + 3x_2 \geq 9$$

Solution. Because this problem involves only two physical variables, the constraints and lines of constant y can be graphed, as in Fig. 12-7.

For the solution by linear programming, first write the constraint inequalities as equations by introducing the slack variables x_3, x_4, and x_5.

$$5x_1 + x_2 - x_3 = 10$$
$$9x_1 + 13x_2 - x_4 = 74$$
$$x_1 + 3x_2 - x_5 = 9$$

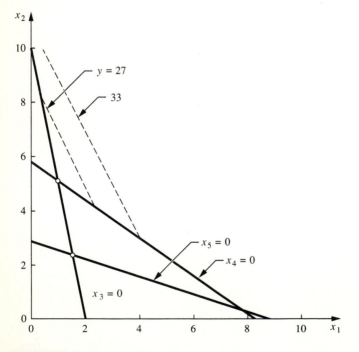

FIGURE 12-7
Minimization in Example 12.2.

Next the artificial variables, x_6, x_7, and x_8, are inserted in each equation

$$5x_1 + x_2 - x_3 + x_6 = 10 \qquad (12.23)$$

$$9x_1 + 13x_2 - x_4 + x_7 = 74 \qquad (12.24)$$

$$x_1 + 3x_2 - x_5 + x_8 = 9 \qquad (12.25)$$

Now the remaining conditions surrounding the artificial variables are specified in that the objective function is also revised as follows:

$$y = 6x_1 + 3x_2 + Px_6 + Px_7 + Px_8 \qquad (12.26)$$

The coefficient P never assumes a numerical value but is only considered to be extremely large. The existence of the products of P and the artificial variables in Eq. (12.26) is so penalizing to the minimization attempt that no satisfactory minimization will occur until the values of the artificial variables have been driven to zero.

The linear-programming process can now start at the origin in Fig. 12-7 with all the slack variables equal to zero but all the artificial variables having positive values. The simplex algorithm applies, but one further operation is performed on the objective function before writing the first tableau. From Eqs. (12.23) to (12.25)

$$x_6 = 10 - 5x_1 - x_2 + x_3$$

$$x_7 = 74 - 9x_1 - 13x_2 + x_4$$

$$x_8 = 9 - x_1 - 3x_2 + x_5$$

Substituting these values of the artificial variables into Eq. (12.26) and grouping gives

$$y = (6 - 15P)x_1 + (3 - 17P)x_2 + Px_3 + Px_4 + Px_5 + 93P$$

The first tableau can now be constructed. In the minimization operation the variable with the *largest positive* difference coefficient is chosen to be programmed. In tableau 1 (Table 12.10) this choice will be x_2 because P appears with the largest coefficient and the P values dominate over any purely numerical terms.

TABLE 12.10
Tableau 1 of Example 12.2

		x_1	x_2	x_3	x_4	x_5	
10	x_6	5	1	−1	0	0	10
74/13	x_7	9	13	0	−1	0	74
⇒ 3	x_8	1	3	0	0	−1	9
		$15P - 6$	$17P - 3$	$-P$	$-P$	$-P$	$93P$

TABLE 12.11
Tableau 2 of Example 12.2

		x_1	x_3	x_4	x_5	x_8	
⇒ 3/2	x_6	14/3	-1	0	1/3	$-1/3$	7
105/14	x_7	14/3	0	-1	13/3	$-13/3$	35
9	x_2	1/3	0	0	$-1/3$	1/3	3
		$\dfrac{28P-15}{3}$	$-P$	$-P$	$\dfrac{14P-3}{3}$	$\dfrac{-17P+3}{3}$	$42P+9$

TABLE 12.12
Tableau 3 of Example 12.2

		x_3	x_4	x_5	x_6	x_8	
21	x_1	$-3/14$	0	1/14	3/14	$-1/14$	3/2
⇒ 7	x_7	1	-1	4	-1	-4	28
-7	x_2	1/14	0	$-5/14$	$-1/14$	5/14	5/2
		$\dfrac{14P-15}{14}$	$-P$	$\dfrac{56P-9}{14}$	$\dfrac{-28P+15}{14}$	$\dfrac{-70P+9}{14}$	$\dfrac{56P+33}{2}$

TABLE 12.13
Tableau 4 of Example 12.2

	x_3	x_4	x_6	x_7	x_8	
x_1	$-13/56$	1/56	13/56	$-1/56$	0	1
x_5	1/4	$-1/4$	$-1/4$	1/4	-1	7
x_2	9/56	$-5/56$	$-9/56$	5/56	0	5
	$-51/56$	$-9/56$	$-P+51/56$	$-P+9/56$	$-P$	21

The next question is this: What is the limit to which x_2 can be increased? Just as in the maximization process, the variable x_2 is increased to the most limiting constraint, which in this case is the third one. The procedure in transforming to tableau 2 (Table 12.11) is the standard simplex algorithm. The difference coefficient with the largest positive value is $(28P - 15)/3$, so x_1 is programmed next until limited by the point where x_6 starts to go negative in the first constraint. In tableau 3 (Table 12.12) x_5 has the largest positive difference coefficient, so x_5 is increased until it reaches the limiting value of 7, beyond which x_7 would become negative.

In tableau 4 (Table 12.13) all difference coefficients are negative, so no further reduction in the objective function is possible. The solution is

$$x_1^* = 1 \qquad \text{and} \qquad x_2^* = 5$$

at which point $y^* = 21$.

12.18   REVIEW OF MINIMIZATION CALCULATION

Now that the minimization process in Example 12.2 has been completed, a reexamination of the successive tableaux in the problem will present a more complete picture of the operation. The introduction of the artificial variables in the objective function and the constraint equations permits a temporary violation of the constraints but only at the expense of an enormously large value of the objective function. The solution will certainly not be satisfactory until all the P terms are removed from the expression for the objective function.

In contrast to the maximization problem, e.g., Example 12.1, where the constraints were like solid walls that could not be surmounted, in the minimization problem the constraints are like stiff rubber bands that can be violated temporarily, but with a severe penalty in the magnitude of the objective function.

Looking first at tableau 1, we see that the position represented in Fig. 12-8 is the origin, because $x_1 = x_2 = 0$. The slack variables x_3, x_4, and x_5 are also zero, but the artificial variables have nonzero values: $x_6 = 10$, $x_7 = 74$, and $x_8 = 9$. The value of the objective function is $93P$, which is prohibitively large.

In tableau 2 the nonzero values of variables are $x_2 = 3$, $x_6 = 7$, and $x_7 = 35$. In moving from point 1 to point 2 in Fig. 12-8, the magnitude of the objective function drops from $93P$ to $42P + 9$. In other words, the pure numerical value increases by 9, but the staggering P term decreases from $93P$ to $42P$.

The next shift is to point 3, represented by tableau 3, where the nonzero values are $x_1 = \frac{3}{2}$, $x_2 = \frac{5}{2}$, $x_7 = 28$. The magnitude of the objective function is $(56P = 33)/2$, showing a continued increase of the

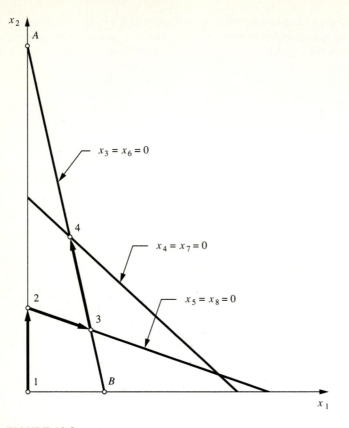

FIGURE 12-8
Points represented by successive tableaux in Example 12.2.

numerical portion and a decrease of the P coefficient. At this point x_3 and x_6 are zero, as are x_5 and x_8. Point 3 requires a positive value of the $-x_4 + x_7$ combination which is achieved by a zero value of x_4 and a positive value of x_7.

The final move is from point 3 to point 4, where the nonzero values are $x_1 = 1$, $x_2 = 5$, and the slack variable $x_5 = 7$. All the artificial variables are now zero, so the objective function in tableau 4 is free of P terms and has the value of 21. Since there are no remaining positive difference coefficients, the minimum value has been reached.

The minimization process thus consisted of starting at the origin, which was made permissible but highly uncomfortable by the introduction of artificial variables, and then moving through the thicket of constraint lines until finally the light of day was seen at point 4. In this problem, the first feasible point encountered was also the minimum, but it is possible for the process to move around several corners after it breaks into the feasible region.

12.19 APPEARANCE OF NEGATIVE NUMBER FOR MOST RESTRICTIVE CONSTRAINT

One further comment should be made regarding tableau 3, where, follow-ing the decision to program x_5, the limitation for increasing x_5 set by the first equation was 21, by the second equation 7, and by the third equation -7. Strict consistency would have required us to choose the third as the controlling equation. Geometrically, in Fig. 12-8, the test to find the con-trolling equation consists of moving along the $-x_3 + x_6 = 0$ line. Beyond point A, where $x_5 = 21$, the value of x_1 would go negative. Beyond 4, where $x_5 = 7$, x_7 would go negative. The limitation of $x_5 = -7$ is repre-sented by point B, which would result in a step backward, so the limitations with negative values are ignored.

12.20 EQUALITITES AND SENSES OF INEQUALITIES

The two classes of linear-programming problems considered so far are (1) maximization with less-than constraints and (2) minimization with greater-than constraints. Physical problems sometimes require mixtures of senses of the constraints, i.e., some less-than and the remainder greater-than constraints. One or more equality constraints may also appear in combina-tion with the inequalities. The next several sections explore these combina-tions and certain other situations introduced by such physical requirements as mass balances.

12.21 MIXTURES OF EQUALITIES AND INEQUALITIES AND OF SENSES OF INEQUALITIES

A few general guidelines are applicable to setting up problems where the constraints are mixed. These guidelines control the introduction of slack and artificial variables. In the two classes of problems treated so far, where the constraints were all inequalities with the same sense, the insertion of slack and artificial variables accomplished two functions: (1) converting all inequalities into equalities and (2) permitting programming to begin at the origin of the physical variables. The same objective prevails if the constraints are mixed. One additional variation is that when an artificial variable is introduced, it must appear in the objective function as $+Px_i$ if the objective function is being minimized and $-Px_i$ if the function is being maximized.

Example 12.3. Find the maximum of y, where

$$y = 3x_1 + 2x_2 + 4x_3$$

subject to

$$3x_1 + 4x_2 + 5x_3 \le 40 \qquad (12.27)$$

$$x_1 + x_2 + x_3 = 9 \qquad (12.28)$$

$$7x_1 + 4x_2 + 4x_3 \ge 42 \qquad (12.29)$$

Solution. Equation (12.27) requires only the insertion of a positive slack variable x_4 in order to convert the inequality into an equality and also permit $x_1 = x_2 = x_3 = 0$ with $x_4 \ge 0$. Equation (12.27) thus becomes

$$3x_1 + 4x_2 + 5x_3 + x_4 = 40$$

Equation (12.28) is already an equality, but the equation will not permit programming to start at the origin. An artificial variable x_5 will be introduced into Eq. (12.28) which retains the equality and also permits $x_1 = x_2 = x_3 = 0$,

$$x_1 + x_2 + x_3 + x_5 = 9 \qquad (12.30)$$

The objective function must also be revised so that x_5 will ultimately be driven to zero to satisfy Eq. (12.28),

$$y = 3x_1 + 2x_2 + 4x_3 - Px_5 \qquad (12.31)$$

The penalty term in the objective function is assigned a negative value because the goal is to maximize the function and the maximum should not be reached until x_5 has shrunk to zero.

The final constraint, Eq. (12.29), requires both a slack and an artificial variable. The slack variable $-x_6$ converts the inequality into an equality. The artificial variable x_7 allows programming to start at the origin. Equation (12.29) then becomes

$$7x_1 + 4x_2 + 4x_3 - x_6 + x_7 = 42 \qquad (12.32)$$

Introducing the penalty term associated with x_7 into the objective function revises Eq. (12.31) further

$$y = 3x_1 + 2x_2 + 4x_3 - Px_5 - Px_7 \qquad (12.33)$$

As the final step before writing the first tableau, substitute x_5 from Eq. (12.30) and x_7 from Eq. (12.32) into the objective function, Eq. (12.33), to get

$$y = -51P + (8P + 3)x_1 + (5P + 2)x_2 + (5P + 4)x_3 - Px_6$$

The first tableau is shown in Table 12.14. Since programming begins at the origin, x_1, x_2, and x_3 are assigned the value of zero. The slack variable x_6 is also zero because the nonzero value of x_7 permits x_1, x_2, and x_3 in Eq. (12.32) to go to zero.

Since the goal is the maximum, the variable with the largest negative difference coefficient, which is x_1, will be programmed first. The succession of tableaux is shown in Tables 12.15 to 12.18.

The P terms have disappeared from the difference conefficients associated with the physical variables x_1, x_2, and x_3, and so the position indicated by

TABLE 12.14
  Tableau 1 of Example 12.3

$\Downarrow$

		x_1	x_2	x_3	x_6	
40/3	x_4	3	4	5	0	40
9	x_5	1	1	1	0	9
$\Rightarrow$ 6	x_7	7	4	4	-1	42
		$-8P - 3$	$-5P - 2$	$-5P - 4$	P	$-51P$

TABLE 12.15
Tableau 2 of Example 12.3

$\Downarrow$

		x_2	x_3	x_6	x_7	
$\Rightarrow$154/23	x_4	16/7	23/7	3/7	$-3/7$	22
7	x_5	3/7	3/7	1/7	$-1/7$	3
42/4	x_1	$-4/7$	4/7	$-1/7$	1/7	6
		$\dfrac{-3P - 2}{7}$	$\dfrac{-3P - 16}{7}$	$\dfrac{-P - 3}{7}$	$\dfrac{8P + 3}{7}$	$-3P + 18$

TABLE 12.16
Tableau 3 of Example 12.3

$\Downarrow$

		x_2	x_4	x_6	x_7	
154/16	x_3	16/23	7/23	3/23	$-3/23$	154/23
$\Rightarrow$ 1	x_5	3/23	$-3/23$	2/23	$-2/23$	3/23
25/2	x_1	4/23	$-4/23$	$-5/23$	5/23	50/23
		$\dfrac{-3P + 30}{23}$	$\dfrac{3P + 16}{23}$	$\dfrac{-2P - 3}{23}$	$\dfrac{25P + 3}{23}$	$\dfrac{-3P + 766}{23}$

TABLE 12.17
Tableau 4 of Example 12.3

		x_4	x_5	x_6	x_7	
-18	x_3	1	$-16/3$	$-1/3$	$1/3$	6
$\Rightarrow 3/2$	x_2	-1	$23/3$	$2/3$	$-2/3$	1
-6	x_1	0	$-4/3$	$-1/3$	$1/3$	2
		2	$\dfrac{3P-30}{3}$	-1	$P+1$	32

TABLE 12.18
Tableau 5 of Example 12.3

	x_2	x_4	x_5	x_7	
x_3	$1/2$	$1/2$	$-3/2$	0	$13/2$
x_6	$3/2$	$-3/2$	$23/2$	-1	$3/2$
x_1	$1/2$	$-1/2$	$5/2$	0	$5/2$
	$3/2$	$1/2$	$\dfrac{2P+3}{2}$	P	33.5

tableau 4 is now in the feasible region. The optimal solution, then, is $x_1^* = 2.5, x_2^* = 0, x_3^* = 6.5$, and $y^* = 33.5$.

12.22 MATERIAL BALANCES AS CONSTRAINTS

In the application of optimization methods, including linear programming, to thermal systems, material balances often impose constraints. The input-output material balance often results in a unique situation during the simplex procedure that should be recognized in order to provide the interpretation that allows the procedure to continue. The situation will be illustrated by an example.

Example 12.4. In the processing plant shown in Fig. 12-9 the operation is essentially one of concentrating material A. The concentrator receives a raw material consisting of 40 percent A by mass and can supply two products of 60 and 80 percent A, respectively. The flow rate of the raw material is designated x_1 metric tons per day and the 60 percent and 80 percent products are designated x_2 and x_3, respectively. The prices are:

FIGURE 12-9
Discharge
Concentrator in Example 12.4.

Amount	x_1	x_2	x_3
Price per metric ton	$40	$80	$120

The capacity of the loading facility imposes the constraint

$$2x_2 + 3x_3 \leq 60$$

Determine the combination of raw material and products that results in maximum profit for the plant.

Solution. The objective function is the difference between the income of the products and the cost of the raw materials

$$y = 80x_2 + 120x_3 - 40x_1$$

In addition to the constraint of the loading facility, the mass balance of material A imposes another constraint that can be expressed by either Eq. (12.34) or (12.35)

$$0.6x_2 + 0.8x_3 = 0.4x_1 \qquad (12.34)$$

$$0.6x_2 + 0.8x_3 \leq 0.4x_1 \qquad (12.35)$$

Equation (12.34) is a strict equality for material, while Eq. (12.35) allows for the possibility of dumping some of material A during the processing. In some rare instances the most economical solution can be achieved by dumping some material of value, so in the solution presented below the possibility of dumping will be used and Eq. (12.35) will be chosen. The mathematical statement of the problem becomes:

Maximize $\qquad\qquad y = 80x_2 + 120x_3 - 40x_1$

subject to $\qquad\qquad\qquad 2x_2 + 3x_3 \leq 60$

$$-0.4x_1 + 0.6x_2 + 0.8x_3 \leq 0$$

The first tableau is shown in Table 12.19. The variable to be programmed first is x_3 for which the most restrictive constraint is the second one. Using the simplex algorithm gives tableau 2 (Table 12.20).

TABLE 12.19
   Tableau 1 of Example 12.4

		x_1	x_2	x_3	
20	x_4	0	2	3	60
$\Rightarrow 0/0.8 = 0$	x_5	−0.4	0.6	0.8	0
		40	−80	−120	0

The difference coefficient of x_1 in tableau 2 has the largest negative value, so x_1 is programmed next. The straightforward procedure would indicate that the second constraint controls, because $0 < 40$. The key feature in solving this class of problems, however, is the interpretation of the indicators of control. We shall interpret the zero as a "negative zero" and, as is the practice, ignore any constraint associated with a negative number in the left column. The first constraint is used as the controlling one and results in tableau 3 (Table 12.21). No negative difference coefficients remain, so the maximum has been reached: $x_1^* = 40$, $x_3^* = 20$, and $y^* = 800$.

Let us now return to tableau 2 and explore what would have occurred if the second equation had been chosen as the controlling one. That tableau would have been as shown in Table 12.22. The values indicated by the

TABLE 12.20
Tableau 2 of Example 12.4

		x_1	x_2	x_5	
$\Rightarrow 40$	x_4	1.5	−0.25	−3.75	60
$0/-0.5 = 0$	x_3	−0.5	0.75	1.25	0
		−20	10	150	0

TABLE 12.21
Tableau 3 of Example 12.4

	x_2	x_4	x_5	
x_1	−0.1667	0.6667	−2.5	40
x_3	0.833	0.333	0	20
	13.33	13.33	100	800

TABLE 12.22
Incorrect Tableau 3 of Example 12.4

	x_2	x_3	x_5	
x_4	2	3	0	60
x_1	−1.5	−2	−2.5	0
	−10	−40	100	0

tableau in Table 12.22 are $x_1 = x_2 = x_3 = x_5 = 0$, and $x_4 = 60$. These are precisely the starting conditions of tableau 1, so the process is in an infinite loop. The interpretation of the negative zero breaks open the loop.

The physical explanation of what occurs between tableaux 1 and 2 is as follows. Most profit can be made by selling x_3, so this variable is understandably the one to be increased first. But tableau 2 show that x_3 is still zero, and the reason is that the second constraint, the material balance, will not allow x_3 to increase until some raw material x_1 has also been brought in. By the shift of the slack variable x_5 to zero in the column heading, the next programming operation will assure that x_1 is increased in proper proportion to the product x_3.

12.23 PRIMAL-DUAL PROBLEMS

The primal/dual pattern will be illustrated by two examples, Examples 12.5 and 12.6.

Example 12.5. Maximize $\qquad y = 4x_1 + 3x_2$

subject to $\qquad\qquad\qquad 2x_1 + x_2 \le 8$

and $\qquad\qquad\qquad\qquad 2x_1 + 3x_2 \le 12$

Solution. The initial and final (third) tableaux are shown in Table 12.23 yielding the optimal results of $x_1^* = 3, x_2^* = 2$, and $y^* = 18$.

Example 12.6. Minimize $\qquad Y = 8w_1 + 12w_2$

subject to $\qquad\qquad\qquad 2w_1 + 2w_2 \ge 4$

and $\qquad\qquad\qquad\qquad w_1 + 3w_2 \ge 3$

Solution. The first tableau is shown in Table 12.24, and the final tableau in Table 12.25. The optimal values are $w_1^* = 3/2$, $w_2^* = 1/2$, and $Y^* = 18$.

The links between the two problem statements are the following: (1) the coefficient matrices of the constraints are transposes, and (2) the

TABLE 12.23
Initial and final tableaux in Example 12.5

	x_1	x_2			x_3	x_4	
x_3	2	1	8	x_1	3/4	−1/4	3
x_4	2	3	12	x_2	−1/2	1/2	2
	−4	−3	0		3/2	1/2	18

TABLE 12.24
Initial tableau in Example 12.6

	w_1	w_2	w_3	w_4	
w_5	2	2	−1		4
w_6	1	3		−1	3
	$3P - 8$	$5P - 12$	$-P$	$-P$	$7P$

where w_5 and w_6 are artificial variables.

TABLE 12.25
Initial tableau in Example 12.6

	w_3	w_4	w_5	w_6	
w_1	−3/4	1/2	3/4	−1/2	3/2
w_2	1/4	−1/2	−1/4	1/2	1/2
	−3	−2	$-P + 3$	$-P + 2$	18

right-side constants of one problem are the coefficients in the objective function of the other.

There are three relationships between the two solutions: (1) the optimal values of the objective function are the same, (2) the solution of the x's in the primal are the numerical values of the difference coefficients under the artificial variables of the dual, and (3) the solutions of the w's in the dual are the difference coefficients in the primal.

The primal and dual problems in Examples 12.5 and 12.6 were two-variable, two-constraint optimizations in which both the constraints were active. There was no ambiguity in identifying the solution of the dual based on the final tableau of the primal. When the number of constraints differs from the number of variables, some interpretation is needed to select the applicable numbers. To the maximization problem in Example 12.5, introduce an additional constraint.

Example 12.7. Maximize $y = 4x_1 + 3x_2$

subject to

$$x_1 + 4x_2 \leq 12$$

$$2x_1 + x_2 \leq 8$$

$$2x_1 + 3x_2 \leq 12$$

The new constraint turns out to be inactive which permits tracing the terms that appeared in Table 12.23.

Solution. The first and third tableaux appear in Table 12.26.

Example 12.8. The dual of the problem in Example 12.7 is

minimize

$$Y = 12w_1 + 8w_2 + 12w_3$$

subject to

$$w_1 + 2w_2 + 2w_3 \geq 4$$

and

$$4w_1 + w_2 + 3w_3 \geq 3$$

Solution. The initial and final tableaux are shown in Tables 12.27 and 12.28, respectively.

TABLE 12.26
Initial and final tableaux in Example 12.7

	x_1	x_2			x_4	x_5	
x_3	1	4	12	x_3	5/4	−7/4	1
x_4	2	1	8	x_1	3/4	−1/4	3
x_5	2	3	12	x_2	−1/2	1/2	2
	−4	−3	0		3/2	1/2	18

TABLE 12.27
Initial tableau in Example 12.8

	w_1	w_2	w_3	w_4	w_5	
w_6	1	2	2	−1		4
w_7	4	1	3		−1	3
	$5P - 12$	$3P - 8$	$5P - 12$	$-P$	$-P$	$7P$

where w_6 and w_7 are artificial variables.

TABLE 12.28
Final tableau in Example 12.8

	w_1	w_4	w_5	w_6	w_7	
w_2	$-5/4$	$-3/4$	$1/2$	$3/4$	$-1/2$	$3/2$
w_3	$7/4$	$1/4$	$-1/2$	$-1/4$	$1/2$	$1/2$
	-1	-3	-2	$-P + 3$	$-P + 2$	18

Let us propose using the results of the primal in Example 12.7 to predict the optimal solution in Example 12.8. For the three w's there are only two values, 3/2 and 1/2, available in Example 12.7. One of the w's will be zero because there are three variables and two constraints in Example 12.8, but which one? The natural order is followed: $x_3 \neq 0$, so $w_1 = 0$. The difference coefficient of $x_4 = 3/2 = w_2$, and the difference coefficient of $x_5 = 1/2 = w_3$.

12.24 POST-OPTIMAL ANALYSIS

Post-optimal analysis is the activity that begins after the original problem has been solved, and asks the following question: "What if a constant or coefficient is changed slightly in the statement of the original problem?" Three of the sensitivity coefficients applicable to linear programming are to evaluate the change in y^* with respect to (1) a coefficient in the objective function, (2) a right-side constant in a constraint, and (3) a coefficient in one of the constraint equations. The first two of these sensitivity coefficients will be examined now.

The sensitivity coefficient with respect to a coefficient in the objective function is readily available. If, for example, the objective function is

$$y = Ax_1 + Bx_2$$

then at the optimum

$$y^* = Ax_1^* + Bx_2^*$$

so the sensitivity coefficients are: $\partial y^*/\partial A = x_1^*$ and $\partial y^*/\partial B = x_2^*$.

The primal/dual relationship offers the key to expressing the sensitivity coefficient with respect to a right-side term in a constraint. In the maximization problem of Example 12.5 the constraints are

$$2x_1 + x_2 \leq 8 = C$$
$$2x_1 + 3x_2 \leq 12 = D$$

The dual of this problem is Example 12.6 having an objective function

$$Y = Cw_1 + Dw_2$$

and at the optimum,

$$y^* = Y^* = Cw_1^* + Dw_2^*.$$

so $\partial y^*/\partial C = w_1^*$ and $\partial y^*/\partial D = w_2^*$. These values are 3/2 and 1/2, respectively.

12.25 EXTENSIONS OF LINEAR PROGRAMMING

The chapter illustrated hand calculations to solve linear-programming problems, and there is some benefit to understanding the techniques. Hereafter library routines will certainly be used, particularly for large-scale problems.

Two classes of sensitivity coefficients were explored; one remaining is that with respect to a coefficient in a constraint equation. The solution to a body of problems opens when an optimization not completely linear is linearized and subjected to linear programming.

PROBLEMS

12.1. A farmer wishes to choose the proportions of corn and soybeans to plant to achieve maximum net return. He has 240 hectares available for planting although he may elect to operate less than 240 hectares. The capital and labor requirements and the net return on each of the two crops are

Crop	Capital cost per hectare	Labor, worker-hours per hectare	Net return per hectare
Corn	$120	5	$50
Soybeans	40	10	75

The farmer has a maximum of 2000 worker-hours of labor available and maximum capital of $19,200.
(a) Set up the objective function and constraints.
(b) Use the simplex algorithm of linear programming to determine the optimal number of hectares of corn and soybeans and the maximum net return.

   Ans.: Maximum net return = $16,000.

12.2. During a summer session a student enrolls in two courses, psychology for four hours and engineering for three hours. The student wants to use the available time to amass the largest number of grade points (numerical grade multiplied by the number of hours). Students who have taken these courses previously indicate that probable grades are functions of the time spent:

$$G_e = \frac{x_1}{5} \qquad \text{and} \qquad G_p = \frac{x_2}{8}$$

where G = numerical grade (4.0 = A, 3.0= B, etc.)
x_1 = number of hours per week spent outside of class studying engineering
x_2 = number of hours per week spent outside of class studying psychology

The total number of hours available for outside study per week cannot exceed 39. Furthermore, most engineering students can tolerate the two courses in a combination such that

$$x_1 + 0.5x_2 \leq 27$$

Also it is fruitless to spend more time studying beyond that necessary to earn an A.

Use the simplex algorithm to determine how to distribute study time, subject to the constraints, in order to acquire the maximum number of grade points.

Ans.: B in both courses.

12.3. A food-processing firm is planning construction of a plant that could man-ufacture a combination of three frozen food products, pot pies, TV dinners, and pizzas. The investment costs consist of plant costs and machinery costs, and the credit rating of the firm will permit loans of $4.8 million or less for building construction and $1.6 million or less for machinery. The building and machinery costs for each of the proposed products are

Food product	Building cost	Machinery cost
Pot pies	$600,000$x_1$	$100,000$x_1$
TV dinners	500,000x_2	400,000x_2
Pizzas	400,000x_3	200,000x_3

where x_1, x_2, and x_3 represent the hourly production rate of pot pies, TV dinners, and pizzas, respectively, in thousands of units.

The hourly profit from the manufacture of each of the products in dollars in 25x_1, 30x_2, and 40x_3, respectively.

The values of x_1, x_2, and x_3 are to be determined such that the profit is maximum within the prevailing constraints.

(*a*) Set up the objective function and constraints.

(*b*) Use the simplex algorithm of linear programming to determine optimal values of x_1, x_2, x_3, and the hourly profit.

Ans.: Optimum profit = $340 per hour.

12.4. In the manufacture of cement the basic operations are to grind limestone, mix with clay or shale, and then heat the mixture in a rotary kiln, as shown in Fig. 12-10. A certain cement plant can produce three ASTM types of cement:

Type I. Standard portland
Type II. Sulfate-and alkaline-resistant
Type III. High early strength

FIGURE 12-10
Cement plant in Prob. 12.4.

The profit of each type and the capabilities of the grinder and kiln in processing these cements are shown in the table.

Cement type	Profit, per megagram	Grinder capacity, Gg/day	Kiln capacity Gg/day
I	$ 6	Coarse 10	8
II	10	Fine 5	4.8
III	9	Fine 5	6

The grinder capacity shown in the table of 10 Gg/day for type I means, for example, that the grinder could grind the limestone for 10 Gg of type I if it operated all day solely on limestone for type I. The grinder and kiln operate 24 h/day and can switch from one cement type to another instantaneously. The limestone storage space and mixer capacity are more than adequate for any rates that the grinder and kiln will permit.

Use the simplex algorithm of linear programming to determine what daily production of the various types of cement will result in maximum profit.

Ans.: Maximum daily profit = $51,000.

12.5. Three materials *A, B,* and *C* of varying thicknesses are available for combining into a building wall, as shown in Fig. 12-11. The characteristics and costs of the materials are

Material	Thermal resistance, units per centimeter thickness	Load bearing capacity, units/cm	Cost per centimeter
A	30	7	$8
B	20	2	4
C	10	6	3

The total thermal resistance of the wall must be 120 or greater, and the total load-bearing capacity must be 42 or greater. The minimum-cost wall is sought.

FIGURE 12-11
Composite wall in Prob. 12.5.

(a) Set up the objective function and constraints.
(b) Use the simplex algorithm of linear programming to determine the optimal thicknesses of each material.
   Ans.: $30 minimum cost.

12.6. The optimization of the combined gas- and steam-turbine plant in Prob. 7.4 resulted in a linear objective function and three linear constraints. Use the simplex algorithm to determine the optimum value of q_1 and q_2. To simplify mathematical manipulation, use the following equations instead of those in Prob. 7.4:

Objective function: $\qquad q = q_1 + q_2$

Subject to $\qquad q_1 + 1.2q_2 \geq 28\text{MW}$

$\qquad\qquad q_1 + 0.4q_2 \geq 19\text{MW}$

$\qquad\qquad q_1 + 1.7q_2 \geq 32\text{MW}$

   Ans.: Optimal $q = 25.75$ MW.

12.7. The furnace serving a certain steam-generating plant is capable of burning coal, oil, and gas simultaneously. The heat-release rate of the furnace must be 2400 kW, which with the 75 percent combustion efficiency of this furnace requires a combined thermal input rate in the fuel of 3200 kW. Ordinances in certain cities impose a limit on the average sulfur content of the fuel mixture, and in the city where this plant is located the limit is 2 percent or less. The sulfur contents, costs, and heating values of the fuels are shown in the table.

Fuel	Sulfur content, %	Cost per megagram	Heating value, kJ/kg
Coal	3.0	$24	35,000
Oil	0.4	36	42,000
Gas	0.2	42	55,000

Using the simplex algorithm of linear programming, determine the combination of fuel rates that results in minimum costs and yet meets all constraints.
   Ans.: Minimum cost = $0.00231 per second.

12.8. A manufacturer of cattle food mixes a combination of wheat and soybeans to form a product which has minimum requirements of 24 percent protein and 1.2 percent minerals by mass. The composition and prices of the wheat and soybeans are given in the table.

	Percent composition by mass		
Constituent	Protein	Minerals	Price per 100 kg
Wheat	16	1.5	$10
Soybeans	48	1.0	20

Use the simplex algorithm of linear programming to determine the mass of each of the constituents of 100 kg of product such that the cost of the raw materials is minimum and the nutritional requirements are met.

   Ans.: 75 kg of wheat and 25 kg of soybeans.

12.9. A wax concentrating plant (Fig. 12-12) receives feedstock with a low concentration of wax and refines it into a product with a high concentration of wax. The selling prices of the products are x_1, $8 per megagram and x_2, $6 per megagram. The raw material costs are x_3, $1.5 per megagram and x_4, $3 per megagram. The plant operates under the following constraints:

1. No more wax leaves the plant than enters.
2. The receiving facilities of the plant are limited to a total of 1600 Mg/h.
3. The packaging facilities can accommodate a maximum of 1200 Mg/h of x_2 or 1000 Mg/h of x_1 and can switch from one to the other with no loss of time.

If the operating cost of the plant is constant, use the simplex algorithm of linear programming to determine the purchase and production plan that results in the maximum profit.

   Ans.: Profit = $3650 per hour.

12.10. A dairy operating on the flow diagram shown in Fig. 12-13 can buy raw milk from either or both of two sources and can produce skim milk, homogenized milk, and half-and-half cream. The costs and butterfat contents of the sources and products are

Item	Designation, L/day	Sale or purchase cost	Butterfat content, vol %
Source 1	x_1	$0.23 per liter	4.0
Source 2	x_2	0.24 per liter	4.5
Half-and-half	x_3	0.48 per half-liter	≥ 10.0
Skim milk	x_4	0.60 per two-liter	≥ 1.0
Homogenized milk	x_5	0.68 per two-liter	≥ 3.0

The daily quantities of sources and products are to be determined so that the plant operates with maximum profit. All units in the dairy can operate for a maximum of 8 h/day.

(a) Write the equation for the profit and the constraint equations in terms of the x variables.

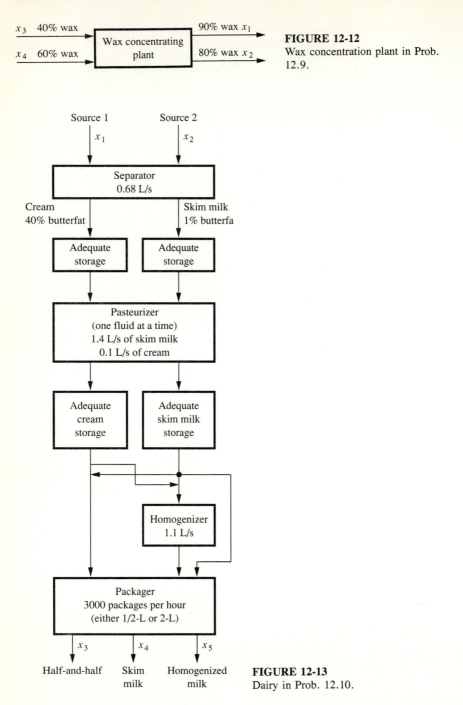

x_3 40% wax

x_4 60% wax

Wax concentrating plant

90% wax x_1

80% wax x_2

FIGURE 12-12
Wax concentration plant in Prob. 12.9.

Source 1

x_1

Source 2

x_2

Separator
0.68 L/s

Cream
40% butterfat

Skim milk
1% butterfa

Adequate storage

Adequate storage

Pasteurizer
(one fluid at a time)
1.4 L/s of skim milk
0.1 L/s of cream

Adequate cream storage

Adequate skim milk storage

Homogenizer
1.1 L/s

Packager
3000 packages per hour
(either 1/2-L or 2-L)

x_3 x_4 x_5

Half-and-half   Skim milk   Homogenized milk

FIGURE 12-13
Dairy in Prob. 12.10.

(b) Use linear programming to solve for the plan that results in maximum profit.

    Ans.: (a).

$$y = -0.23x_1 - 0.24x_2 + 0.96x_3 + 0.30x_4 + 0.34x_5$$

subject to

Separator:	$x_1 + x_2 \leq 19,584$
Pasteurizer:	$x_1 + 1.08x_2 \leq 20,150$
Homogenizer:	$x_5 \leq 31,680$
Packager:	$4x_3 + x_4 + x_5 \leq 48,000$
Butterfat:	$-4x_1 - 4.5x_2 + 10x_3 + x_4 + 3x_5 \leq 0$
Total Mass:	$-x_1 - x_2 + x_3 + x_4 + x_5 \leq 0$

    (b) Maximum profit = $5908 per day.

12.11. A chemical plant whose flow diagram is shown in Fig. 12-14 manufactures ammonia, hydrochloric acid, urea, ammonium carbonate, and ammonium chloride from carbon dioxide, nitrogen, hydrogen, and chlorine. The x values in Fig. 12-14 indicate flow rates in moles per second.

    The costs of the feed stocks are c_1, c_2, c_3, and c_4 dollars per mole, and the values of the products are p_5, p_6, p_7, and p_8 dollars per mole, where the subscript corresponds to that of the x value. In reactor 3 the ratios of molal flow rates are $m = 3x_7$ and $x_1 = 2x_7$, and in the other reactors straightforward material balances apply. The capacity of reactor 1 is equal to or less than 2 mol/s of NH_3, and the capacity of reactor 2 is equal to or less than 1.5 mol/s.

(a) Develop the expression for the profit.

(b) Write the constraint equations for this plant.

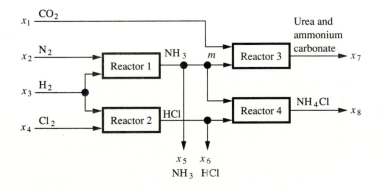

FIGURE 12-14

Flow diagram of chemical plant in Prob. 12.11.

Ans.: (*b*)

$$x_1 - 2x_7 = 0$$

$$2x_2 - x_5 - 3x_7 - x_8 = 0$$

$$2x_3 - 3x_5 - x_6 - 9x_7 - 4x_8 = 0$$

$$2x_4 - x_6 - x_8 = 0$$

$$x_6 + x_8 \leq 1.5$$

$$x_5 + 3x_7 + x_8 \leq 2.0$$

12.12. When large fabric filter installations, called *baghouse filters*, filter high-temperature gases, e.g., from a smelter, the temperature of the gases must be reduced to 265°C or less, even when using glass-fiber filters. Three methods[1] of reducing the temperature may be used singly or in combination, as shown in Fig. 12-15: (1) reject heat to ambient air through the use of a heat exchanger, (2) dilute the hot gas with ambient air that is at a temperature of 25°C, and (3) inject water for evaporative cooling. If either dilution air or water injection is used, the baghouse must be enlarged to accommodate the additional mass flow. Designate

$$x_1 = \text{area of heat exchanger}, \text{m}^2$$

$$x_2 = \text{mass flow rate of dilution air}, \text{kg/s}$$

$$x_3 = \text{mass flow rate of spray water}, \text{kg/s}$$

Data

Cost of baghouse is $2000 for each kilogram per second of capacity. The heat-transfer surface costs $15 per square meter.

There is no cost for the dilution air and spray water other than that due to the enlargement of the baghouse.

FIGURE 12-15
Cooling a gas before it enters a baghouse filter.

The entering flow rate of gas $= 16$ kg/s at 500°C.

Each square meter of heat exchanger reduces the temperature of gas by 2°C.

The evaporation of spray water cools the gas-air mixture equivalent to a sensible-heat removal from the gas entering the water spray of 2000 kJ/kg of spray water.

The specific heat of the gas, dilution air, and gas-air mixture is 1.0 kJ/(kg · K).

To avoid corrosion, the mass rate of flow of spray water must be 5 percent or less of the combined flow of gas and dilution air.

(a) Set up the mathematical statement of the optimization problem in terms of x_1 to x_3 to minimize the cost, subject to the appropriate constraints.

(b) Develop the first tableau of the simplex algorithm of the linear-programming solution of this optimization and indicate which variable to program first.

Ans.: (a) $y = 32,000 + 15x_1 + 2000x_2 + 2000x_3$

subject to $\qquad 2x_1 + 15x_2 + 125x_3 \geq 235$

$$- \quad x_2 + \quad 20x_3 \leq 16$$

12.13. Some petrochemical plants take a large flow rate from a natural-gas transmission line, remove ethane and propane from it, and return the methane to the pipeline, as shown in Fig. 12-16.

Cost data

	Cost per cubic meter
Feed	$0.06
Price of x_1 returned to pipeline†	0.0595
x_2	0.08
x_3	0.10
Operating cost:	
Extractor, per cubic meter of total flow	0.001
Separator, per cubic meter of pure propane	0.003

† Since the methane has a lower heating value than the original feed, it is less valuable.

Restrictions

The composition of the feed if 90% methane, 8% ethane, and 2% propane.

The maximum capacity of the extractor is 200 m³/s of feed.

The maximum capacity of the separator is 3 m³/s of pure propane.

Set up the objective function and constraints to maximize the profit of this plant.

Pipeline

FIGURE 12-16
Removal of propane and ethane from a natural-gas pipeline. The units of the flow rates, x_1, and x_2, and x_3, are in cubic meters per second.

Ans.: Constraints are

$$x_3 \leq 3$$

$$x_2 + x_3 \leq 20$$

$$x_1 - 9x_2 - 9x_3 = 0$$

$$x_2 - 4x_3 \geq 0$$

12.14. Consider as the primal problem:

minimize $\qquad y = 2x_1 + 3x_2 + 2x_3$

subject to $\qquad 3x_1 + 5x_2 + 6x_3 \geq 21$

$\qquad\qquad 4x_1 + 2x_2 + 5x_3 \geq 16$

$\qquad\qquad 2x_1 + 2x_2 + \quad x_3 \geq 9$

(a) Construct the dual, and
(b) by solution of the dual determine the optimal values of the x's in the
  primal.
    Ans.: $y^* = 10.667$

12.15. (a) Maximize $\qquad y = 2x_1 + 4x_2$

subject to $\qquad x_1 + 3x_2 \leq 21$

$\qquad\qquad 2x_1 + 3x_2 \leq A$, where $A = 24$

$\qquad\qquad x_1 + \quad x_2 \leq 10.5$

(b) From information in the final tableau, determine $\partial y^*/\partial A$.
    Ans.: part (b), 2/3.

12.16. A simplified version of a certain oil refinery is shown in Figure 12-17 and
consists of one distillation unit which separates the three constituents into

napthas, oils, and bottoms; an oil cracking unit that converts some of the oils into naptha; a bottom cracking unit that converts some of the bottoms into oils; a gasoline blender; and a fuel oil blender.

The composition of the feed and products and their prices are

	Naptha, percent	Oils, percent	Bottoms, percent	Price $/barrel
Crude 1	10	75	15	20.00
Crude 2	15	65	20	24.00
Fuel oil		80	20	25.00
Gasoline	30	70		36.00

The output of Cracking Unit 1 is 40% naptha and 60% oils. The output of Cracking Unit 2 is 70% oils and 30% bottoms. The following capacity limitations apply:

$$\text{naptha output from distillation unit} \leq 3{,}000 \text{ barrels/day}$$

$$\text{Cracking Unit 1 input} \leq 8{,}000 \text{ barrels/day}$$

$$\text{Cracking Unit 2 input} \leq 10{,}000 \text{ barrels/day}$$

In terms of x_1, x_2, x_3, and x_4, where the x's have units of thousands of barrels per day,

(a) write the objective function of the difference in income from the products and costs of the crude.

(b) construct the constraints.

(c) solve for the optimal values.

   Ans.: part (c) $377,333 per day.

12.17. A plant in a process industry is capable of generating steam at 4,000 kPa and at 1,100 kPa. This steam is available for heat exchangers or for

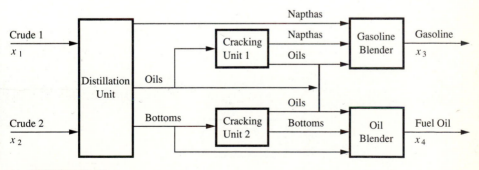

FIGURE 12-17
Oil refinery in Prob. 12.16.

generating power in turbines. Power also may be purchased from the local electric utility. The plant requires 30 MW of shaft power which may be obtained by combination of the three sources shown in Fig. 12-18. There are three process heating requirements and the flow rates of the steam required for these needs are shown in Fig. 12-19.

The flow rates of the 4000 and 1100 kPa steam may be combined linearly to meet the requirements, for example, 15 kg/s of 4000 kPa steam plus 20 kg/s of 1100 kPa steam could serve the 175°C need. Because of a limitation on the boiler capacity, a maximum of 70 kg/s of 4000 kPa steam can be generated. The investment cost of the heat exchangers is

$$\text{investment cost, dollars per s} = \frac{(0.1)(\text{capacity in kg/s})}{(c - t)}$$

where t = outlet fluid temperature, °C
c = 250°C for 4000 kPa steam
c = 185°C for 1100 kPa steam
c = 150°C for 480 kPa steam

The costs of the utilites are

4000 kPa steam, $3.20 per Mg

1100 kPa steam, $2.30 per Mg

Purchased electric power, $0.012 per MJ

In terms of the x's which are rates per second, (a) write the expression for total operating and investment cost per second of operation, (b) write the constraints, and (c) use linear programming to determine the operating conditions that result in minimum cost.

Ans.: (c) minimum cost is $0.694 per second.

FIGURE 12-18
Means of providing shaft power in Prob. 12.17.

4,000 kPa

x_1

Condensate   175°C

Full load handled by
30 kg/s of 4,000 kPa or
40 kg/s of 1,100 kPa steam or
linear combinations thereof.

1,100 kPa

x_2

4,000 kPa

x_3

1,100 kPa   140°C

x_4

Full load handled by
20 kg/s of 4,000 kPa or
24 kg/s of 1,100 kPa steam or
linear combinations thereof.

1,100 kPa

x_5

480 kPa   100°C

x_6

Full load handled by
24 kg/s of 1,100 kPa or
27 kg/s of 480 kPa steam or
linear combinations thereof.

Condensate

FIGURE 12-19
Combinations of steam flow rates to meet the three process heating requirements in Prob. 12.17.

REFERENCE

1. P. Vandenboeck, "Cooling Hot Gases before Baghouse Filtration," *Chem. Eng.*, vol. 79, pp. 67-70, May 1, 1972.

ADDITIONAL READINGS

Charnes, A., W. W. Cooper, and A. Henderson: *An Introduction to Linear Programming*, Wiley, New York 1953.
Dano, S.: *Linear Programming in Industry*, 3d ed., Springer, New York, 1965.
Dantzig, G. B.: *Linear Programming and Extension*, Princeton University Press, Princeton, N. J., 1963.
Garvin, W. W.: *Introduction to Linear Programming*, McGraw-Hill, New York, 1960.
Greenwald, D. U.: *Linear Programming*, Ronald Press, New York, 1957.
Hadley, G.: *Linear Programming*, Addison-Wesley, Reading, Mass., 1962.
Llewellyn, R. W.: *Linear Programming*, Holt, New York, 1964.

CHAPTER
13

MATHEMATICAL MODELING— THERMODYNAMIC PROPERTIES

13.1 NEED FOR MATHEMATICAL MODELING

A necessary preliminary step before a simulation or optimization of a thermal system can begin is almost invariably that of modeling some characteristic(s) of the equipment or processes. The simulation and optimization operations almost always use data in equation form, and the conversion of data to equation form is called mathematical modeling. The groundwork for modeling has been laid in Chapters 4 and 5, and in many cases the techniques covered there are adequate for our needs. Chapter 4, on equation fitting, approached the task as a number-processing operation and showed how the method of least squares could apply to polynomial and other forms of equations. Chapter 5, on the modeling of thermal equipment, took advantage of physical relationships that prevail for heat exchangers, distillation towers, and turbomachines to choose effective forms of the equations.

This chapter carries on the work of mathematical modeling by concentrating on non-linear as well as linear regression, and on expressing thermodynamic properties in equation form. Before settling on those topics the chapter briefly introduces several criteria for the fidelity of the model and comments on the choice of the form of the equation.

The ability to express thermodynamic properties in equation form is valuable in work with thermal systems. It is true that property equations abound, but the accurate ones are usually complex. Our objective will be to develop equations that are simpler, although slightly less accurate. In many cases it will be possible to use some classical thermodynamic property relationships to suggest an additional term or two that can be added to a simple or ideal relation.

13.2 THE FORM OF THE EQUATION

Section 4.12 described the "art of equation fitting," and suggested that having some insight into the relationship may result in a choice of equation form that gives a good representation with a simple equation. Problem 13.1 illustrates two forms of equation that both require the determination of two coefficients. One of the representations is a far better match of the data however. While there are some expectations of how given individual functions behave, it is difficult to set rules for choosing the form of the equation. When no insight exists, polynomial representations might be chosen. Transcendental functions should frequently be explored, and certain techniques of nonlinear regression allow the determination of exponents other than integers which provide a better fit of the data to the equation.

13.3 CRITERIA FOR FIDELITY OF REPRESENTATION

There needs to be some measure of the effectiveness of the equation fitting the data in order to have some basis for selecting the constants and coefficients in a given form of equation and also in choosing between available forms of equations. There are numerous criteria in use, but we shall limit ourselves to three: (1) sum of the deviations squared (SDS), (2) average percent absolute deviation (APD), and (3) goodness of fit (GOF).

The SDS criteria, used in the method of least squares and first presented in Sec. 4.10, is the sum of the squares of the deviations,

$$\text{SDS} = \sum_{i=1}^{n} (y_i - Y_i)^2 \tag{13.1}$$

where y_i = value of the dependent variable computed from the equation
Y_i = value of the dependent variable from the original data
n = total number of data points

Squaring the deviations prohibits the negative deviations from canceling out the positive ones and giving a low value of the summation for what would be a poor representation. A characteristic of the SDS is that since the deviations are squared, the data points that are far off of the ultimate equation contribute more than a proportional influence to the summation. The minimization therefore works toward reducing the large deviations.

The second criterion proposed is the APD. While the SDS has some virtues, it gives equal weights to deviations (squared) regardless of the magnitude of the variable. Thus, a deviation of 0.1 in the SDS criterion is just as important when the dependent variable has a magnitude of 5, for example, as it has when the variable has a magnitude of 5000. In many cases we are interested in the percentage deviation being minimized, which suggests using the APD,

$$APD = \frac{100}{n} \sum_{i=1}^{n} \sqrt{\left(\frac{Y_i - y_i}{Y_i}\right)^2}$$

(13.2)

The deviation expressed as a fraction of the original data value is squared and then the square root extracted to obtain the absolute value of the fraction of deviation.

The third criterion, the GOF, is

$$GOF, \% = 100\left(1 - \frac{SDS}{G}\right)^{0.5}$$

(13.3)

where G = sum of the squares of the deviations of Y_i from

the mean value of Y, $\sum_{i=1}^{n}(Y_i - y_{mean})^2$

The highest GOF results with a zero value of SDS, and a perfect fit of all points gives a GOF of 100%. The GOF is more demanding of flat curves than of steep curves. In Fig. 13-1, for example, the same deviations of the fitted curve from the original points exist for both the nearly-horizontal and the steep lines. The value of G, however, is greater for the steep curve which reduces the SDS/G term in Eq. (13.3). The GOF criterion is less demanding when y varies over a wide span.

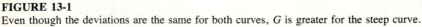

FIGURE 13-1
Even though the deviations are the same for both curves, G is greater for the steep curve.

13.4 LINEAR REGRESSION ANALYSIS

Regression analysis, as the term will be used here, applies to the process of determining constants and coefficients (called parameters) in equations that represent a dependent variable as a function of one or more independent variables. In "linear regression analysis," the partial derivatives of the fidelity criterion with respect to all the parameters are independent of the other parameters. Another way of identifying a linear regression situation is to ask whether the parameters can be solved using a set of linear simultaneous equations. Nonlinear regression will require the solution of one or more nonlinear equations.

The method of least squares presented in Sec. 4.10 is an example of linear regression. Values of the constants and coefficients in the equation $y = f(x)$ are sought that minimize the sum of the squares of the deviations, the SDS. Section 4.11 emphasized that the designation *linear* does not refer to the form in which the independent variable(s) appear, but rather that the parameters appear in linear form. In the minimization process of the SDS with respect to the parameters, the partial derivatives are extracted and equated to zero. Even non-linear forms of the independent variable can provide a set of linear equations with respect to the parameters. Some of the example of forms soluble by linear regression are:

$$y = a + bx$$
$$y = a + bx + cx^2$$
$$y = a + b/x + cx^{1.5}$$
$$y = ae^{-x} + b\sin(2x)$$
$$y = a + bx^2 - c\ln(x)$$

13.5 NONLINEAR REGRESSION ANALYSIS

When the parameters do not appear in a linear form in the equation, determining their values usually consists of an optimization process, perhaps using a search method. The task is to find in the unconstrained optimization problem the optimal values of the parameters, $a_0, a_1, \ldots a_n$ that provide the most favorable value of the fidelity criteria, C. One possibility is the method of steepest descent (Sec. 9.13) which starts with trial values of the parameters and revises them in each step according to the relation

$$\frac{\Delta a_0}{\partial C/\partial a_0} = \frac{\Delta a_1}{\partial C/\partial a_1} = \ldots = \frac{\Delta a_n}{\partial C/\partial a_n} \tag{13.4}$$

The partial derivatives must be extracted numerically,

$$\frac{\partial C}{\partial a_i} = \frac{C(a_0, \ldots a_i + \delta, \ldots a_n) - C(a_0, \ldots, a_i, \ldots a_n)}{\delta} \tag{13.5}$$

because C is a summation, not a single mathematical equation.

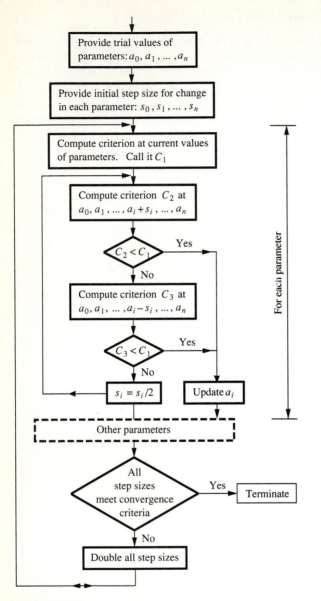

FIGURE 13-2
Flow diagram of a search method for a nonlinear regression. Criterion is being minimized.

While the steepest-descent method is efficient in principle, the choice of the step size and of δ in extracting the partial derivatives may require some care when writing the computer program. It is possible that a more reliable technique that is also easier to program is the one shown in the flow diagram in Fig. 13-2. The process is essentially one of working with one parameter at a time, carrying out the following operations:

TABLE 13.1
Search for Parameters in Example 13.1

Iteration	a	b	Average Percent Deviation	a-step	b-step
0	3.00000	0.40000	54.8306	0.10000	0.020000
1	3.10000	0.42000	49.5006	0.10000	0.020000
2	3.20000	0.44000	44.4117	0.10000	0.020000
3	3.30000	0.46000	38.4853	0.10000	0.020000
. . . .					
42	2.58753	0.6006298	0.1360213	7.6294E–7	1.90735E–8
43	2.58753	0.6006298	0.1360212	9.5367E–8	1.19209E–9

1. Increase the parameter by the initially-assigned step size. If the fidelity criterion is improved, go on to the next parameter.
2. If the criterion was not improved in (1), decrease the parameter by the step size. If the criterion is improved, go on to the next parameter.
3. If neither (1) nor (2) improve the criterion, decrease the step size and return to (1).

A technique that often increases the speed of convergence is to increase the step size after each operation, because there could have been an occasion where the step size had to be reduced, but a larger step size may work for some future operations.

When the step sizes of the parameters have been reduced to preassigned levels, the search terminates.

Example 13.1. The following pairs of (x, y) points are to be fit to the equation, $y = ae^{bx}$: (1, 4.7), (2, 8.6), (3, 15.7), (4, 28.7), (5, 52.2), (6, 95.2), (7, 173.4), (8, 316), (9, 576), and (10, 1049). Use a search technique to determine the a and b parameters that result in the minimum average-percent deviation.

Solution. Trial values of $a = 3$ and $b = 0.4$ are chosen, and the initial step sizes selected are 0.1 and 0.02 for a and b, respectively. The early and final values are shown in Table 13.1. For purposes of comparison, the same search was conducted except that the regular increases in step size were omitted. That search required more than 1600 iterations to arrive at the final values shown in Table 13.1.

13.6 THERMODYNAMIC PROPERTIES

The need for equations to represent thermodynamic properties arises continually in work with thermal systems. Those properties of particular interest are the temperature, pressure, enthalpy, density or specific volume, entropy,

FIGURE 13-3
Some widely used property charts.

and occasionally the internal energy. The approach that will be followed in the remainder of this chapter is a compromise between exhaustive effort to achieve precise representations on the one hand, and rough idealizations on the other. We will be satisfied if the accuracy of the equation can be improved from 5%, for example, to an accuracy of, say, 1 or 2%.

Representation of the thermodynamic properties requires both property relations from classical thermodynamics and experimental data. There are a number of possible starting points and paths in arriving at desired property equations, but a frequently used technique[1,2] is to seek from experimental data the following four equations:

1. p-v-T equations for vapor
2. c_{p0}, the specific heat at zero pressure
3. p-T relation for saturated conditions
4. ρ_f, the density of liquid.

Based on the above four equations, it is possible, by the judicious use of classical thermodynamics, to compute two nonmeasurable properties that are of great importance in thermal system work—enthalpy and entropy. It is no accident that some of the most useful property charts, such as those shown in skeleton form in Fig. 13-3, relate the enthalpy and entropy to other properties. The next several sections will introduce some relations from classical thermodynamics which will later build on the four equations to compute other properties.

13.7   INTERNAL ENERGY AND ENTHALPY

Computation of rates of energy flow is a key operation in analyzing thermal systems. This requirement contributes to the importance of the internal energy and enthalpy properties. The irony is that there are no meters and instruments to measure these properties; instead they have to be deduced from measurable properties, such as pressure, temperature, and specific volume.

The first law, when applied to a closed system consisting of a piston-cylinder, such as Fig. 13-4, yields the equation,

$$dq = du + dw \qquad (13.6)$$

where q = rate of heat transfer per unit mass in vessel, kJ/kg
u = internal energy, kJ/kg
w = external work per unit mass in vessel, kJ/kg

If the heat is transferred to the system in a thermodynamically reversible process,

$$dq\big)_{\text{rev}} = T \, ds \qquad (13.7)$$

where T = temperature, K
s = entropy, kJ/kg · K

and if the work is performed reversibly

$$dw\big)_{\text{rev}} = p \, dv \qquad (13.8)$$

where p = pressure, kPa
v = specific volume, m³/kg

Substitute Eqs. (13.7) and (13.8) into (13.6) to arrive at

$$T \, ds = du + p \, dv \qquad (13.9)$$

Even though Eq. (13.9) was arrived at through the use of a nonflow process with the assumption of reversibility, thermodynamic logic permits us to state that Eq. (13.9) is perfectly general and is applicable to all processes whether flow or nonflow, reversible or irreversible. The reason is that Eq. (13.9) is exclusively a relationship of properties, and the procedure by which we arrived at the equation is immaterial.

Classical thermodynamic is replete with partial derivatives of properties, and some of these relationships will prove to be valuable in building a full set of thermodynamic properties. Equation (13.9) offers the basis for the first two of these relationships. If v is held constant, $dv = 0$, and the

FIGURE 13-4
Energy balance in a closed system.

equation may be written,

$$\frac{\partial u}{\partial s}\bigg)_v = T \qquad (13.10)$$

and if s is held constant,

$$\frac{\partial u}{\partial v}\bigg)_s = -p \qquad (13.11)$$

The enthalpy is also the basis of some relationships of thermodynamic properties. By definition,

$$h = u + pv$$

so

$$dh = du + p\,dv + v\,dp \qquad (13.12)$$

In combination with Eq. (13.9)

$$dh = T\,ds + v\,dp \qquad (13.13)$$

Equation (13.13) is the basis of two property relationships

$$\frac{\partial h}{\partial s}\bigg)_p = T \qquad (13.14)$$

$$\frac{\partial h}{\partial p}\bigg)_s = v \qquad (13.15)$$

Example 13.2. In the Mollier diagram with $h - s$ coordinates as shown in Fig. 13-5, what is the slope of the constant-pressure line in the liquid-vapor mixture and the superheated vapor regions?

Solution. The slope of a constant pressure line on $h - s$ coordinates is $\partial h/\partial s)_p$, which from Eq. (13.14) equals T. In the liquid-vapor mixture

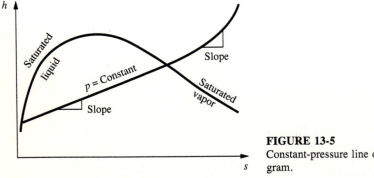

FIGURE 13-5
Constant-pressure line on an h-s diagram.

region when p is constant the temperature is also constant, so in this region the line is straight. When the line moves farther into the superheated vapor region T progressively increases, so the line curves upward.

13.8 THE CLAPEYRON EQUATION

A useful expression for relating some thermodynamic liquid and vapor properties at saturated conditions is the Clapeyron equation. This equation can be developed from purely thermodynamic principles and is applicable to all pure substances. The development begins with the expression of the Gibbs function F :

$$F, \text{ kJ/kg} = h - Ts \qquad (13.16)$$

Because it is based on properties, F is also a property. In differential form,

$$dF = dh - T\,ds - s\,dT \qquad (13.17)$$

and substituting Eq. (13.12),

$$dF = du + p\,dv + v\,dp - T\,ds - s, dT \qquad (13.18)$$

Next apply Eq. (13.18) to a change between saturated liquid and saturated vapor at the same pressure. In the mixture region the temperature remains constant in a constant-pressure process, so both dT and dp in Eq. (13.18) are zero,

$$dF = du + p\,dv - T\,ds \qquad (13.19)$$

Since from Eq. (13.9) $T\,ds = du + pdv$,

$$dF = 0$$

so the Gibbs functions of saturated liquid and saturated vapor are equal at the same temperature and pressure, $F_f = F_g$.

The next operation will be to change slightly the pressure and temperature of a sample of liquid/vapor at saturation as shown in Fig. 13-6. Since $F_{f,1} = F_{g,1}$, and $F_{f,2} = F_{g,2}$,

$$dF_g = dF_f$$

$$dh_g - T\,ds_g - s_g\,dT = dh_f - T\,ds_f - s_f\,dT$$

or $\qquad v_g\,dp - s_g\,dT = v_f\,dp - s_f\,dT$

and $\qquad dp(v_g - v_f) = dT\,(s_g - s_f)$

so $\qquad \dfrac{dp}{dT} = \dfrac{s_g - s_f}{v_g - v_f} = \dfrac{T(s_g - s_f)}{T(v_g - v_f)} = \dfrac{h_{fg}}{T(v_g - v_f)} \qquad (13.20)$

Equation (13.20) is the Clapeyron equation and relates the derivative of the

FIGURE 13-6
Changing the pressure and temperature of a liquid/vapor mixture.

saturation pressure with respect to the saturation temperature to certain other properties at saturated conditions.

> **Example 13.3.** At 50°C and 51°C, respectively, water has the following properties at saturated conditions, $v_f = 0.0010121$, 0.0010126 m³/kg, $v_g = 12.05$, 11.50 m³/kg; $h_f = 209.26$, 213.44 kJ/kg; and $h_g = 2592.2$, 2593.9 kJ/kg. Use the Clapeyron equation to predict the change in saturation pressure between 50 and 51°C.
>
> **Solution.** Substitute the property values at 50.5°C into Eq. (13.20)
>
> $$\Delta p = (1K)\frac{(2593.1 - 211.4)}{(50.5 + 273.15)(11.775 - 0.0010124)}$$
>
> $$= 0.625 \text{ kPa}$$
>
> The steam tables give the following change in the saturation pressure,
>
> $$12.961 - 12.335 = 0.626 \text{ kPa}$$

13.9 PRESSURE-TEMPERATURE RELATIONSHIPS AT SATURATED CONDITIONS

By making two bold approximations the Clapeyron equation can be used to suggest the form of an equation to relate the saturation pressure to its corresponding saturation temperature. Assume (1) that $v_f \ll v_g$, and (2) that the ideal gas equation applies, even for saturated vapor, $pv_g = RT$. Equation (13.20) then becomes

$$\frac{dp}{dT} = \frac{h_{fg}}{T(RT/p)}$$

and

$$\frac{dp}{p} = \frac{dT}{T^2}\frac{h_{fg}}{R}$$

Integrating,

$$\ln p = -h_{fg}/(RT) + \text{constant of integration}$$

or
$$\ln p = A + B/T \tag{13.21}$$

where A and B are positive and negative constants, respectively. The form suggested by Eq. (13.21) is widely used and indicates that on a $\ln p$ vs $1/T$ graph a straight line represents the characteristics of a given substance. Figure 13-7 shows such a graph for several fluids.

A goal of this chapter is to identify means of making slight refinements to the form of an equation to achieve a noticeable improvement in the fidelity of representation. An example of this process is the Antoine equation (1888) which introduces one additional parameter,

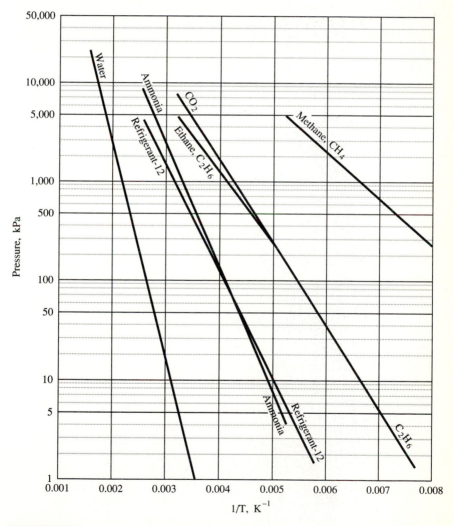

FIGURE 13-7
Display of terms in Eq. (13.21) for several substances.

TABLE 13.2
Thermodynamic Properties of Water

Temperature		Pressure p, kPa	Liquid Density kg/m³	Sp. Vol. of Vapor m³/kg	Enthalpy, kJ/kg		Entropy, kJ/kg · K	
t, °C	T, K				h_f	h_g	s_f	s_g
0	273.15	0.6108	999.800	206.3	−0.04	2501.6	−0.00002	9.1577
20	293.15	2.337	998.303	57.84	83.86	2538.2	0.2963	8.6684
40	313.15	7.375	992.260	19.55	167.45	2574.4	0.5721	8.2583
60	333.15	19.920	983.187	7.679	251.09	2609.7	0.8310	7.9108
80	353.15	47.360	971.628	3.409	334.92	2643.8	1.0753	7.6132
100	373.15	101.33	958.130	1.673	419.06	2676.0	1.3069	7.3554
120	393.15	198.54	942.863	0.8915	503.72	2706.0	1.5276	7.1293
140	413.15	361.40	925.840	0.5085	589.10	2733.1	1.7390	6.2984
160	433.15	618.10	907.276	0.3068	675.47	2756.7	1.9425	6.7475
180	453.15	1002.7	886.918	0.1938	763.12	2776.3	2.1393	6.5819
200	473.15	1554.9	864.678	0.1272	852.37	2790.9	2.3307	6.4278
220	493.15	2319.8	840.336	0.08604	943.67	2799.9	2.5178	6.2817
240	513.15	3347.8	813.603	0.05965	1037.6	2802.2	2.7020	6.1406
260	533.15	4694.3	783.945	0.04213	1134.9	2796.4	2.8848	6.0010
280	553.15	6420.2	750.525	0.03013	1236.8	2780.4	3.0683	5.8586
300	573.15	8592.7	712.200	0.02165	1345.0	2751.0	3.2552	5.7081
320	593.15	11289.	666.889	0.01548	1462.6	2703.7	3.4500	5.5423
340	613.15	14605.	610.240	0.01078	1595.5	2626.2	3.6616	5.3427
360	633.15	18675.	528.011	0.00694	1764.2	2485.4	3.9210	5.0600
374.15	647.30	22120.	315.457	0.00317	2107.4	2107.4	4.4429	4.4429

$$\ln p = A + B/(T - C) \qquad (13.22)$$

A proposal for the estimate[3] of C is

$$C, \text{ K} = -18 + 0.19T$$

For the saturation pressures of water shown in Table 13.2, the average percent deviation (from Prob. 13.4) using the Antoine form is 3.39%, which may be compared to 6.53% for the form of Eq. (13.21) used in Prob. 13.2.

13.10 THE MAXWELL RELATIONS

A set of equations that assists in achieving the goals of developing convenient, yet reasonably accurate, property relations are the Maxwell relations.
The Maxwell relations are

$$\left. \frac{\partial T}{\partial v} \right)_s = - \left. \frac{\partial p}{\partial s} \right)_v \qquad (13.23)$$

$$\left. \frac{\partial T}{\partial p} \right)_s = \left. \frac{\partial v}{\partial s} \right)_p \qquad (13.24)$$

$$\left. \frac{\partial v}{\partial T} \right)_p = - \left. \frac{\partial s}{\partial p} \right)_T \qquad (13.25)$$

$$\left. \frac{\partial p}{\partial T} \right)_v = - \left. \frac{\partial s}{\partial v} \right)_T \qquad (13.26)$$

The Maxwell relations provide a link between entropy and the properties in the p-v-T equation of state and are derived from four differential equations:

$$T \, ds = du + p \, dv \qquad (13.9)$$

$$T \, ds = dh - v \, dp \qquad (13.13)$$

$$dF = dh - T \, ds - s \, dT \qquad (13.17)$$

and from the definition of the Helmholtz function

$$A = u - Ts$$

the differential form is

$$dA = du - T \, ds - s \, dT \qquad (13.27)$$

The same pattern is used in developing each of the Maxwell relations, and Eq. (13.9), which is the basis for Eq. (13.23), will be used as an illustration. The internal energy u can be specified by two independent properties, s and v, so

$$u = f(s, v)$$

and
$$du = \left. \frac{\partial u}{\partial s} \right)_v ds + \left. \frac{\partial u}{\partial v} \right)_s dv \qquad (13.28)$$

Equating the two expressions for du from Eqs. (13.9) and (13.28) gives

$$T \, ds - p \, dv = \left. \frac{\partial u}{\partial s} \right)_v ds + \left. \frac{\partial u}{\partial v} \right)_s dv$$

The coefficients of ds and dv must be equal, so

$$T = \left. \frac{\partial u}{\partial s} \right)_v \qquad \text{and} \qquad -p = \left. \frac{\partial u}{\partial v} \right)_s$$

The second derivatives of u with respect to s and v from each of the two above relations are

$$\frac{\partial^2 u}{\partial s \, \partial v} = \left. \frac{\partial T}{\partial v} \right)_s = - \left. \frac{\partial p}{\partial s} \right)_v$$

which is the first Maxwell relation.

13.11   SPECIFIC HEATS

One of the tools listed in Sec. 13.6 for determining a complete set of thermo-dynamic properties is the specific heat. The two specific heats of interest are those at constant volume and at constant pressure, c_v and c_p, respectively. Their definitions are

constant volume, $\qquad\qquad c_v = \left. \frac{\partial u}{\partial T} \right)_v$ $\qquad\qquad$ (13.29)

constant pressure, $\qquad\qquad c_p = \left. \frac{\partial h}{\partial T} \right)_p$ $\qquad\qquad$ (13.30)

Specific heats are means of computing enthalpies, entropies, and internal energies. Arbitrary base values are always chosen for those properties, so the task is more precisely defined as computing the differences in enthalpy, entropy, and internal energy between two state points.

In general, specific heats vary both with respect to pressure and temperature, although we shall find that the temperature has the dominant influence. The computation of a change in property occurring in a change of pressure and temperature thus entails an integration, for which an equation for the specific heat as a function of p and T would be valuable. The strategy to be followed is to first express the so-called "specific heat at zero pressure," c_{p0}, as a function of temperature. Such an expression would come from experimental data, possibly in conjunction with spectographic measurements. In combination with p-v-T relations, c_{p0} can be extended into both c_p and c_v at various temperatures and pressures.

An important relation is the partial derivative of c_p with respect to pressure at constant temperature. This relation can be developed by solving for ds in Eq. (13.13) and showing for the constant pressure process that

$$\frac{1}{T}\frac{\partial h}{\partial T}\bigg)_p = \frac{\partial s}{\partial T}\bigg)_p$$

Substituting c_p from Eq. (13.30) and differentiating with respect to p at constant T gives

$$\frac{1}{T}\frac{\partial c_p}{\partial p}\bigg)_T = \frac{\partial^2 s}{\partial T\partial p} = \frac{\partial}{\partial T}\left[\frac{\partial s}{\partial p}\bigg)_T\right]_p$$

Substitution of the Maxwell relation, Eq. (13.25) results in

$$\frac{\partial c_p}{\partial p}\bigg)_T = -T\frac{\partial^2 v}{\partial T^2}\bigg)_p \tag{13.31}$$

An important feature of Eq. (13.31) is that $\partial c_p/\partial p)_T$ is expressed in terms of the p-v-T properties, which are assumed to be available in equation form. The first use of Eq. (13.31) is to apply it to a perfect gas which has the equation of state

$$pv = RT \tag{13.32}$$

The second derivative, $\partial^2 v/\partial T^2)_p$, is zero, so with a perfect gas there is no change of c_p with respect to pressure. It is acknowledged, however, that even for a perfect gas c_p can vary with temperature. The lack of influence of pressure on c_p of a perfect gas encourages the concept of c_p at zero pressure, c_{p0}, because all gases behave as perfect gases as their pressure approaches zero. Thus, if c_{p0} can be expressed in terms of temperature, and if the equation of state is known, with the help of Eq. (13.31) c_p can be expressed over a range of temperatures and pressures. Values of c_{po} determined from experimental data can be expressed as functions of temperature using simple polynomials. An example of these representations for several common gases is presented in Table 13.3.

Some of the enthusiasm for the use of Eq. (13.31) will be blunted by attempts to apply the equation. Most useful equations of state implicitly incorporate the specific volume v, so the expression for $\partial^2 v/\partial T^2)_p$ is complex. The ultimate use of c_p is likely to be in computing changes in enthalpy or entropy over a change in pressure and temperature which then requires an integration of some complex expressions. All of this represents a departure from the overriding objective of this chapter of achieving an improvement over the ideal representations with a modest increase in complexity. Returning to this objective, we may wish to proceed no further than expressing c_p in a form shown graphically in Fig. 13-8. The zero-pressure values of a given substance are functions of temperature, such as from Table 13.3. The slope of any curve is available from Eq. (13.31) using the equation of state, perhaps by numerical rather than analytical differentiation (see Example 13.4). If the curve for a given temperature is

TABLE 13.3
Equations for zero-pressure c_p in the range of 300 to 3500 K, where $\Theta = T/100$ (Ref. 3).

Gas	c_{po}, kJ/(kg $\cdot$ K)
H_2	$28.253 - 351.37\Theta^{-0.75} + 582.5\Theta^{-1} - 280.35\Theta^{-1.5}$
CH_4	$-42.054 + 27.483\Theta^{0.25} - 2.228\Theta^{0.75} + 11.423\Theta^{-3}$
H_2O	$7.947 - 10.20\Theta^{0.25} + 4.597\Theta^{0.5} - 0.2055\Theta$
N_2	$1.395 - 18.314\Theta^{-1.5} + 38.31\Theta^{-2} - 29.3\Theta^{-3}$
O_2	$1.1698 - 0.000628\Theta^{1.5} - 5.580\Theta^{-1.5} + 7.4025\Theta^{-2}$
CO_2	$-0.0849 + 0.6938\Theta^{0.5} - 0.09326\Theta + 0.000550\Theta^2$

approximated by a second-degree equation, this equation can be developed knowing (1) c_p at zero pressure, (2) that the slope is zero at zero pressure, and (3) the slope at some nonzero pressure from Eq. (13.31).

Example 13.4. Compute the equation for c_p of oxygen as a function of pressure from 0 to 2000 kPa at a temperature of 600 K if the van der Waals equation of state applies,

$$p, \text{kPa} = \frac{RT}{v - b} - \frac{a}{v^2}$$

where $R = 0.2598$ kJ/(kg $\cdot$ K)
$a = 0.1346$ (kPa)(m^6/kg^2)
$b = 0.000995$ m^3/kg

Solution. From Table 13.3, c_{p0} at 600 K for $O_2 = 0.9865$ kJ/(kg $\cdot$ K). Next compute $\partial c_p / \partial p)_T$ at 2000 kPa using Eq. (13.31). Extract the second derivative numerically by computing v at 595, 600, and 605 K.

FIGURE 13-8
Expression of c_p as a function of temperature and pressure.

$$v_{595} = 0.0774274693033$$

$$v_{600} = 0.0780840914874$$

$$v_{605} = 0.0787405951740$$

The second derivative with respect to temperature at $p = 2000$ kPa is

$$\frac{\left[\dfrac{(v_{605} - v_{600})}{5} - \dfrac{(v_{600} - v_{595})}{5}\right]}{5} = 0.470 \times 10^{-8}$$

Then from Eq. (13.31)

$$\left.\frac{\partial c_p}{\partial p}\right|_T = 2.844 \times 10^{-6} \quad \text{kJ/(kg} \cdot \text{K) per kPa}$$

At 600 K, $\qquad\qquad\qquad \left.\dfrac{\partial c_p}{\partial p}\right|_T = 0$ at $p = 0$,

and $\qquad\qquad\qquad \left.\dfrac{\partial c_p}{\partial p}\right|_T = 2.844 \times 10^{-6}$ at $p = 2000$ kPa

Proposing a second degree variation with pressure as suggested in Fig. 13-8,

At 600 K, $\qquad c_p - c_{p0} = a_1 p + a_2 p^2$

where $a_1 = 0$
$$a_2 = (2.844 \times 10^{-6})/2p$$
$$= 1.422 \times 10^{-9}$$

Thus c_p at 600 K and 2000 kPa is

$$c_p = 0.9865 + (1.422 \times 10^{-9})(2000)^2$$

$$= 0.9865 + 0.0057$$

$$= 0.992 \text{ kJ/(kg} \cdot \text{K)}$$

In Example 13.4 the correction for pressure revises c_p by 0.6%. In general, the influences of pressure on c_p are much less than those of temperature. The tools are now available to assess whether a correction for pressure on c_p needs to be made at all.

13.12 p-v-T EQUATIONS

Many forms of equations are in use for p-v-T relations, and there are still more versions when the various modifications of the original forms are included. This section identifies several widely used equations of state that require knowledge of only a small number of constants. Furthermore, in several of the equations presented below these constants can be computed knowing only the molecular weight, and the critical temperature and pressure of the substance.

The simplest and probably most valuable equation of state is the perfect gas equation,

$$pv = RT \tag{13.32}$$

in which the specific gas constant R for the substance can be determined from the universal gas constant,

$$R, \text{ kJ/(kg} \cdot \text{K)} = 8.314/(\text{molecular weight})$$

A refinement of the perfect gas equation incorporates the compressibility factor, Z,

$$pv = ZRT \tag{13.33}$$

If Z is known, Eq. (13.33) gives a precise representation, but unfortunately Z is a function of T and p. Graphs are available providing values of $Z = f(T,p)$ for specific substances. For a wide range of temperatures and pressures the calculation reverts to a cumbersome expression for Z.

An example of a form based heavily on empirical data is the Beattie-Bridgeman equation.

$$p = \frac{RT}{v} + \frac{\alpha}{v^2} + \frac{\beta}{v^3} + \frac{\gamma}{v^4} \tag{13.34}$$

where the values of α, β and γ are determined from best fits of experimental data.

Critical temperatures and pressures of substances are readily available data, and the advantage of two equations of state—the van der Waals and the Redlich-Kwong equations—is that the constants are derivable from the critical properties. At the critical point both the first and second derivatives of p with respect to v are zero.

$$\left.\frac{\partial p}{\partial v}\right|_{T_c} = 0 \qquad \text{and} \qquad \left.\frac{\partial^2 p}{\partial v^2}\right|_{T_c} = 0 \tag{13.35}$$

where T_c = critical temperature, K

When Eq. (13.35) is imposed on the van der Waals equation

$$p = \frac{RT}{v - b} - \frac{a}{v^2} \tag{13.36}$$

$$a = \frac{27}{64}\frac{R^2 T_c^2}{p_c} \qquad \text{and} \qquad b = \frac{RT_c}{8p_c} \tag{13.37}$$

where p_c = critical pressure, kPa

The development of Eq. (13.37) is addressed in Problem 13.7. The values of the a and b constants in Example 13.4 were computed from Eq. (13.37). The van der Waals equation uses two constants in addition to R, and its accuracy extends over a wider range than the perfect gas equation. A still more accurate two-constant equation is one proposed by Redlich and Kwong:[4]

$$p = \frac{RT}{v - b} - \frac{a}{T^{1/2}v(v + b)} \tag{13.38}$$

Soave[5] made even further improvements in the accuracy by expressing the a constant as a function of temperature. The simplicity of the Redlich-Kwong equation of state in Eq. (13.38) is, however, consistent with the approach of this chapter. Applying Eq. (13.35) to the Redlich-Kwong equation yields

$$a = 0.42748\frac{R^2 T_c^{5/2}}{p_c} \tag{13.39}$$

$$b = 0.08664\frac{RT_c}{p_c} \tag{13.40}$$

Example 13.5. Compute the specific volume of superheated water vapor at 5000 kPa and 350°C using (a) the perfect gas equation, (b) the van der Waals equation, and (c) the Redlich-Kwong equation.

Solution.

(a) R of water vapor $= 8314/18 = 0.461889 \text{kJ}/(\text{kg} \cdot \text{K})$

$$v = \frac{(0.461889)(350 + 273.15)}{5000 \text{ kPa}}$$

$$= 0.057565 \text{ m}^3/\text{kg}$$

(b) Using the values of critical pressure and temperature from Table 13.2, the van der Waals constants can be computed,

$$a = 1.7048 \quad \text{and} \quad b = 0.0016895$$

The van der Waals equation,

$$5000 = \frac{(0.461889)(623.15)}{v - 0.0016895} - \frac{1.7048}{v^2}$$

can be solved for v by an iterative process, such as Newton-Raphson, to find

$$v = 0.05303 \text{ m}^3/\text{kg}$$

(c) For water, the Redlich-Kwong constants from Eqs. (13.39) and (13.40) are

$$a = 43.951 \quad \text{and} \quad b = 0.001171$$

Substituting these constants into Eq. (13.38) and solving for v yields

$$v = 0.05230 \text{ m}^3/\text{kg}$$

The values computed in Example 13.5 may be compared with the specific volume from steam tables[6] at 5000 kPa and 623.15 K of 0.05194 m³/kg. The van der Waals equation showed an appreciable improvement over the perfect gas equation, and the Redlich-Kwong equation yielded a result deviating only 0.7% from the tabular value.

13.13 BUILDING A FULL SET OF DATA

This chapter concludes with a summary of how a complete set of thermo-dynamic properties can be computed using the four tools listed in Sec. 13.6, namely, (1) the p-v-T equation of state, (2) the zero-pressure specific heat as a function of temperature, (3) the pressure-temperature relationship at saturated conditions, and (4) the density of saturated liquid. The p-v-T properties of superheated and saturated vapor have already been explored; the properties that remain are enthalpy and entropy.

One of the several approaches to developing a full set of property data is represented symbolically in Fig. 13-9. Base conditions of enthalpy and entropy are always arbitrary choices which usually are in reference to convenient values assigned to saturated liquid at a chosen temperature. The Clapeyron equation then permits jumping across to the enthalpy and entropy of saturated vapor. From this saturated vapor condition the enthalpies and entropies of saturated and superheated vapor can be computed using the

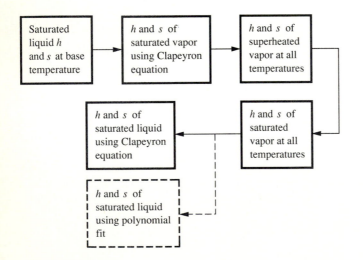

FIGURE 13-9
A possible calculation sequence to develop enthalpies and entropies.

equations outlined below. The enthalpy and entropy of saturated liquid at one temperature have already been assumed, and the values at other temperatures could be computed in one of two ways. It would be possible to jump back from saturated vapor to saturated liquid using the Clapeyron equation, or it might be as simple to perform a polynomial fit of tabular data for the enthalpy and entropy of saturated liquid.

Figure 13-9 implies that specific values of enthalpy and entropy are to be computed. In many such cases a calculation will not be needed, however, because only a difference of enthalpy (during a process) needs to be computed. Similarly, equal entropies before and after a compression may need to be assured so the specific value of entropy is of no consequence.

The change in enthalpy in the vapor region can be found by integrating an expression for dh which is developed as follows. For vapor, h can be represented as a function of p and T, and then the expression for c_p substituted.

$$dh = \left.\frac{\partial h}{\partial T}\right)_p dT + \left.\frac{\partial h}{\partial p}\right)_T dp = c_p\, dT + \left.\frac{\partial h}{\partial p}\right)_T dp$$

Also, since $T\,ds = dh - v\,dp$

$$\left.\frac{\partial h}{\partial p}\right)_T = v + T\left.\frac{\partial s}{\partial p}\right)_T$$

Substitute the expression for $\partial s/\partial p)_T$ from the Maxwell relation, Eq. (13.25), to obtain

$$dh = c_p\, dT + \left[v - T\left.\frac{\partial v}{\partial T}\right)_p\right]dp \qquad (13.41)$$

The expression for a differential change in entropy comes from a parallel development:

$$ds = \left.\frac{\partial s}{\partial T}\right)_p dT + \left.\frac{\partial s}{\partial p}\right)_T dp$$

and from $\qquad\qquad\qquad T\,ds = dh - v\,dp$

$$\left.\frac{\partial s}{\partial T}\right)_p = \frac{1}{T}\left.\frac{\partial h}{\partial T}\right)_p = \frac{c_p}{T}$$

Then

$$ds = c_p\frac{dT}{T} - \left.\frac{\partial v}{\partial T}\right)_p dp \qquad (13.42)$$

Equations (13.41) and (13.42) can be integrated through a change in p and T to find the change in enthalpy and entropy, respectively, with c_p

treated as a variable with respect to temperature, and possibly pressure. In Eq. (13.42) $\partial v / \partial T)_p$ is variable with respect to p. With an ideal gas, for example, $\partial v / \partial T)_p = \frac{R}{p}$, and the integral is $R \ln(p_2/p_1)$. The Redlich-Kwong equation may be used to modify R/p for real gases for the integration.

Example 13.6. Starting with a base enthalpy of $h = 0$ kJ/kg for saturated liquid water at 0°C, compute the enthalpy of superheated vapor at 4000 kPa and 500°C using the following equations: (1) Redlich-Kwong equation of state in Example 13.5, (2) c_{p_0} from Table 13.3, and (3) the p-T equation for saturation conditions at low pressure, $\ln p = 19.335 - 5416/T$.

Solution. The first stage of the calculation is to move from the base enthalpy to that of saturated vapor at 0°C. Needed for the Clapeyron equation

$$\frac{dp}{dT} = \frac{h_{fg}}{T(v_g - v_f)}$$

are dp/dT, v_g, and v_f. The p-T equation at saturation yields p of 0.6108 kPa, and

$$\frac{dp}{dT} = p\frac{5416}{T^2} = 0.04434 \text{ kPa/K}$$

From the Redlich-Kwong equation, when $p = 0.6108$ and $T = 273.15$ K the specific volume $v_g = 206.54$ m³/kg. With respect to this magnitude of v_g, the specific volume of saturated liquid v_f is negligible. Substituting into the Clapeyron equation yields

$$h_{fg} = (0.0434)(273.15)(206.54) = 2501.5 \text{ kJ/kg}$$

so $$h_g \text{ at } 0°C = 0 + 2501.5 = 2501.5 \text{ kJ/kg}$$

The process of moving from saturated vapor at 0°C to superheated vapor at 4000 kPa and 500°C comes from an integration of Eq. (13.41). First find the enthalpy at 0.6108 kPa and 500°C by integrating $c_p \, dT$. At this low pressure the specific heat at zero pressure applies, so use Table 13.3 to compute c_{p0} at three temperatures, say 0, 250, and 500°C. Those values are 1.8703, 1.9603, and 2.1319 kJ/(kg · K), respectively, which translate to a second degree equation,

$$c_{p0} = 1.8703 + 0.0001968t + 0.6528 \times 10^{-6}t^2$$

where t is in °C

The integration from 0 to 500°C at 0.6108 kPa of $c_{p0} \, dT$ is 986.95, so h at 500°C and 0.6108 kPa $= 2501.5 + 986.95 = 3488.5$ kJ/kg.

The other term to be integrated in Eq. (13.41) is

$$\left[v - T\frac{\partial v}{\partial T}\Big)_p \right] dp \tag{13.43}$$

using the Redlich-Kwong equation. Unfortunately v cannot be solved explicitly in this equation, so a numerical computation of $\partial v/\partial T)_p$ must be made. For example, at 4000 kPa the following specific volumes can be computed:

$$t = 499.9°C \qquad v_a = 0.085963245748$$

$$t = 500.0 \qquad v_b = 0.085975713311$$

$$t = 500.1 \qquad v_c = 0.085988180541$$

Thus

$$v - T\left.\frac{\partial v}{\partial T}\right)_p = v_b - \frac{773.15(v_c - v_a)}{0.2}$$

$$= -0.010 \text{ m}^3/\text{kg}$$

The bracketed term in Eq. (13.43) is essentially constant at a value of -0.010 throughout the pressure range of 0.6108 to 4000 kPa so the integral is $(-0.010)(4000 - 0.61) = -40.0$ kJ/kg. Finally, h at 4000 kPa and 500°C is $3488.5 - 40.0$ kJ/kg $= 3448.5$ kJ/kg.

The agreement of the calculations of Example 13.6 with tabular values is good, as Table 13.4 shows.

One action that helped provide accuracy was to use a p-T equation for saturation applicable to the low pressure range, rather than the one developed in Prob. 13.4, which is a best fit to data of the wide range of 0°C to the critical point.

When the techniques presented in this chapter are used, it is unusual to start at one point, as in Example 13.6, and to build property values over the wide range of conditions based on equations alone. Instead, the tools provided here are most often used for more confined regions, such as condensation, evaporation, or heating or cooling a vapor. Using several tabular values of properties as reference values at various positions on the property map will prevent the accumulation of errors.

When simulating or optimizing systems the need often arises to perform property calculations. This chapter has presented several tools which

TABLE 13.4
Comparison of Example 13.6 calculations with tabular values[6]

		Vapor enthalpies, kJ/kg	
t, °C	p, kPa	Table	Example 13.6
0	0.6108	2501.6	2501.5
500	0.6108	3489.2	3488.5
500	4000	3445.0	3448.5

improve the accuracy of these calculations by a modest increase in complexity.

PROBLEMS

13.1. Use the method of least squares to find the three parameters in the following equations that give the best fit of the liquid density of water to the absolute temperature. Use the data from Table 13.2.

(a) density $= a_0 + a_1 T + a_2 T^2$

(b) density $= c_0 + c_1(1 - T/T_c)^{1/3} + c_2(1 - T/T_{T_c})^{2/3}$

where T_c = critical temperature, 647.30 K

Compute the sum of the deviations squared for both representations.

      Ans.: (a) 30,885, (b) 1747.

13.2. Using the 20 temperature-pressure data points for water at saturated conditions from Table 13.2, write a computer program or use one from a computer library to determine the values of A and B in the equation

$$p = e^{(A + B/T)}$$

that give the minimum average-percent deviation. Select the trial values of A and B by substituting two different p-T points into the equation.

(a) What are the values of A and B?

(b) What is the minimum average-percent deviation?

      Ans.: (b) 6.61% or 6.53% depending on trial values.

13.3. A vapor passes through a venturi-type flow measuring device, as in Fig. 13-10, having an area ratio $A_1/A_2 = 1.8$. The process from point 1 to point 2 is usually assumed to be isentropic. The specific volumes are $v_1 = 0.06$ m³/kg, $v_2 = 0.068$ m³/kg; the arithmetic mean of these two values may be used in Eq. (13.15). If $\Delta p_{1-2} = 40$ kPa, what is v_2?

      Ans.: 82.1 m/s.

13.4. Examine the 20 temperature-pressure pairs for water at saturated conditions from Table 13.2.

(a) Determine the values of A, B, and C in the revised form of Eq. (13.22)

$$p = e^{[A + B/(T - C)]}$$

that give the minimum average percent deviation of p.

(b) Determine the minimum average percent deviation, and compare this value with the results from Prob. 13.2.

      Ans.: (a) $A = 16.577$, $B = -4023.05$, and $C = 37.20$
      (b) 1.02%

FIGURE 13-10
Venturi flow meter in Prob. 13.3.

13.5. Starting with the differential of the Gibbs function, Eq. (13.17), develop the Maxwell relation, Eq. (13.25).

13.6. When c_p is known, c_v may be determined if an expression for $c_p - c_v$ can be developed. Using $s = f_1(T, p)$ and $s = f_2(T, v)$, equate the two expressions for ds to develop an equation for dT.

(a) Equate this dT to the expression for dT from $T = f_3(v, p)$, and make appropriate substitutions to develop an equation for $c_p - c_v$.

(b) What is $c_p - c_v$ for a perfect gas?

(c) What is the magnitude of $c_p - c_v$ for water vapor at 300°C and 7000 kPa using the Redlich-Kwong equation in Example 13.5?

$$\text{Ans.: } (a) \; T\left(\frac{\partial v}{\partial T}\right)_p \left(\frac{\partial p}{\partial T}\right)_v$$

$$(c) \; 0.920 \text{ kJ/(kg} \cdot \text{K)}$$

13.7. Show that because of the required conditions at the critical point (expressed in Eq. (13.35)), the a and b constants in the van der Waals equation are as represented in Eq. (13.37).

13.8. Several properties of Refrigerant-12 (dichlorodifluoromethane, CCl_2F_2) are as follows: molecular mass of 120.93, a critical temperature of 112.0°C, and a critical pressure of 4113 kPa. Compute the specific volume of R-12 superheated vapor at 100°C and 2350 kPa using (a) the perfect gas equation, (b) the van der Waals equation, and (c) the Redlich-Kwong equation.

Ans.: property tables show $v = 0.0079888 \text{ m}^3/\text{kg}$.

13.9. Using as the starting point the entropy of saturated water vapor of 8.9767 kJ/(kg · K) at 1 kPa (6.98°C), compute the entropy of superheated water vapor at 400°C and 4500 kPa. Use the Redlich-Kwong equation and the specific heat data from Example 13.6.

Ans.: from tables,

$$\text{at } 400°C \text{ and } 1 \text{ kPa, } s = 10.6711 \text{ kJ/(kg} \cdot \text{K)}$$

$$\text{at } 400° \text{and } 4500 \text{ kPa, } s = 6.7093 \text{ kJ/(kg} \cdot \text{K)}.$$

13.10. Refrigerant-12 has a molecular mass of 120.93, a critical temperature of 112.0°C, and a critical pressure of 4113 kPa. An expression for the zero pressure c_p is $0.64496 + (0.1783 \times 10^{-3})t + (2.1384 \times 10^{-6})t^2$. Starting from saturated vapor at 0°C ($p = 308.61$ kPa) where the enthalpy is 351.477 kJ/kg, compute with the aid of the Redlich-Kwong equation the enthalpy of superheated vapor at 100°C and $p = 800$ kPa.

Ans.: from tables,

$$\text{at } 100°C \text{ and } 308.61 \text{ kPa, } h = 417.591 \text{ kJ/kg}$$
$$\text{at } 100°C \text{ and } 800 \text{ kPa, } h = 413.388 \text{ kJ/kg}.$$

REFERENCES

1. *Handbook of Fundamentals*, Chapter 1, American Society of Heating, Refrigerating, and Air-Conditioning Engineers, Atlanta, GA, 1985.
2. W. C. Reynolds, *Thermodynamic Properties in SI*, Stanford University, Palo Alto, CA, 1979.

3. G. J. Van Wylen and R. E. Sonntag, *Fundamentals of Classical Thermodynamics*, 2d ed., John Wiley, New York, 1978.
4. O. Redlich and J. N. S. Kwong, "On the Thermodynamics of Solutions, V—An Equation of State," *Chemical Review*, vol. 44, pp. 233–244, 1949.
5. G. Soave, "Equilibrium Constants from a Modified Redlich-Kwong Equation of State," *Chemical Engineering Science*, vol. 27, pp. 1197–1203, 1972.
6. E. Schmidt, ed., *Properties of Water and Steam in SI-Units*, Springer-Verlag, New York, 1969.

CHAPTER
14

STEADY-STATE SIMULATION OF LARGE SYSTEMS

14.1 DEVELOPMENTS IN SYSTEM SIMULATION

Chapter 6 defined system simulation as the prediction of the performance of a system when the following information is known: performance characteristics of components, properties of working substances, and conditions imposed by the surroundings. The value of simulating thermal systems lies in exploring the operating status at off-design conditions as a step in optimization and to detect potential operating problems such as excessively high or low pressures, temperatures, and flow rates. The mathematical description of the system simulation process is the simultaneous solution of a set of algebraic equations that may be nonlinear.

Chapter 6 provided several tools for simulating systems including successive substitution and the Newton-Raphson technique. Successive substitution is straightforward to program and does not demand large computer memory. The Newton-Raphson method is powerful and convenient, particularly when using generalized computer programs[1] that extract the partial derivatives numerically. For moderately sized problems (less than 50–100 equations) there is not much need for more sophisticated approaches. The

331

principal challenge lies in setting up the equations to properly represent the system. A major thrust of this chapter is to extend simulation to large problems where computer time and memory requirements are of concern. Since successive substitution is frugal in memory requirements, this method may take on additional importance, but Chapter 6 already pointed out several drawbacks of successive substitution in that sometimes convergence is slow, and in other calculations the sequence is divergent. Successive substitution has been a popular method for some of the early large simulation programs in the chemical and process industries, although it appears that Newton-Raphson and similar techniques are now gaining ground. This chapter will look deeper into the divergence problem, including ways to avoid the problem and correcting it if it does occur.

When moving to large problems, the computer time and memory requirements when using the Newton-Raphson method increase roughly as the square of the number of equations. This chapter presents refinements of the Newton-Raphson method that facilitate the solutions of large problems.

Another concept that is important in large systems is the modular approach, in which the complete system is considered to be a collection of subsystems, each having identifiable performance characteristics.

Two final topics covered in this chapter, applicable to both small and large systems, are (1) influence coefficients and (2) correction of component models when experimental or field data become available to verify the simulation.

14.2 CONVERGENCE AND DIVERGENCE IN SUCCESSIVE SUBSTITUTION

Section 6.8 emphasized that the calculating sequence dictated by the structure of the information-flow diagram determines whether the sequence will converge or diverge. In multi-equation simulations there may be a number of possible information-flow diagrams, and we can hope that some better approach can be found than blindly trying one after the other until a successful one is found. We shall point out that techniques are available to test a sequence without actually performing the successive substitution. Unfortunately, the test often requires more effort than the actual successive substitution. We shall, consequently, devote our effort to identifying qualitative indications of how best to structure the sequence.

A clue to the convergence criterion can be derived from the Gauss-Seidel method of solving a set of linear, simultaneous equations. The Gauss-Seidel method starts with trial values of all except one unknown, uses the first equation to solve for that unknown, the next equation for the next unknown, and so on. Each time a new variable is computed it is used in subsequent calculations. After one set of unknowns has been computed the calculation returns to the first variable, and the iterations proceed until convergence is achieved.

FIGURE 14-1
The Gauss-Seidel method as a successive substitution process.

Example 14.1. Using the Gauss-Seidel method, solve for the x's in the following set of simultaneous linear equations:

$$A: \quad 4x_1 - 3x_2 + x_3 = 12$$

$$B: \quad x_1 - 2x_2 + 2x_3 = 6$$

$$C: \quad 2x_1 + x_2 + 3x_3 = 6$$

Solution. The Gauss-Seidel method is a form of successive substitution, as shown by the information-flow diagram, Fig. 14-1. If trial values of $x_2 = 0$ and $x_3 = 0$ are chosen and the diagram of Fig. 14-1 is executed, the results are as shown in Table 14.1.

The Gauss-Seidel method in Example 14.1 converged to the correct result: $x_1 = 2$, $x_2 = -1$, and $x_3 = 1$. If the order in which the equations were solved changed, however, the calculation sequence may not have converged. If, for example, the A-C-B order were chosen such that the following equations were solved in sequence:

$$A: \quad x_1 = (12 + 3x_2 - x_3)/4$$

$$C: \quad x_2 = 6 - 2x_1 - 3x_3$$

$$B: \quad x_3 = (6 - x_1 + 2x_2)/2$$

the calculation, as shown in Table 14.2, diverges.

TABLE 14.1
Gauss-Seidel Solution of Example 14.1

Cycle	x_1	x_2	x_3
1	3.0	−1.5	0.5
2	1.75	−1.625	1.375
. . .	. . .	. . .	. . .
10	2.045	−1.021	0.977
∞	2	−1	1

TABLE 14.2
Example 14.1 with equations solved in the A-C-B sequence

Cycle	x_1	x_2	x_3
1	3	0	1.5
2	2.625	− 3.75	− 2.0625
3	0.703	10.78	13.43
4	7.29	−49.75	−50.61

For this particular set of equations, of the six possible sequences only the A-B-C sequence converges. One characteristic of the difference between the two sequences is provided by an examination of the coefficient matrices:

A-B-C, Convergent   A-C-B, Divergent

$$\begin{bmatrix} 4 & -3 & 1 \\ 1 & -2 & 2 \\ 2 & 1 & 3 \end{bmatrix} \qquad \begin{bmatrix} 4 & -3 & 1 \\ 2 & 1 & 3 \\ 1 & -2 & 2 \end{bmatrix}$$

In the convergent sequence the large-magnitude coefficients appear in the diagonal position of the matrix.

More precise means are available to test the coefficient matrix to determine whether the sequence that it represents is convergent or divergent. One technique[2] is to multiply the lower-triangular terms by the constant λ, set the determinant of the resulting matrix equal to zero, and solve for the values of λ. If all the λ's have an absolute magnitude less than unity, the solution is convergent. When applied to sequence A-B-C

$$\begin{vmatrix} 4\lambda & -3 & 1 \\ \lambda & -2\lambda & 2 \\ 2\lambda & \lambda & 3\lambda \end{vmatrix} = 0$$

$$\lambda = 0, 0.125 + 0.696i, 0.125 - 0.696i,$$

so A-B-C is convergent

but for sequence A-C-B

$$\begin{vmatrix} 4\lambda & -3 & 1 \\ 2\lambda & \lambda & 3 \\ \lambda & -2\lambda & 2\lambda \end{vmatrix} = 0$$

$$\lambda = 0, 0.2713, -4.146$$

so A-C-B is divergent

The translation of the Gauss-Seidel criterion to nonlinear equations will be approached through an example.

Example 14.2. The power required by a certain automobile is a function of its speed,

$$P = 4.2 + 0.45V + 0.0025V^{2.8} \qquad (14.1)$$

where P = power, kW

V = speed of the auto, m/s

The power delivered by a direct-drive engine at speeds above 12 m/s is

$$P = 60 + 8V - 0.16V^2 \qquad (14.2)$$

By means of successive substitution, determine the speed of which the auto is capable and the power delivered by the engine at that speed.

Solution. Two different information-flow diagrams are possible, as shown in Fig. 14-2. Flow diagram a, with a trial velocity of 50 m/s, yields a set of values abstracted in Table 14.3 and converges to $P = 112.39$ kW and $V = 42.25$ m/s. Solution of Eq. (14.1), which is a nonlinear equation, requires an iterative process.

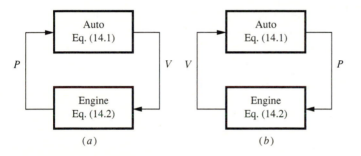

FIGURE 14-2
Information-flow diagrams for Example 14.2.

TABLE 14.3
Abstract of successive substitution calculations with flow diagram a in Example 14.2

Iteration	P	V
0		50.0000
1	60.0000	32.0188
2	151.9144	47.8273
3	76.6263	35.8487
. . . .
78	112.3907	42.2501
79	112.3893	42.2499
80	112.3905	42.2501

TABLE 14.4
Abstract of successive substitution
calculations with flow diagram *b* in
Example 14.2

Iteration	P	V
0		42.0000
1	110.8078	42.5343
2	114.2082	41.9174
. . . .
24	156.8580	29.4314
25	49.8492	51.2382
26	Square root of negative number in Eq. (14.2)	

Flow diagram *b* diverges, however, as Table 14.4 shows, even though the trial value of $V = 42.0$ was close to the solution.

The insight provided by the coefficient matrices in the Gauss-Seidel method will now be translated to the set of nonlinear equations. The elements that make up the coefficient matrix with the set of linear equations are comparable to the elements of a matrix of partial derivatives:

$$\begin{bmatrix} \dfrac{\partial f_1}{\partial x_1} & \cdots & \dfrac{\partial f_1}{\partial x_n} \\ \cdots & \cdots & \cdots \\ \dfrac{\partial f_n}{\partial x_1} & \cdots & \dfrac{\partial f_n}{\partial x_n} \end{bmatrix}$$

and we propose that the partial derivatives be used for the nonlinear equations. The arrangement is that if variable x_i is computed from f_j, the $\partial f_j / \partial x_i$ must be one of the diagonal elements. At the converged values of $V = 42.25$ and $P = 112.39$, the matrices for the two flow diagram are as shown in Fig. 14-3. For convergence we seek a coefficient matrix that has large magnitudes down the diagonal. Figure 14-3 shows that Flow Diagram a, which is convergent, best meets that requirement.

$$\begin{array}{cc} & V \quad P \\ \text{Eq. (14.1)} & \begin{bmatrix} 6.36 & -1 \\ -5.52 & -1 \end{bmatrix} \\ \text{Eq. (14.2)} & \end{array}$$

Flow diagram *a* (convergent)

$$\begin{array}{cc} & P \quad V \\ \text{Eq. (14.1)} & \begin{bmatrix} -1 & 6.36 \\ -1 & -5.52 \end{bmatrix} \\ \text{Eq. (14.2)} & \end{array}$$

Flow diagram *b* (divergent)

FIGURE 14-3
Matrices of partial derivatives in Example 14.2.

The test of the absolute values of λ described previously is valid for Example 14.2 and conceivably could be used for calculation loops with many equations. With large numbers of equations, however, the method becomes cumbersome and is quite impractical. Our approach will be to use the concept in a qualitative manner in trying to arrange the calculation sequence so that the diagonal elements are the large-magnitude partial derivatives. Another way of stating the same objective is to try to structure the sequence so that the variable computed through use of an equation changes only a small amount for large changes in the other variables in that equation.

It will not always be possible to choose a calculating sequence such that the largest-magnitude partial derivatives appear in the diagonal position for all equations. Choosing the variable to be solved by one equation may force into the diagonal position of another equation a partial derivative that is not the largest for that equation. A crude rule is to resolve the conflict by placing in the diagonal position the partial derivative that has the greatest ratio in comparison to the other derivatives of its equation.

14.3 PARTIAL SUBSTITUTION IN SUCCESSIVE SUBSTITUTION

Example 14.2 illustrated two undesirable characteristics of successive substitution. Flow diagram *a* produced slow convergence, while flow diagram *b* produced divergence. Partial substitution may be effective in correcting both of these problems. The equation that defines the partial substitution process[3] to be applied to each of the variables is

$$x_{j,\,i+1} = \beta \, x_{j,\,i+1*} + (1 - \beta) \, x_{j,i} \qquad (14.3)$$

where
β = partial substitution factor
x_j = variable being computed
i = subscript indicating previous value
$i + 1$ = subscript indicating new value
$i + 1*$ = subscript indicating new value computed directly from the equation

Choosing $\beta = 1$ gives the unaltered successive substitution process and for a set of linear equations defines the Gauss-Seidel method. Reducing β shortens the steps during the iterations and in general drives the calculation toward a more convergent process. No general statement can be made about the influence of β on the speed of convergence. Figures 14-4*a* and 14-4*b* show the effect of β on the number of iterations needed for convergence of flow diagram *a* and flow diagram *b*, respectively, of Example 14.2.

FIGURE 14-4
Number of iterations for convergence (defined as the fraction of change of both variables less than 0.0000001) for (*a*) flow diagram *a* with trial $V = 50$, and (*b*) flow diagram *b* with trial $V = 48$ in Example 14.2.

Table 14.3 showed that when the value of β was unity that the changes in V and P during an iteration were too large, resulting in oscillations around the ultimate solution. When β is set at 0.55, convergence to the criterion listed in the caption of Fig. 14-4 occurred in 8 iterations. Increasing β beyond 1.06 results in divergence. Flow diagram b with $\beta = 1$ is divergent, but, as Fig. 14-4b shows, choosing β less than 0.95 permits the simulation to converge.

14.4 EVALUATION OF THE NEWTON-RAPHSON TECHNIQUE WHEN IMPLEMENTED AS A GENERALIZED PROGRAM

Two methods for solving a set of simultaneous, nonlinear algebraic equations (the mathematical description of system simulation) that are presented in this book are successive substitution and Newton-Raphson. Some relative advantages of these two methods are as follows:

Successive substitution. Straightforward to program for the computer, and sparing in the use of computer memory.

Newton-Raphson. More reliable and more rapidly convergent. Also, it is not necessary to list the equations in any special order.

The above comparison probably translates into the following approach. If a person forsees the need of simulating a small system with no likelihood of simulating other systems, successive substitution is probably the best choice. The flow diagram should be constructed following the recommendations listed in Sec. 14.2. Should the sequence diverge or converge too slowly, partial substitution as described in Sec. 14.3 can be applied. If the worker anticipates a frequent need of performing simulations, gaining access and developing a familiarity with a computer library routine for solving simultaneous nonlinear equations is recommended. Some library routines have extensive capabilities and can accommodate a broad range of assignments, but this characteristic may also be a disadvantage. The additional and nonused capabilities may only complicate and add to the expense of using the program when applied to the single purpose of system simulation.

An alternate approach is to write one's own generalized program which offers the possibility of building in such conveniences as input requirements matching the system simulation assignment, providing intermediate printouts following each iteration, and building in diagnostics for the types of problems anticipated in simulating thermal systems. The generalized system simulation program cited in Reference 1 is less than 200 lines of FORTRAN code, and is reproduced in Appendix II. Certain operations are standard in the Newton-Raphson solution such as the computation of the f-functions that are to be driven to zero, the computation of partial derivatives of all functions with respect to all variables, and the solution of a set of simultaneous linear equations to compute the corrections to each of the variables. The user has the responsibility, and indeed will demand the right, to describe the system by specifying the equations and providing trial values of the variables.

The possible structure of the generalized system simulation program listed in Appendix II is shown in Fig. 14-5. The user must prepare the MAIN program and the subroutine containing the equations, called EQNS. In the MAIN program the user provides the required information indicated in Fig. 14-5 and finally calls the simulation routine, here named SIMUL. SIMUL in turn calls EQNS, PARDIF, and GAUSSY. Subroutine EQNS computes the values of the f-functions, PARDIF numerically extracts partial derivatives, and GAUSSY solves for the changes to be made in all the variables. After the values of the variables are updated, the convergence test determines whether to terminate or to return to EQNS for another cycle.

FIGURE 14-5
Structure of a generalized program for steady-state simulation of systems.

A generalized program saves considerable user time by extracting the partial derivatives numerically, rather than requiring mathematical expressions for the derivatives. The partial derivative of function f_i with respect to variable x_j, for example, is

$$\frac{\partial f_i}{\partial x_j} =$$

$$\frac{f_i(x_1, x_2, \ldots, x_j + \Delta, \ldots, x_n) - f_i(x_1, x_2, \ldots, x_j, \ldots, x_n)}{\Delta}$$

(14.4)

Subroutine PARDIF makes a call of each equation in EQNS n times, so $(n \times n)$ equations are calculated.

A generalized program such as this is useful for frequent or even occasional uses. The program can be executed either on a mainframe computer or a microcomputer, and the execution time and memory requirements are not severe for simulations of systems represented by, for example, 50 equations. If the system to be simulated is of this size or smaller, it is questionable whether effort in developing a program of greater speed or sophistication can be justified.

14.5 SOME CHARACTERISTICS OF THE NEWTON-RAPHSON TECHNIQUE

In comparison to successive substitution, the Newton-Raphson technique has several advantages. Newton-Raphson is not subject to divergence that results from an improper choice of the information-flow diagram or calculation sequence. In fact, the sequence of equations listed in Newton-Raphson is immaterial, and the probability of convergence is high. The Newton-Raphson technique can diverge, however, if the trial values are too far from the final solution. If the iterations drive the values of the variables too far from the physical solution, the computer program may blow up, or it may also be possible that the calculation converges to a nonphysical solution. This nonphysical solution may satisfy all the equations, but one or more of them may be in a range beyond that intended for their use.

One of the characteristics of Newton-Raphson is that the variables may change a large amount on the first iteration. It is also possible that the variables following the first iteration are farther from the solution than the trial values, as illustrated for one equation in Fig. 14-6. Often after the big jump in the first iteration the process then converges to the correct solution. On some other occasions the variables are thrown so far from realistic values that the program eventually fails. It is possible to damp the first iteration

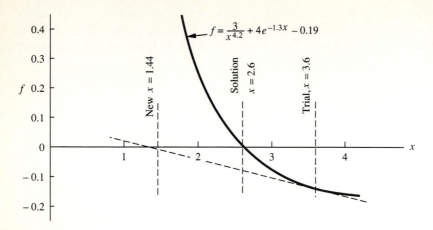

$$f = \frac{3}{x^{4.2}} + 4e^{-1.3x} - 0.19$$

New $x = 1.44$

Solution $x = 2.6$

Trial, $x = 3.6$

FIGURE 14-6
First Newton-Raphson iteration may move the values of the variables further from the solution than the trial values.

by changing the variables only a fraction of that indicated by pure Newton-Raphson.

Since one or more heat exchangers are likely components of any thermal system, it is appropriate to point out several oddities that may occur.

Example 14.3. Water enters the heat exchanger in Fig. 14-7 at 50°C and is heated by steam that is at a constant temperature of 120°C. The UA value of the heat exchanger remains constant at 2.13 kW/K. What flow rate of water, w, is required in order to transfer 100 kW in the heat exchanger?

Solution. The combination of the rate equation and energy balance yields one equation,

$$f = 1 - e^{-0.508/w} - 0.341/w = 0 \qquad (14.5)$$

The intent of this example is to show some consequences of the choice of trial value of w for a Newton-Raphson solution. To illustrate, the behavior

Steam 120°C

Water   50°C

w kg/s $c = 4.19$ kJ/kg•K

$q = 100$ kW

$UA = 2.13$ kW/K

Condensate 120°C

FIGURE 14-7
Water heater in Example 14.3.

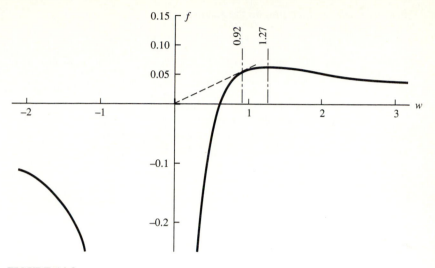

FIGURE 14-8
Behavior of f in Eq. (14.5) over a range of values of the water flow rate w.

of f in Eq. (14.5) is shown in Fig. 14-8, which is information not normally available when performing a simulation. The solution is: $w = 0.6$ kg/s.

The Newton-Raphson technique converges to the solution if the trial value w_t is

$$0 < w_t < 0.92$$

If $0.92 < w_t < 1.27$, the first iteration jumps to a negative value of w and from there drives to $-\infty$. Trial values of w greater than 1.27 drive w to $+\infty$.

An example will illustrate several other pitfalls that in rare instances may arise when simulating systems that include heat exchangers.

Example 14.4. A counterflow heat exchanger shown in Fig. 14-9 has a UA value of 6.5 kW/K and uses a stream of water to cool oil from 80°C to a temperature regulated by a proportional controller set at 60°C. The temperature of the entering water is 35°C and the controller regulates the

FIGURE 14-9
Oil cooler simulated in Example 14.4.

flow of water w_w according to the equation,

$$w_{w'} \text{ kg/s} = 0.3(t_o - 60)$$

where t_o = outlet oil temperature, °C. Simulate this system to determine t_o and w_w.

Solution. The solution is $t_o = 62.70°C$ and $w_w = 0.8095$ kg/s under which conditions t_w, the outlet water temperature, is 69.21°C. The purpose of this example, however, is to show several experiences that occur with different equation formulations and trial values. First use the equations:

$$f_1 = 0.3(t_o - 60) - w_w \tag{14.5}$$

The logarithmic-mean-temperature difference (LMTD) may be computed by the equation,

$$\text{LMTD} = [(80 - t_w) - (t_o - 35)]/\ln[(80 - t_w)/(t_o - 35)]$$

$$f_2 = 6.5(\text{LMTD}) - (3.2)(2.1)(80 - t_o) \tag{14.6}$$

$$f_3 = (3.2)(2.1)(80 - t_o) - w_w(4.19)(t_w - 35) \tag{14.7}$$

A simulation attempt with the trial values shown in Table 14.5 fails. One approach to circumventing the problem of the negative logarithm is to express Eq. (14.6) in terms of an exponent of e which has no limitation on its argument. The execution may be able to pass temporarily through the physically impossible situation that terminated the execution in Table 14.5. Revise Eq. (14.6) by taking the antilog,

$$f_2 = \frac{(80 - t_w)}{(t_o - 35)} - \text{EXP}\left\{\frac{[6.5(80 - t_w) - (t_o - 35)]}{[2.1(3.2)(80 - t_o)]}\right\} \tag{14.8}$$

and use in combination with Eqs. (14.5) and (14.7). As Table 14.6 shows, the simulation is successful.

Lest it be concluded that the exponential form of the rate equation is a panacea, choose trial values as shown in Table 14.7.

The form converges, but to incorrect values. The simulation has driven the water flow rate to a value such that $wc_{\text{water}} = wc_{\text{oil}}$, with the consequence

TABLE 14.5
Simulation of Example 14.4 using log-mean-temperature difference in the rate equation

	w_w	t_o	t_w
Trial values	5.000	70.00	48.00
Iteration 1	−0.667	57.78	56.82
Iteration 2	1.807	66.02	82.26
	Execution terminated, negative log		

TABLE 14.6
Successful simulation of Example 14.4 using
the exponential form of the rate equation

	w_w	t_o	t_w
Trial values	5.000	70.00	48.00
Iteration 1	3.507	71.69	41.50
Iteration 2	−1.000	56.67	53.98
Iteration 3	4.563	75.20	132.87
Iteration 4	2.958	69.86	72.85
. . . .
Iteration 8	0.8095	62.70	69.21

TABLE 14.7
Unsuccessful attempt at simulating
Example 14.4 using the exponential form
of the rate equation

	w_w	t_o	t_w
Trial values	5.000	61.00	51.00
Iteration 1	1.184	63.95	53.31
Iteration 2	1.577	65.25	49.18
. . . .
Iteration 5	1.604	65.35	49.60

that the rise in water temperature equals the drop in oil temperature (this situation was first explored in Sec. 5.4). The mean temperature difference is $80 - 49.6 = 30.4°C = 65.4 - 35$. Thus the rate of heat transfer = $(2.1)(3.2)(80 - 65.35) = 98.45$ kW. The required UA value is $98.45/30.4 = 3.24$ kW/K which does not agree with the specified value of 6.5 kW/K.

The source of the difficulty lies in Eq. (14.8), where the numerator of the argument of the exponent drives to zero, so that the control exerted by the UA value in the numerator is wiped out.

Incidentally, the use of the logarithmic form of the rate equation is successful with the trial values of Table 14.7.

A third form of the rate equation is available—that of Eq. (5.6), which, along with Eq. (14.5), provides two independent equations for w_w and t_o. Unfortunately, with either set of trial values used so far, (5 kg/s, 61°C) and (5 kg/s, 70°C) the simulation converges to impossible values of $w_w = 3.24$ kg/s and $t_o = 49.14°C$.

Example 14.4 showed some difficulties that can arise when simulating systems that contain heat exchangers, and most thermal systems incorporate heat exchangers. To place the situation in perspective, it should be pointed out that the trial values chosen in Example 14.4 were rather far from the correct solution. When the trial values are close to the final solution, any of the formulations should work. The occurrence of any of the problems cited in Example 14.4 is rare, but can be troublesome, so it is beneficial to be alert to this possibility.

14.6   ACCELERATING THE SOLUTIONS OF LINEAR EQUATIONS

Because the Newton-Raphson technique possesses many strengths, it can serve as the basis for further improvements. The targets for these refinements are the reduction of execution time and memory requirements when simulating large systems. The first means of addressing these targets that will be described in this section is to improve the procedure for solving the set of linear simultaneous equations. A faster method for extracting partial derivatives is explained in Sec. 14.7, and Sec. 14.8 treats a change in the basic technique—the modified Newton method.

Improvements should be possible in the standard Gaussian elimination routine for solving the set of linear simultaneous equations, since most of the elements in the coefficient matrix are zero. One specific example of this class of sparse matrix techniques is a routine called XGAUSS[4] along with an associated routine NZERO, both of which are listed in Appendix III.

Sparse matrix techniques[5] often consist of storing and handling elements outside of the matrix to which the elements belong. The row and column numbers and the value of each nonzero element are identified in the subroutine NZERO. Three arrays, A( ), IROW( ), and JCOL(,) are set up by a call of NZERO for each nonzero element in the coefficient matrix. Suppose that there are 6 nonzero elements in the set of equations,

$$x_1 + 2x_2 = 7$$
$$2x_1 - x_3 = 8 \qquad\qquad (14.9)$$
$$3x_2 + x_3 = 1$$

In matrix form the equations are:

$$\begin{bmatrix} 1 & 2 & 0 \\ 2 & 0 & -1 \\ 0 & 3 & 1 \end{bmatrix} \begin{bmatrix} x_1 \\ x_2 \\ x_3 \end{bmatrix} = \begin{bmatrix} 7 \\ 8 \\ 1 \end{bmatrix} \qquad\qquad (14.10)$$

or

$$AX = B \qquad\qquad (14.11)$$

The first step is to proceed through the matrix, row by row, numbering each nonzero element in sequence, as in Fig. 14-10. Call the encircled number the "designator." The A array expresses the values of the elements in the sequence of the designator, $A(1) = 1$, $A(2) = 2$, $A(3) = 2$,

1 ① $\quad 2$ ② $\quad 0$

2 ③ $\quad 0 \quad -1$ ④

$0 \quad\quad 3$ ⑤ $\quad 1$ ⑥

FIGURE 14-10
Numbering the nonzero elements in the matrix.

A(4) $= -1$, A(5) $= 3$, and A(6) $= 1$. Instead of the A array being two-dimensional as it is in conventional Gaussian elimination, it has now become a one-dimensional array of nonzero values. The IROW array provides the designator of the first nonzero element in each row. Thus, IROW(1) $= 1$, IROW(2) $= 3$, and IROW(3) $= 5$.

There are two JCOL arrays: JCOL(1,) and JCOL(2,). The JCOL(1,designator) indicates the column number of each nonzero element. Thus JCOL(1,1) $= 1$, JCOL(1,2) $= 2$, JCOL(1,3) $= 1$, JCOL(1,4) $= 3$, JCOL(1,5) $= 2$, and JCOL(1,6) $= 3$. The JCOL(2,) array specifies the designator of the next element in a row after the first nonzero element. In the first row, for example, the first element is designated 1 and the next element in that row is designated 2, so JCOL(2,1) $= 2$. Designator 2 is the last one in that row, so there is no element after 2 and JCOL(2,2) $= 0$. Continuing, JCOL(2,3) $= 4$, JCOL(2,4) $= 0$, JCOL(2,5) $= 6$, and JCOL(2,6) $= 0$.

After the subroutine NZERO has established the A, IROW, and JCOL arrays, these are fed to the XGAUSS subroutine which is a modified version of the standard Gaussian elimination. The structure of XGAUSS is as follows:

I. Formation of the upper triangular matrix.
   A. The outer loop operates on each row, designated K, from 1 to the total number of equations N. For each K the rows, designated I, are processed from K to N.
   Form the upper triangular matrix by first determining which row, called IMAX, contains this largest coefficient. (If all coefficients in this column are zero, the equations are dependent). Then exchange the K row terms with those in IMAX.
   B. Produce zeros in the K column below the diagonal term by subtracting from all terms in the I equation the quantity $A(I,K)/A(K,N)$, where I runs from $K + 1$ to N.

II. Back substitution.
   Back substitution is performed to solve for the X values.

The operations performed by XGAUSS on the set of equations (14.10) are summarized in Table 14.8. The basic Gaussian elimination processes are identifiable, but most operations involving a zero are omitted.

TABLE 14.8
Operations performed by XGAUSS on the set of equations, (14.10)

Revision	Matrix	Array	n 1	2	3	4	5	6	7
Original	1^1 2^2 $\;0\;$ 7	IROW(n)	1	3	5				
	2^3 0 -1^4 8	JCOL(1,n)	1	2	1	3	2	3	0
	0 $\;3^5\;$ 1^6 1	JCOL(2,n)	2	0	4	0	6	0	0
		A(n)	1	2	2	−1	3	1	0
		B(n)	7	8	1				
Switch	2^3 0 -1^4 8	IROW(n)	3	1	5				
rows 1	1^1 2^2 $\;0\;$ 7	JCOL(1,n)	1	2	1	3	2	3	0
and 2	0 $\;3^5\;$ 1^6 1	JCOL(2,n)	2	0	4	0	6	0	0
		A(n)	1	2	2	−1	3	1	0
		B(n)	8	7	1				
Add	2^3 0 -1^4 8	IROW(n)	3	2	5				
$-(1/2)$(Row 1)	0 2^2 $1/2^7$ 3	JCOL(1, n)	0	2	1	3	2	3	3
to Row 2	0 $\;3^5\;$ 1^6 1	JCOL(2, n)	0	7	4	0	6	0	0
		A(n)	0	2	2	−1	3	1	1/2
		B(n)	8	3	1				
Switch	2^3 0 -1^4 8	IROW(n)	3	5	2				
Rows 2 and 3	0 $\;3^5\;$ 1^6 1	JCOL(1, n)	0	2	1	3	2	3	3
	0 2^2 $1/2^7$ 3	JCOL(2, n)	0	7	4	0	6	0	0
		A(n)	0	2	2	−1	3	1	1/2
		B(n)	8	1	3				
Add	2^3 0 -1^4 8	IROW(n)	3	5	7				
$-(2/3)$(Row 2)	0 $\;3^5\;$ 1^6 $\;1\;$	JCOL(1, n)	0	0	1	3	2	3	3
to Row 3	0 0 $-1/6^7$ $7/3$	JCOL(2, n)	0	0	4	0	6	0	0
		A(n)	0	0	2	−1	3	1	−1/6
		B(n)	8	1	7/3				

By back substitution, $x(3) = -14$, $x(2) = 5$, and $x(1) = -3$.

Some elements that were originally nonzero may reduce to zero, and some that were originally zero may take on nonzero values. Revision 2 in the third line of Table 14.8 reduces A(1) to zero at which value it remains. Revision 2 also introduced a new nonzero element designated 7.

The economies provided by XGAUSS, while not obvious in the small example shown in Table 14.8, occur in the following ways:

1. In searching down a column for the largest element, only the nonzero values indicated by JCOL(1,_) are checked.
2. When rows are interchanged, only the nonzero elements are switched.
3. In the triangulation process of producing zeros in the column below the diagonal term, computations are performed on the nonzero elements.
4. In the back substitution process only the nonzero terms in a row are considered.

A test of XGAUSS was conducted on a 295×295 matrix[4] in which most equations contained only 2 or 3 nonzero variables. The memory requirements on a mainframe computer to solve this set of simultaneous linear equations using conventional Gaussian elimination was 360,000 bytes, and the execution time was in excess of six minutes. Using XGAUSS on the same problem, the requirements were 27,000 bytes and 3.5 seconds, respectively.

14.7   FAST PARTIAL DERIVATIVE ROUTINE

A time-consuming process in the generalized Newton-Raphson simulation is the computation of the partial derivatives, which are computed numerically using Eq. (14.4). Each equation is computed once for each variable, thus $n \times n$ equations. Most of the equations do not contain a given variable, so the result of most of the computations is zero.

An approach that saves computer time is to execute the following steps: (1) compute the complete set of partial derivatives in the usual fashion, (2) identify the nonzero partial derivatives, and (3) when computing partial derivatives for succeeding iterations, compute only those equations associated with nonzero partial derivatives. Applying the computed-go-to command is one approach to executing step 3. Application of a fast partial derivative routine on an 81-equation simulation[6] resulted in a 15% increase in compilation time, but in a 40% decrease in execution time.

14.8   QUASI-NEWTON METHOD

In the Newton-Raphson technique the corrections of the variables, the Δx's, are found by solution of the set of linear equations whose coefficient matrix J is composed of the partial derivatives, as in Eq. (6.31). In the quasi-Newton technique with the Broyden update[7]

$$X = -H \, F \tag{14.11}$$

where $X =$ the column vector of the Δx's that are additive to correct the variables. X is thus the negative of the Δx's used in Sec. 6.11.

$H =$ the inverse of the partial derivative matrix $= J^{-1}$

$F =$ the column vector of the values of the functions f_1, f_2, etc.

Furthermore, H is updated for each new iteration not by recomputation of the partial derivatives and inverting the matrix, but by an operation[7] expressed symbolically as

$$H_{k+1} = H_k + \frac{(X_k - H_k \, Y_k) X_k^T H_k}{X_k^T H_k Y_k} \tag{14.12}$$

where $Y_k = F_{k+1} - F_k$

subscript k indicates current values
subscript $k + 1$ indicates values for next iteration
T indicates the transpose.

The simulation of a simple system will illustrate the quasi-Newton method and provide a comparison with the conventional Newton-Raphson technique. The system consists of two components—a fan and a duct, as shown in Fig. 14-11a, with pressure-flow characteristics shown in Fig. 14-11b. The component equations are

duct $$f_1 = 0.0625 + 0.653Q^{1.8} - P$$

fan $$f_2 = 0.3 - 0.2\,Q^2 - P$$

Table 14.9 shows the Newton-Raphson solution including the matrix of partial derivatives and the inverse of that matrix at each iteration. The solution is $P = 0.25$ kPa and $Q = 0.5$ m^3/s.

The quasi-Newton method starts with trial values and a temporary inverse H. For this illustration we use the same trial values of $P = 0.1$ and $Q = 1.0$, compute the partial derivative matrix, and invert it to obtain the initial inverse. With the above assumptions,

$$F_k = \begin{bmatrix} \text{duct equation} \\ \text{fan equation} \end{bmatrix} = \begin{bmatrix} 0.6155 \\ 0 \end{bmatrix}$$

and

$$X_k = -\begin{bmatrix} -0.25392 & -0.74608 \\ 0.63449 & -0.63449 \end{bmatrix}\begin{bmatrix} 0.6155 \\ 0 \end{bmatrix} = \begin{bmatrix} 0.15629 \\ -0.39053 \end{bmatrix}$$

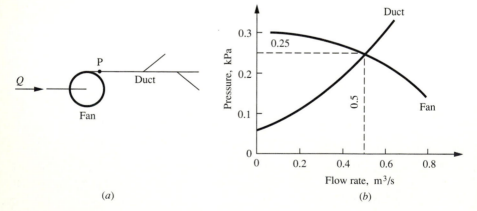

(a)   (b)

FIGURE 14-11
(a) A fan-duct system, and (b) the pressure-flow characteristics.

TABLE 14.9
Newton-Raphson simulation of fan-duct system

Iteration	Variables	Matrix of Partial Derivatives		Inverse of Matrix	
At trial values	$P = 0.1$ $Q = 1.0$	-1.0 -1.0	1.1759 -0.4000	-0.2539 0.6345	-0.7461 -0.6345
1	$P = 0.256$ $Q = 0.608$	-1.0 -1.0	0.7902 -0.2435	-0.2356 0.9674	-0.7644 -0.9674
2	$P = 0.250$ $Q = 0.508$	-1.0 -1.0	0.6836 -0.2032	-0.2291 1.1276	-0.7709 -1.1276
3	$P = 0.250$ $Q = 0.500$	-1.0 -1.0	0.6748 -0.1999	-0.2285 1.1432	-0.7715 -1.1432
4	$P = 0.250$ $Q = 0.500$	-1.0 -1.0			

so

$$V_{k+1} = \begin{bmatrix} 0.1 \\ 1.0 \end{bmatrix} + \begin{bmatrix} 0.15629 \\ -0.39053 \end{bmatrix} = \begin{bmatrix} 0.25629 \\ 0.60947 \end{bmatrix}$$

where V_{k+1} = new values of the variables.

At these values of the variables the new function matrix is

$$F_{k+1} = \begin{bmatrix} 0.07402 \\ -0.03058 \end{bmatrix} \text{ and } Y_k = \begin{bmatrix} 0.07402 \\ -0.03058 \end{bmatrix} - \begin{bmatrix} 0.6155 \\ 0 \end{bmatrix} = \begin{bmatrix} -0.54148 \\ -0.03058 \end{bmatrix}$$

Applying Eq. 14.12 yields the new inverse

$$H_{k+1} = \begin{bmatrix} -0.24630 & -0.74956 \\ 0.76030 & -0.69190 \end{bmatrix}$$

The complete set of quasi-Newton calculations for this example is shown in Table 14.10.

TABLE 14.10
Quasi-Newton simulation of the fan-duct system

Iteration	Variables, V	Functions, F	Inverse, H	
1	0.25629 0.60947	0.074024 -0.030579	-0.24630 0.76030	-0.74956 -0.69190
2	0.25160 0.53203	0.020603 -0.008211	-0.23200 1.04286	-0.76370 -0.97139
3	0.25011 0.50257	0.001656 -0.000624	-0.22850 1.13130	-0.76713 -1.05802
4	0.25001 0.50004	0.000042 -0.000016	-0.22744 1.16053	-0.76816 -1.08636

The quasi-Newton algorithm expressed in Eq. (14.12) is a rather clever procedure that uses the experience of a previous inverse with the results of its corrections on the variables and functions to arrive at an improved inverse. Comparison of the inverses computed in Table 14.10 with that of the Newton-Raphson in Table 14.9 shows that the quasi-Newton technique drives toward the correct inverse and the correct solution. The example might have showed favoritism toward the quasi-Newton method, because the first inverse was provided by what is essentially a Newton-Raphson concept. The fact is that quasi-Newton has some tolerance to errors in the initial inverse, and an approximate inverse is often adequate to set the process in motion.

What advantages, if any, does the quasi-Newton method offer in comparison to Newton-Raphson for the large simulations which are the focus of this chapter? The extraction of the partial derivatives and the solution of a set of linear, simultaneous equations that are integral operations in Newton-Raphson are eliminated. In their places the quasi-Newton method substitutes a number of matrix additions, subtractions, and multiplications. The potential benefit of quasi-Newton in large systems appears not so much in reducing computation time as in allowing greater tolerance in selecting the trial values of the variables. This advantage could be important in large systems where selecting trial values close to the solutions may be difficult. Henley and Rosen[8] concluded that the Newton-Raphson method is rapid when convergence does occur. The quasi-Newton technique is reasonably rapid and may require many iterations if the trial values are not good, but will converge to a solution in some situations with trial values that do not permit Newton-Raphson to converge.

14.9   INFLUENCE COEFFICIENTS

Influence coefficients seek to answer the following question: "What happens if . . . ?" For example, "What happens to the capacity of a system if a heat exchanger is enlarged by 5 percent?" Influence coefficients have been used in some form for many years. For example, a traditional rule of thumb applicable to refrigeration systems is shown in Table 14.11. The table indicates the effect on the refrigeration capacity of a 10 percent increase in capacity of one component at a time while the capacities of the other two components remain fixed. The dominating influence is exerted by the compressor. If the capacities of all the components are increased by 10 percent the system is simply 10 percent larger, and so is the system capacity.

Influence coefficients provide insight about the system as part of simulation, but may also be a partial derivative that provides data for optimization. Influence coefficients, therefore, build a bridge between system simulation and optimization. The direct approach to finding an influence coefficient is to rerun a simulation program with appropriate changes

TABLE 14.11
Effect of 10% increases in component capacities on refrigeration capacity

Component whose capacity is increased 10%	Percent increase in refrigeration capacity
compressor	6
evaporator	2^+
condenser	2^-
all	10

in the equation(s) representing the performance of the component(s) being examined. The value of the variable of interest in the rerun is then compared to the value in the base run.

When numerous influence coefficients are to be computed, it is probably more efficient to apply the principles outlined below, which use the matrix of partial derivatives developed in the final iteration of the Newton-Raphson solution. Most influence coefficients fall into one of two forms which require somewhat different treatment. Form 1 asks for the change in a certain variable of the simulation occurring with a given change in a constant that appears in one or more equations. In Form 2 the change in a variable of the simulation is sought when the capacity of a certain component is altered by a given fraction. The influence coefficients applicable to these two forms will be treated individually in the next two sections.

14.10 INFLUENCE COEFFICIENT WITH RESPECT TO A PARAMETER—FORM 1

An example of Form 1 is the computation for a steam power plant of the change in power P when the area of the condenser A changes. The influence coefficient sought is $\partial P/\partial A$ which has units of kW/m^2.

The Newton-Raphson computation consists of the repeated application of the matrix equation,

$$J \ \Delta = -F$$

where

$$J = \begin{bmatrix} \dfrac{\partial f_1}{\partial x_1} & \cdots & \dfrac{\partial f_1}{\partial x_n} \\ \vdots & & \vdots \\ \dfrac{\partial f_n}{\partial x_1} & \cdots & \dfrac{\partial f_n}{\partial x_n} \end{bmatrix}, \ \Delta = \begin{bmatrix} x_{1,new} - x_{1,old} \\ \vdots \\ x_{n,new} - x_{n,old} \end{bmatrix}, \text{ and } f = \begin{bmatrix} f_1 \\ \vdots \\ f_n \end{bmatrix}$$

At the solution, $F = 0$, $\Delta = 0$, and J is known. An approach applicable to both Form 1 and Form 2 influence coefficients is to make a slight change in a function f_i following convergence of the Newton-Raphson simulation, and to call the change δ_i.

$$J \ \Delta = - \begin{bmatrix} 0 \\ 0 \\ \cdot \\ \delta_i \\ \cdot \\ 0 \\ 0 \end{bmatrix} \leftarrow i\text{th row} \qquad (14.13)$$

The change in a particular variable, Δx_j, can be found through the application of Cramer's rule:

$$\Delta x_j = x_{j,\text{new}} - x_{j,\text{old}} = - \frac{\det \begin{bmatrix} \frac{\partial y_1}{\partial x_1} & & \\ & 0 & \\ & 0 & \\ & \delta_i & \\ & 0 & \\ & 0 & \frac{\partial y_n}{\partial x_n} \end{bmatrix}}{\det J}$$

jth column ——→

$\leftarrow i$th row

Expanding the determinant in the numerator about the jth column yields

$$\Delta x_j = - \frac{\delta_i \, J_{ij}}{\det J} \qquad (14.14)$$

where J_{ij} is the cofactor of the $i - j$ term in J.

Equation (14.14) can be translated into an influence coefficient of Form 1 with respect to a parameter. A parameter is a constant in an equation, for example, c in function i, designated c_i. The first step is to express δ_i in terms of a small change in c_i, Δc_i.

$$\delta_i = \frac{\partial f_i}{\partial c_i} \Delta c_i \qquad (14.15)$$

Substituting δ_i from Eq. (14.15) into Eq. (14.14) yields

$$\Delta x_j = - \frac{\partial f_i}{\partial c_i} \frac{J_{ij}}{\det J} \Delta c_i$$

The influence coefficient, $\partial x_j / \partial c_i$, is the limit as Δc_i approaches zero, or

$$\frac{\partial x_j}{\partial c_i} = \frac{\Delta x_j}{\Delta c_i} \Big|_{\Delta c_i \to 0} = - \frac{\partial f_i}{\partial c_i} \frac{J_{ij}}{\det J} \qquad (14.16)$$

Example 14.5. Air passes in series through the two heat exchangers shown in Fig. 14-12. The air flow rate is 0.8 kg/s, the c_p of the air is 1.0 kJ/kg · K, and its entering temperature is $-5°C$. The steam coil receives saturated steam at 230°C, and the enthalpy of evaporation at 230°C is 1812 kJ/kg. The condensate leaves the steam coil at 230°C, which is a high enough temperature to preheat the air in a condensate coil that is assumed to operate as a counterflow heat exchanger. The U-values and areas of the coils are:

Coil	U, kW/m² · K	Area, m²	UA, kW/K
Condensate	0.04	12.5	0.5
Steam	0.055	22	1.21

(a) Simulate this system to determine the steam flow rate w, and the air temperatures leaving the condensate coil t_1 and leaving the steam coil t_2.
(b) If there is the possibility of increasing the area of either coil by a given amount, which coil should be increased in size to provide the greater increase in t_2?

Solution. The three equations that describe the system are as follows:

Energy balance on the steam coil,

$$f_1 = (0.8)(1.0)(t_2 - t_1) - w(1812)$$

Rate equation, steam coil,

$$f_2 = -(t_2 - t_1) + (230 - t_1)[1 - e^{-(0.055)(A_s)/0.8}]$$

Rate equation, condensate coil, from Eq. (5.6),

$$f_3 = -[t_1 - (-5)] + [230 - (-5)] \left[\frac{1 - e^{0.04A_c(1/0.8 - 1/4.19w)}}{\frac{0.8}{4.19w} - e^{0.4A_c(1/0.8 - 1/4.19w)}} \right]$$

where $A_s = 22$ m² and $A_c = 12.5$ m²

FIGURE 14-12
Series of heat exchangers in Example 14.5 for heating air.

Solving by Newton-Raphson yields $w = 0.0603$ kg/s, $t_1 = 54.92°C$, and $t_2 = 191.4°C$. Calling these variables x_1, x_2, and x_3, respectively, the matrix of partial derivatives from the final Newton-Raphson simulation is

$$\begin{bmatrix} -1812 & -0.8 & 0.8 \\ 0 & 0.220 & -1 \\ 572.6 & -1 & 0 \end{bmatrix}$$

whose determinant is 2169.

Two derivatives needed in Eq. (14.16) are

$$\partial f_2 / \partial A_s = 2.652 \text{ and } \partial f_3 / \partial A_c = 1.681$$

Substitution into Eq. (14.16) yields

$$\frac{\partial t_2}{\partial A_s} = -2.652 \ \frac{\det \begin{bmatrix} -1812 & -0.8 \\ 572.6 & -1 \end{bmatrix}}{2169} = 2.776 \ °C/m^2$$

$$\frac{\partial t_2}{\partial A_c} = -1.681 \ \frac{\det \begin{bmatrix} -1812 & -0.8 \\ 0 & 0.22 \end{bmatrix}}{2169} = 0.309 \ °C/m^2$$

So a given amount of heat-transfer area added to the steam coil is ten times as effective as if added to the condensate coil.

14.11 INFLUENCE COEFFICIENT WITH RESPECT TO A VARIABLE OF THE SIMULATION—FORM 2

The previous section explored influence coefficients with respect to a parameter that was a constant appearing in one or more equations. Changing the magnitude of a parameter normally changes the performance of a component, but another way in which the change in performance must sometimes be expressed is with respect to one of the variables of the simulation. Suppose, for example, that a pump is one of the components in the system having a pressure-rise/flow rate curve as illustrated in Fig. 14-13. What is meant by a 1 percent increase in capacity of the pump? Does it mean a 1 percent increase in pressure rise at a given flow rate, or does it mean a 1 percent increase in flow rate at a given pressure rise? Either definition is valid, but whichever is chosen must be specified. The pressure rise and flow rate are doubtless variables of the simulation in contrast to Form 1 where the parameter was a constant in one of the equations. The term $p_{k,i}$ will be defined as the fraction increase in x_k in equation i while other variables and parameters remain constant. If the pump performance in Fig. 14-13 is given by the f_9 equation, x_5 representing the pressure rise, and x_{11} the flow rate, $p_{11,9} = 0.005$ indicates a 0.5 percent increase in flow rate at constant values of the other variable(s) in the equation. If $p_{k,i}$ can be related to δ_i

FIGURE 14-13
Two interpretations of a 1 percent
increase in pump capacity.

in Eq. (14.13), the matrix of partial derivatives from the Newton-Raphson simulation will once again provide the additional data for calculating the influence coefficients. What, then, is the expression for δ_i if the function f_i is revised so that variable x_k increases by the fraction p? To revise f_i so that x_k increases by the fraction p, divide x_k wherever it appears by $(1+p)$. An alternate procedure is to multiply x_k by $(1-p)$ wherever it appears. The results are essentially the same for small values of p, since from the geometric series

$$\frac{1}{1+p} = 1 - p \overset{\longrightarrow \text{ neglect}}{\boxed{+p^2 - p^3 \cdots}}$$

A positive p indicates that x_k increases by a fraction p. Multiplying x_k at all appearances by $(1-p)$ distorts the original f_i equation by δ_i,

$$f_{i,\text{new}} - f_{i,\text{orig}} = \delta_i$$

Express f_i as a series of p,

$$f_{i,\text{new}} = f_i\Big|_{p=0} + \frac{\partial f_i}{\partial p}\Big|_{p=0}(p) \overset{\longrightarrow \text{ neglect}}{\boxed{+ \frac{1}{2}\frac{\partial^2 f_i}{\partial p^2}\Big|_{p=0}(p^2)}}$$

$$\delta_i = \frac{\partial f_i}{\partial p}\Big|_{p=0} \quad p = p\left[\frac{\partial f_i}{\partial[x_k(1-p)]}\frac{\partial[x_k(1-p)]}{\partial p}\right]_{p=0}$$

$$\frac{\partial[x_k(1-p)]}{\partial p} = -x_k$$

and, since all terms in the following differentiation disappear upon differentiation and setting p to 0,

$$\frac{\partial f_i}{\partial[x_k(1-p)]}\Big|_{p=0} = \frac{\partial f_i}{\partial x_k}$$

Then

$$\delta_i = -p\, x_k\, \frac{\partial f_i}{\partial x_k}$$

Substituting into Eq. (14.14) yields

$$\Delta x_j = p\, x_k\, \frac{\partial f_i}{\partial x_k}\, \frac{J_{ij}}{\det J} \tag{14.17}$$

Example 14.6. In the fan-duct system shown in Fig. 14-11 the fan capacity is increased by 1 percent (1 percent greater flow at a given static pressure). What is the increase in system flow?

Solution. The equations are

duct
$$f_1 = 0.0625 + 0.653\, x_2^{1.8} - x_1$$

fan
$$f_2 = 0.3 - 0.2\, x_2^2 - x_1$$

where x_1 = the pressure P
$\quad\ \ x_2$ = the flow rate Q

At the final Newton-Raphson iteration

$$x_1 = 0.25$$

$$x_2 = 0.5$$

$$J = \begin{bmatrix} -1 & 0.675 \\ -1 & -0.2 \end{bmatrix}$$

The component in which the capacity increase occurs is the fan, so $i = 2$. The nature of the capacity increase is that Q increases while other variables remain constant, so $k = 2$. Finally, the effect on the system flow is sought, so $j = 2$. Applying Eq. (14.17),

$$\Delta x_2 = p x_2\, \frac{\partial f_2}{\partial x_2}\, \frac{J_{22}}{\det J}$$

$$\Delta x_2 = (0.01)(0.5)(-0.2)\, \frac{-1}{0.875} = 0.00114 \text{ m}^3/\text{s}$$

So a 1 percent increase in fan capacity results in a 0.22 percent increase in flow rate when the fan is a part of a larger system.

14.12   CONTINUING DEVELOPMENTS IN SIMULATING LARGE SYSTEMS

One of the directions taken in recent years by developers of computer programs for simulating large systems is the modular approach. In this concept numerous subroutines are provided which the user can call in the application to a given system. A property routine would be provided, for example, that relates pressure, temperature, and enthalpy. The user could call this subroutine, providing two of the properties in order to calculate the third. The user selects from the large assortment of subroutines the component performance and property functions needed, organizing the variables in an appropriate manner for the system being simulated.

This chapter has explored some techniques that are useful, and sometimes even necessary, when simulating large systems. These topics might be considered an extension of techniques of solution of the set of simultaneous nonlinear equations and can be very useful. It should be emphasized, however, that describing the system (setting up the equations) may continue to be the significant challenge in simulating large systems.

PROBLEMS

14.1. The system consisting of the two air-heating coils in Example 14.5 is to be simulated using successive substitution. The three equations are to be solved for the three unknowns t_1, t_2, and w.

(a) Construct all the possible information-flow diagrams for a successive substitution solution of the equations (there are three different diagrams).

(b) Using the partial derivatives from Example 14.5 and the test on the matrix of partial derivatives described in Sec. 14.2, determine which of the information-flow diagrams of part (a) are convergent and which are divergent.

(c) Solve by successive substitution the flow diagram(s) that are shown to be convergent.

14.2. One of the methods by which some of the water from the blowdown from a boiler can be recovered is to throttle it and condense the steam that flashes into vapor, as shown in Fig. 14-14. The system is to be simulated, primarily to compute the flow rate recovered, w_c.

FIGURE 14-14
Recovering a portion of water blown down from a boiler.

The equations for h_f and h_g as functions of the temperature t, which is in °C, are:

$$h_f, \text{ kJ/kg} = 4.19t \quad \text{and} \quad h_g, \text{ kJ/kg} = 2502 + 1.815t$$

The product of the overall heat-transfer coefficient U and its related area A is a function of the loading of the condenser and expressed by the equation:

$$UA = 335/(w_c)^{0.2}$$

Three equations that describe this system in terms of the unknowns t, w_c, and UA are as follows:

Heat transfer rate and heat balance on the condenser,

$$f_1 = w_c(2502 + 1.815t - 4.19t) - (70)(4.19)(t - 25)[1 - e^{-UA/(70)(4.19)}]$$

Expression for UA,

$$f_2 = UA - \frac{335}{w_c^{0.2}}$$

Heat balance about flash tank,

$$f_3 = w_c(2502 + 1.815t) + (2.8 - w_c)(4.19t) - (2.8)(1185)$$

The system is to be simulated by successive substitution.
(a) Construct the three different flow diagrams that are candidates for specifying the calculation sequence.
(b) Near the solution that is $t = 40°C$, $w_c = 1.2$ kg/s, and $UA = 320$ kW/K, the matrix of partial derivatives is

$$
\begin{array}{c}
f_1 \\
\\
f_2 \\
\\
f_3
\end{array}
\begin{array}{ccc}
w_c & t & UA \\
\left[\begin{array}{ccc}
2407 & -197.7 & -5.03 \\
53.85 & 0 & 1 \\
2407 & 8.88 & 0
\end{array}\right]
\end{array}
$$

Use the test on the matrix of partial derivatives described in Sec. 14.2 to determine which of the information-flow diagrams in part (a) are convergent and which are divergent.

14.3. The pressure p, temperature t, and the flow rate w of ammonia are to be determined by successive substitution for the compressor-condenser combination in Fig. 14-15a using the information-flow diagram of Fig. 14-15b. These are applicable equations:

Compressor: $\qquad w = 0.92 - 0.00234(p/50)^{1.4}$

Saturation relation: $\quad \ln(p) = 16.008 - 2711/(t + 273.15)$

Condenser: $\qquad \ln[(t - 30)/(t - 30 - 16.7w)] = 0.512$

The sequence in Fig. 14-15b is divergent when using direct successive substitution. (a) Determine the value of β in partial substitution that results

FIGURE 14-15
(a) Ammonia compressor-condenser (b) information-flow diagram in Problem 14.3.

in borderline stability, and (b) determine the value of β that provides the most rapid convergence.

14.4. A counterflow heat exchanger having a UA value of 12.3 kW/K and shown in Fig. 14-16 is part of a system being simulated. Within the simulation the heat exchanger is described by two equations:

$$f_n = \frac{(t_i - t_2)}{(t_o - t_1)} - \text{EXP}\left\{ \frac{12.3 \ [(t_i - t_2) - (t_o - t_1)]}{[w_1(3.9)(t_i - t_o)]} \right\}$$

$$f_{n+1} = w_2(3.6)(t_2 - t_1) - w_1(3.9)(t_i - t_o)$$

Depending upon the trial values chosen for the simulation, one of the solutions shown in Table 14.12 results.
(a) Do f_n and $f_{n+1} = 0$ for both solutions?
(b) Which of the solutions is valid?

FIGURE 14-16
Heat exchanger in Problem 14.4.

TABLE 14.12
Two different solutions to the heat exchanger simulation in Prob. 14.4

Solution	w_1	w_2	t_i	t_o	t_1	t_2
1	2.95	2.52	38.60	29.83	20.39	31.51
2	2.63	2.85	37.15	26.51	19.86	30.50

14.5. The following linear equations are to be solved by the sparce-matrix routine XGAUSS (Sec. 14.6):

$$2x_1 - x_3 = -3$$

$$3x_1 + x_2 = 0$$

$$2x_2 - 2x_3 = 4$$

Prepare a table in the style of Table 14.8 showing the designators and values of the arrays for IROW(n), JCOL(1,n), JCOL(2,n), A(n), and B(n).
   Ans.: Just prior to back substitution,

				n	1	2	3	4	5	6	7
2①	0	-1②	-3	IROW(n)	1	4	6				
0	1④	3/2⑦	9/2	JCOL(1,n)	1	3	0	2	0	3	3
0	0	-5⑥	-5	JCOL(2,n)	2	0	0	7	0	0	0
				A(n)	2	-1	0	1	0	-5	3/2
				B(n)	-3	9/2	5				

14.6. Equations (14.1) and (14.2) are to be used for simulating the system and solving for S and P using the quasi-Newton method with the Broyden update, as described in Sec. 14.8. With functions arranged as

$$f_1 = 4.2 + 0.45S + 0.0025S^{2.8} - P$$

$$f_2 = 60 + 8S - 0.16S^2 - P$$

and trial values of speed $S = V(1) = 50$, and power $P = V(2) = 70$, $f_1 = 99.61$ and $f_2 = -10$. Also, the inverse of the partial derivative matrix H is

$$H_1 = \begin{bmatrix} 0.06078 & -0.06078 \\ -0.4862 & -0.5138 \end{bmatrix}$$

Using the Broyden update, what is H_2?
   Ans.: applying the updated inverse yields additive corrections of -0.797 for S and -0.644 for P.

14.7. A heat pump, shown schematically in Fig. 14-17, delivers energy from a low-temperature fluid to a high-temperature one. The condenser has a UA value of 13.31 kW/K, and the fluid being heated enters at 30°C with a product of mass flow rate and specific heat of 19.2 kW/K. The evaporator has a UA value of 7.621 kW/K and the fluid being cooled enters at −5°C with a product of mass flow rate and specific heat of 10 kW/K. The pumping capacity of the compressor is such that the equation for the heat-transfer rate at the evaporator as a function of the condensing temperature t_c and the

FIGURE 14-17
Heat pump in Prob. 14.7.

evaporating temperature t_e is

$$q_e = 204 + 2.4t_e + 0.02t_e t_c - 1.5t_c$$

The ratio of heat rejected at the condenser to that absorbed at the evaporator is 1.2. The system may be be simulated as a four-variable problem of q_e, q_c, t_e, and t_c and described by the following equations:

$$f_1 = (-5 - t_e)(1 - e^{-UA/wc}) - q_e/wc$$

$$f_2 = (t_c - 30)(1 - e^{-UA/wc}) - q_c/wc$$

$$f_3 = 204 + 2.4t_e + 0.02t_e t_c - 1.5t_c - q_e$$

$$f_4 = q_c - 1.2q_e$$

which yield the solution of $t_c = 40°C$, $t_e = -20°C$, $q_e = 80$ kW, and $q_c = 96$ kW.

Using Eq. (14.17) to compute the influence coefficient, determine the percent increase in q_e if the compressor capacity is increased 1.0 percent. The interpretation of the increase in compressor capacity is that q_e increases 1.0 percent for constant values of t_e and t_c.

Ans.: from a simulation, the new $q_e = 80.4334$.

14.8. A steam power plant, shown schematically in Fig. 14-18, consists of a boiler, turbine, condenser, and feed-water pump. The designation of variables and the component characteristics are shown in the figure. The following equations describe the system and are sufficient to determine the values of the unknowns, t_b, t_c, w, q_b, q_c, and q_t.

Boiler heat-transfer rate

$$f_1 = (90.0)(1.1)(650 - t_b)(1 - e^{-190/90/1.1}) - q_b$$

Condenser heat-transfer rate

$$f_2 = (716)(4.19)(t_c - 20)(1 - e^{-963/716/4.19}) - q_c$$

Flow rate through the turbine

$$f_3 = -0.89 + 0.04t_b - 0.12t_c - w$$

FIGURE 14-18
Steam power plant in Prob. 14.8.

Power delivered by turbine

$$f_4 = (280)(w)\left(\frac{t_b + 273.1}{t_c + 273.1} - 0.5\right) - q_t$$

Enthalpy change

Enthalpy change through the boiler is 1900 kJ/kg, so

$$f_5 = 1900 w - q_b$$

Energy balance

$$f_6 = q_t + q_c - q_b$$

The solution is $t_b = 407.7°C$, $t_c = 38.69°C$, $w = 10.7743$ kg/s, $q_b = 20,471$ kW, $q_c = 15,393$ kW, and $q_t = 5078.6$ kW.

The turbine is enlarged 1.0 percent in the sense that the flow rate w in f_3 is increased 1.0 percent for given values of t_b and t_c. Use Eq. (14.17) to compute the influence coefficients and determine the new w and the new q_t.

Ans.: From a simulation new $w = 10.8242$ kg/s, and the new $q_t = 5089.02$ kW.

14.9. A helium liquefier operating on the flow diagram shown in Fig. 14-19 receives high pressure helium vapor, liquefies a fraction of the vapor, and returns the remainder to be recycled. The conditions of helium vapor entering the liquefier at point 1 are as follows: $p = 1,000$ kPa, $w = 4.6$ g/s, $T = 18$ K, and $h = 101.8$ kJ/kg. The separator and the entire low-pressure side operate at 100 kPa. Some helium properties are

At 100 kPa saturation temperature $= 4.2$ K
    enthalpy of saturated liquid $= 9.74$ kJ/kg
    enthalpy of saturated vapor $= 30.31$ kJ/kg
  enthalpy of superheated vapor, kJ/kg $= 2.81 + 6.79 T - 0.0578 T^2$

At 1,000 kPa
  enthalpy of superheated vapor, kJ/kg $= -29.67 + 8.79 T - 0.0826 T^2$

The UA value of both heat exchangers is 70 W/K.

FIGURE 14-19
Helium liquefier in Prob. 14.9.

Turbine characteristics: $\qquad T_6 = 4.6 + 0.1 T_3$

and $\qquad\qquad\qquad w, \text{g/s} = 3.75 - 0.125\, T_3$

Simulate this system and compute the following quantities:

$w_3 = w_6$, g/s; $w_4 = w_5 = w_7$, g/s; w_8 g/s;

w_9, g/s; $w_{10} = w_{11}$, g/s; $T_2 = T_3 = T_4$ K;

T_5 K; T_6 K; T_{10} K; and T_{11} K.

Ans.: $w_3 = 2.00$ g/s, $w_8 = 0.782$ g/s, $T_{11} = 11.28$ K.

14.10. Figure 14-20 shows a diagram of a central chilled water plant that consists of a primary loop and two secondary loops, each with its own secondary pump.
Pump characteristics:

$$A\!-\!B\ \Delta p, \text{kPa} = 300 + 2.464Q - 0.04648Q^2$$

$$C\!-\!D\ \Delta p, \text{kPa} = 180 + 3.294Q - 0.1163Q^2$$

$$F\!-\!G\ \Delta p, \text{kPa} = 180 + 2.107Q - 0.0131Q^2$$

where $Q =$ flow rate in m^3/hr.

FIGURE 14-20
Chilled-water distribution system in Prob. 14.10.

The pressure drop in each of the piping or coil sections is given by the equation: Δp, kPa $= C Q^2$.

The values of C for the various sections are:

Section	C
B–C	0.017
D–E	0.0447
E–C	0.546
E–I	0.089
C–F	0.0359
G–H	0.0302
H–I	0.0089
I–A	0.0199
H–F	0.162

Simulate this system, specifying the pressures in kPa absolute at each lettered point and the flow rates in m³/hr in each of the pipe sections.

Ans.: $p_D = 401.8$, and $Q_{EI} = 27.84$ m³/kg

14.11. The purpose of the multiple-flash evaporator shown in Fig. 14-21 is to concentrate a solution of NaOH that enters with a 20 percent NaOH concentration by mass. The entering temperature of the solution is 20°C, and steam is supplied to the coil of the first stage at a temperature of 115°C. The performance characteristics of several of the components are

UA of heat exchangers in both stages I and II $= 65$ kW/K
Condenser capacity, m_2 kg/s $= 0.03(t_{s2} - 25)$

Equations for the enthalpy of saturated liquid and saturated water vapor are

$$h_f \text{ kJ/kg} = 4.184\, t$$

$$h_v \text{ kJ/kg} = 2524 + 1.68\, t$$

where $t =$ water temperature in °C.

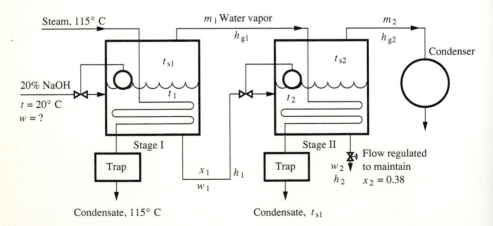

FIGURE 14-21
Multiple-flash evaporator in Prob. 14.11.

The enthalpy of the NaOH solution has the characteristic curves shown in Fig. 14-22. An approximate equation representing these curves is

$$h \text{ kJ/kg} = 111.3 + 77.49x - 2556x^2 + 9592x^3 - 6579x^4$$
$$+ (4.249 - 7.932x + 38.97x^2 - 85.15x^3 + 62.38x^4)(t - 26.6)$$

where x = NaOH mass concentration expressed as a fraction.

The relationship of saturation temperature to solution temperature is shown by the Duhring lines in Fig. 14-23 and expressed by the equation

$$t = (1 + 0.05x)t_s + 58.3x$$

where t = solution temperature
t_s = saturation temperature of water, both in °C.

Simulate this system, specifying the following:

solution flow rates w, w_1, and w_2, kg/s
water vapor flow rates m_1 and m_2, kg/s
saturation temperatures t_{s1} and t_{s2}, °C
solution temperatures t_1 and t_2, °C
solution concentration x_1

FIGURE 14-22
Enthalpy of NaOH solutions.

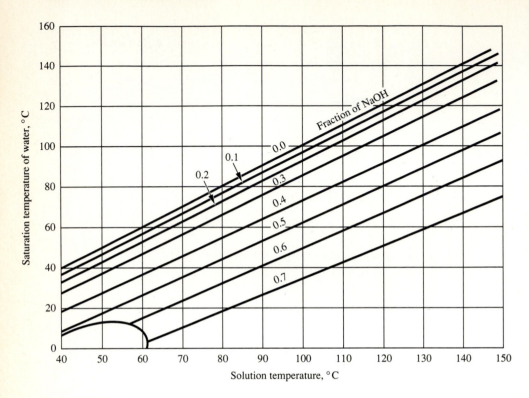

FIGURE 14-23
Relationship of saturation temperature to solution temperature for NaOH solutions.

solution enthalpies h_1 and h_2, kJ/kg
vapor enthalpies h_{g1} and h_{g2}, kJ/kg

Ans.: $w = 1.783$ kg/s.

REFERENCES

1. W. F. Stoecker, "A Generalized Program for Steady-State System Simulation," *ASHRAE Transactions*, vol. 77, Part I, pp. 140–148, 1971.
2. B. Carnahan, H. A. Luther, and J. O. Wilkes, *Applied Numerical Methods*, John Wiley, New York, 1969.
3. R. W. Hornbeck, *Numerical Methods*, Quantum Publishers, Inc., New York, 1975.
4. J. J. Anselmino, "Computer Program to Simulate Central Chilled Water Systems," Master of Science Thesis, University of Illinois at Urbana-Champaign, 1977.
5. R. P. Tewarson, *Sparse Matrices*, Academic Press, New York, 1973.
6. W. F. Stoecker, "Computer Simulation of the Performance of an Aqua-Ammonia Absorption Refrigeration System," Paper B2.30, International Congress of Refrigeration, (International Institute of Refrigeration), Moscow, 1975.
7. C. G. Broyden, "A Class of Methods for Solving Nonlinear Simultaneous Equations," *Mathematics of Computation*, vol. 19, pp 577–593, 1965.
8. E. J. Henley and E. M. Rosen, *Material and Energy Balance Computations*, John Wiley, New York, 1969.

CHAPTER
15

DYNAMIC
BEHAVIOR
OF
THERMAL
SYSTEMS

15.1   IN WHAT SITUATIONS IS
DYNAMIC ANALYSIS IMPORTANT?

The previous chapter concentrated on steady-state simulation; this one focuses on dynamic behavior where there are changes with respect to time. Both steady-state and dynamic simulations are important, but for different reasons. Steady-state simulations are needed and performed more frequently than dynamic simulations; in the majority of cases a steady-state simulation can be justified in the design stage of a plant in order to explore part-load efficiency and potential operating problems. Dynamic simulations, on the other hand, address transient problems that could possibly cause shutdown, damage to the plant, or at least imprecise control. Problems related to dynamic behavior of the system may be infrequent, but could be critical when they occur.

All too often in engineering practice it is assumed that there will be no transient problems and a dynamic analysis is omitted. If a transient problem occurs, it is corrected in the field—perhaps by an adjustment of a controller in a direction indicated by a dynamic analysis. In some other cases modifications must be made to the system, in which case the correction is expensive. Several typical situations calling for dynamic analyses are the investigations of (1) the effects of disturbances,[1] (2) startup transients,[2] and (3) stability of the control system.[3]

15.2   SCOPE AND APPROACH OF THIS CHAPTER

Many topics in this chapter are related to automatic control, and it is clear that any of the excellent textbooks in this subject cannot be condensed into this one chapter. It is assumed, therefore, that the reader has already studied automatic control, and the subjects that are standard to that field will be reviewed only briefly. What, then, is the intent of this chapter? Three directions are (1) the concentration on thermal components, (2) the emphasis of behavior in the time domain, and (3) the translation of physical situations into symbolic or mathematical representation.

The underlying objective of this chapter is to apply the principles of dynamic behavior to physical hardware. Many of the texts on automatic controls unfortunately choose few examples from the thermal field—selecting instead mechanical or electrical systems. One objective, then, is to become more comfortable in making dynamic analyses of thermo-fluid components and systems.

Secondly, the field of automatic controls can be a highly mathematical one, with the challenge being the development of elegant and sometimes lengthy equations to represent a process or system. While the field requires some workers gifted in mathematical skills, there is a far greater need of interpreting a somewhat lower level of mathematics into the physical situation. One of the techniques that helps provide this physical interpretation is, as Shinskey contends,[4] representation of the performance in the time domain. This effort contrasts with the practice of many automatic control specialists who work primarily in the transformed domain.

The third possible contribution of this chapter is to provide more experience in translating the physical situation into a control block diagram. We see engineers completing courses in automatic control highly proficient in analyzing a system that is already represented as a block diagram with transfer functions neatly assigned to the boxes. Sometimes the greater challenge is in arriving at that block diagram in the first place—namely, translating physical equipment and processes into the symbolic form. This chapter offers some practice in this exercise.

As a bridge from the steady-state simulation of Chapters 6 and 14 to the dynamic work of the bulk of this chapter, the next section describes a hybrid situation where the major part of the simulation is adequately represented as a steady state one, but one element introduces changes with respect to time.

15.3 ONE DYNAMIC ELEMENT IN A STEADY-STATE SIMULATION

A modest extension from steady-state to dynamic simulation is appropriate for some systems, where, for example, the thermal capacity of all components except one can be neglected. An example will illustrate the approach.[5] The refrigeration system shown in Fig. 15-1 is to maintain a low temperature in the chamber. The evaporator that is located in the chamber removes the heat from the space that is conducted in through the walls and also that heat removed from the thermal capacity of the contents of the chamber.

First consider the case where the conditions have stabilized throughout the system such that a steady-state simulation is appropriate. The controlling equations are

Compressor refrigeration capacity: $q_e = f_1(T_e, T_c)$ (15.1)

Compressor power: $P = f_2(T_e, T_c)$ (15.2)

Condenser heat transfer: $q_c = (W_A)(c_{pa})(T_c - T_{amb})$
$$\{1 - e^{UA_c/[(W_A)(c_{pa})]}\}   (15.3)$$

Evaporator heat transfer: $q_e = (T_s - T_e)(UA_e)$ (15.4)

FIGURE 15-1
System with one dynamic element—refrigeration plant serving a cold room.

Energy balance: $\qquad q_{c} = P + Q_{e}$ (15.5)

Heat transfer to chamber: $q_{e} = q_{tr} = UA_{ch}(T_{amb} - T_{s})$ (15.6)

The set of six algebraic equations is adequate to compute the six operating variables of the system:

$$T_{c} = \text{condensing temperature, }°C$$

$$T_{e} = \text{evaporating temperature, }°C$$

$$T_{s} = \text{space temperature, }°C$$

$$P = \text{power, kW}$$

$$Q_{c} = \text{heat-transfer rate at the condenser, kW}$$

$$Q_{e} = \text{heat-transfer rate at the evaporator, kW}$$

when the functions f_{1} and f_{2} are known and the following parameters are specified:

c_{pa}, specific heat of air, kJ/kg · K
T_{amb}, temperature of ambient air, °C
UA_{c}, product of U value and area of condenser, kW/K
UA_{e}, product of U value and area of evaporator, kW/K
w_{a}, flow rate of air to condenser, kg/s

Suppose that instead of the steady state we were interested in transient performance, for example, in the pull-down of temperature of the refrigerated space from a warm condition. Furthermore, the objects in the space have a thermal capacity that is a summation of products of the masses and the specific heats of the individual contents of the space. Equation (15.6) must now be revised, because q_{e} is no longer equal to q_{tr}, the rate of heat transfer through the walls. The revised Eq. (15.6), called Eq. (15.6a) is

$$q_{tr} = UA_{ch}(T_{amb} - T_{s}) \qquad (15.6a)$$

Also assume that the heat transfer between the space air temperature T_{s} and the objects is good and that the internal conduction is high so that the temperature of these objects is the same as T_{s}. The energy balance within the space provides the final equation—a differential equation.

$$q_{tr} = q_{e} + M_{c}\left(\frac{dT_{s}}{dt}\right) \qquad (15.7)$$

where M_{c} = product of mass and specific heat of contents

From the standpoint of the simulation a difficulty has arisen in that two new variables have been introduced, q_{tr} and (dT_{s}/dt), but only one new equation. To resolve the difficulty, remove T_s as a variable of the steady-state simulation. The initial value of T_s must be known just as is usually required in the solution of differential equations. If $T_s = 30°C$ initially, the set of 7 equations can be solved for the 7 variables, including (dT_s/dt). Suppose that (dT_s/dt) comes out -0.00024 °C/s. If a time step of 1000 s is arbitrarily chosen, the value of T_s after 1000 s would be $30 + (1000 \text{ s})(-0.00024 \text{ °C/s}) = 29.76$ °C. This value of T_s would then be used for the next steady-state simulation.

The foregoing approach of alternating between a steady-state and a dynamic simulation applies to cases where the response of the steady-state processes and components is rapid in comparison to the element(s) treated dynamically. In the above example the new steady-state values of T_c, T_e, P, q_c, and q_e were assumed to respond very quickly to a change in T_s which is the slowly changing variable.

15.4 LAPLACE TRANSFORMS

Because of their special relationship to differential equations, Laplace transforms are a powerful tool in predicting the behavior of dynamic processes and of control systems. For control systems there is the added utility of predicting stability or instability of the control loop by characterization of the loop as a Laplace transform without inverting to the time domain. It is assumed that the reader has already studied Laplace transforms as well as control systems, so this section and the next two should be considered refreshers which can be bypassed by anyone with the subject at their fingertips.

The definition of the Laplace transform of a function of time $F(t)$ is

$$\mathcal{L}\{F(t)\} = \int_0^\infty F(t)e^{-st}dt = f(s) \qquad (15.8)$$

Example 15.1. What is the Laplace transform of the constant c?

Solution

$$\mathcal{L}\{c\} = \int_0^\infty ce^{-st}dt = -\frac{c}{s}e^{-st}\Big|_0^\infty = \frac{c}{s}$$

Example 15.2. What is the Laplace transform of bt?

Solution

$$\mathcal{L}\{bt\} = \int_0^\infty bte^{-st}dt = -b\frac{d}{ds}\int_0^\infty e^{-st}dt = -b\frac{d(1/s)}{ds} = b/s^2$$

The Laplace transform of e^{at} is a building block in developing the transforms of hyperbolic functions.

$$\mathcal{L}\{e^{at}\} = 1/(s - a)$$

Therefore the Laplace transform of $\sinh(at)$ is

$$\mathcal{L}\{\sinh(at)\} = \mathcal{L}\left\{\frac{e^{at} - e^{-at}}{2}\right\}$$

$$= \frac{1}{2}\left(\frac{1}{s - a} - \frac{1}{s + a}\right) = \frac{a}{s^2 - a^2}$$

To extend the hyperbolic relations to trigonometric functions, make use of the identities

$$e^{ix} = \cos x + i \sin x$$

and

$$e^{-ix} = \cos x - i \sin x$$

where $i = \sqrt{-1}$

so

$$\cos x = \frac{e^{ix} + e^{-ix}}{2} = \cosh(ix)$$

and

$$\sin x = \frac{e^{ix} - e^{-ix}}{2i} = \frac{\sinh(ix)}{i}$$

Then

$$\mathcal{L}\{\sin(at)\} = \mathcal{L}\left\{\frac{e^{iat} - e^{-iat}}{2i}\right\}$$

$$= \frac{1}{2i}\left(\frac{1}{s - ia} - \frac{1}{s + ia}\right)$$

$$\mathcal{L}\{\sin(at)\} = \frac{a}{s^2 + a^2}$$

Table 15.1 is a compilation of transforms of some frequently encountered functions of time.

The expression for the transform of a derivative $F'(t)$ can be found by substituting $F'(t)$ as the function in the definition of the Laplace transform, Eq. (15.8).

$$\mathcal{L}\{F'(t)\} = \int_0^\infty F'(t)e^{-st}\, dt$$

Denote $F'(t)dt$ as dx and e^{-st} as y, and integrate by parts:

$$\mathcal{L}\{F'(t)\} = e^{-st}F(t)\Big]_0^\infty - \int_0^\infty F(t)(-s)e^{-st}\, dt$$

TABLE 15.1
Table of Laplace transforms

$f(s)$	$F(t)$	$f(s)$	$F(t)$
$\dfrac{1}{s}$	1	$\dfrac{s}{s^2 - a^2}$	$\cosh(at)$
$\dfrac{1}{s^2}$	t	$\dfrac{s}{s^2 + a^2}$	$\cos(at)$
$\dfrac{1}{s^n}$ $n = 1, 2, 3 ..$	$\dfrac{t^{n-1}}{(n-1)!}$	$\dfrac{b}{(s+a)^2 + b^2}$	$e^{-at} \sin(bt)$
$\dfrac{1}{\sqrt{s}}$	$\dfrac{1}{\sqrt{\pi t}}$		
$\dfrac{1}{s-a}$	e^{at}	$\dfrac{s+a}{(s+a)^2 + b^2}$	$e^{-at} \cos(bt)$
$\dfrac{1}{(s-a)^2}$	$t e^{at}$	$\dfrac{1}{s(s^2 + a^2)}$	$\dfrac{1}{a^2}(1 - \cos at)$
$\dfrac{1}{(s-a)^n}$ $n = 1, 2, \ldots$	$\dfrac{t^{n-1} e^{at}}{(n-1)!}$	$\dfrac{1}{s^2(s^2 + a^2)}$	$\dfrac{1}{a^3}(at - \sin at)$
$\dfrac{1}{(s-a)(s-b)}$	$\dfrac{1}{a-b}\left(e^{at} - e^{bt}\right)$	$\dfrac{1}{(s^2 + a^2)^2}$	$\dfrac{1}{2a^3}(\sin at - at \cos at)$
$\dfrac{1}{(s-a)(s-b)(s-c)}$	$\dfrac{[(b-c)e^{at} + (c-a)e^{bt} - (a-b)e^{ct}]}{(a-b)(b-c)(c-a)}$	$\dfrac{s}{(s^2 + a^2)^2}$	$\dfrac{t}{2a} \sin(at)$
$\dfrac{s}{(s-a)(s-b)}$	$\dfrac{1}{(a-b)}\left(ae^{at} - be^{bt}\right)$	$\dfrac{1}{(s+a)(s^2 + b^2)}$	$\dfrac{1}{a^2 + b^2}\Big[e^{-at}$
$\dfrac{1}{s^2 + a^2}$	$\dfrac{1}{a} \sin(at)$		$+ \dfrac{\sqrt{a^2 + b^2}}{b} \sin(bt - \theta)\Big]$
$\dfrac{1}{s^2 - a^2}$	$\dfrac{1}{a} \sinh(at)$		where $\Theta = \tan^{-1}\left(\dfrac{b}{a}\right)$

so $$\mathcal{L}\{F'(t)\} = sf(s) - F(0) \tag{15.9}$$

For the transform of the second derivative, simply cascade the results of the first derivative,

$$\mathcal{L}\{F''(t)\} = s\mathcal{L}\{F'(t)\} - F'(0)$$
$$= s^2 f(s) - sF(0) - F'(0) \tag{15.10}$$

15.5 INVERSION OF LAPLACE TRANSFORMS

The understanding of a dynamic process or dynamic behavior of a system will be enhanced by determining the response in the time domain. Relationships based on differential equations can be handled more easily in the s domain, but a final step that provides physical insight is the inversion of the transform. The symbol for inversion is $\mathcal{L}^{-1}\{\}$ and the inversion process is a transformation in the opposite direction from the original transform. Thus,

$$\mathcal{L}^{-1}\{f(s)\} = F(t) \tag{15.11}$$

If the transform appears on a table, such as Table 15.1, the inverse is immediately available. For transforms not found on the table, the objective is to decompose complicated transforms into ones that can be found from the table. A key step is to break the denominator of the transform into factors,

$$\frac{c_0 s + c_1}{(s + a)(s + b)} = \frac{A}{s + a} + \frac{B}{s + b}$$

After A and B are determined, the inverses of the simple terms are surely available from tables.

Example 15.3. Invert $s/(s^2 - 3s + 2)$.

Solution

$$\mathcal{L}^{-1}\left\{\frac{s}{s^2 - 3s + 2}\right\} = \mathcal{L}^{-1}\left\{\frac{A}{s - 2} + \frac{B}{s - 1}\right\}$$

To find A and B, equate the numerators.

$$s = A(s - 1) + B(s - 2)$$

which supplies two equations,

constants: $\qquad\qquad 0 = -A - 2B$

and, s: $\qquad\qquad 1 = A + B$

Then $A = 2$, and $B = -1$.

$$\mathcal{L}^{-1}\left\{\frac{s}{s^2 - 3s + 2}\right\} = \mathcal{L}^{-1}\left\{\frac{2}{s - 2} - \frac{1}{s - 1}\right\} = 2e^{2t} - e^t$$

A special situation prevails when there are repeated roots. In that case the *s-r* denominators must be provided in descending powers from the power of the repeated root.

Example 15.4. Invert

$$\frac{s + 10}{(s - 2)^2(s + 1)}$$

Solution. The attempt to find A and B in the partial fraction representation below is doomed:

$$\frac{s + 10}{(s + 1)(s - 2)^2} = \frac{A}{s + 1} + \frac{B}{(s - 2)^2}$$

Instead, use this expression:

$$\frac{s + 10}{(s + 1)(s - 2)^2} = \frac{A}{s + 1} + \frac{B}{(s - 2)^2} + \frac{B'}{s - 2}$$

This supplies three simultaneous equations.

constants: $10 = 4A + B - 2B'$

s: $1 = -4A + B - B'$

s^2: $0 = A + B'$

from which $A = 1$, $B = 4$, and $B' = -1$. Thus,

$$\mathcal{L}^{-1}\left\{\frac{s + 10}{(s + 1)(s - 2)^2}\right\} = e^{-t} + 4te^{2t} - e^{2t}$$

Another technique for finding the A, B, and C constants in the partial fraction decomposition of

$$\frac{N(s)}{D(s)} = \frac{A}{s - a} + \frac{B}{s - b} + \cdots$$

is, for non-repeated roots,

$$B = \frac{N(s)(s - b)}{D(s)}\bigg|_{s \to b}$$

If b, for example, is a repeated root,

$$\frac{N(s)}{D(s)} = \frac{A}{s - a} + \frac{B}{(s - b)^2} + \frac{B'}{s - b}$$

$$B = \frac{N(s)(s - b)^2}{D(s)}\bigg|_{s \to b} \tag{15.13a}$$

and

$$B' = \frac{d}{ds}\left[\frac{N(s)(s - b)^2}{D(s)}\right]_{s \to b} \tag{15.13b}$$

Example 15.5. Rework Example 15.4 using Eqs. (15.13*a*) and (15.13*b*).

Solution

$$A = \frac{s + 10}{(s - 2)^2}\Big|_{s \to -1} = 1 \qquad B = \frac{s + 10}{s + 1}\Big|_{s \to 2} = 4$$

and

$$B' = \frac{d}{ds}\left(\frac{s + 10}{s + 1}\right)\Big|_{s \to 2} = -1$$

15.6 SOLUTION OF ORDINARY DIFFERENTIAL EQUATIONS

The standard steps in solving a differential equation using Laplace transforms are the following: (1) invert the differential equation, (2) solve for $f(s)$, and (3) invert $f(s)$. The boundary condition(s) may be substituted into the transformed equation following step 1 if the specific values of $F(0)$, $F'(0)$, etc., are known. Otherwise constants may be inserted for $F(0)$, $F'(0)$, etc., after step 1 and the boundary condition(s) substituted into the inverted equation following step 3.

Example 15.6. Solve the differential equation

$$Y''(t) + k^2 Y(t) = 0$$

subject to the boundary conditions: $Y(0) = A$, and $Y'(0) = B$.

Solution. Transform the differential equation,

$$s^2 y(s) - s Y(0) - Y'(0) + k^2 y(s) = 0$$

Insert the boundary conditions and solve for $y(s)$,

$$y(s) = \frac{As}{s^2 + k^2} + \frac{B}{s^2 + k^2}$$

Finally, invert $y(s)$

$$Y(t) = A \cos(kt) + (B/k) \sin(kt)$$

Example 15.7. Solve the differential equation

$$\frac{d^2 y}{dx^2} + y = x$$

with the boundary conditions:

$$dy/dx = 1 \text{ when } x = 0; \text{ and } y = 0 \text{ when } x = \pi$$

Solution. Transform the equation,

$$s^2 y(s) - s Y(0) - Y'(0) + y(s) = 1/s^2$$

$Y'(0) = 1$, but $Y(0)$ is not yet known, so assign it the symbol C. Solving for $y(s)$

$$y(s) = \frac{1}{s^2(s^2 + 1)} + \frac{Cs}{s^2 + 1} + \frac{1}{s^2 + 1}$$

Upon inversion

$$Y = x - \sin x + C \cos x + \sin x = x + C \cos x$$

The final boundary condition may now be substituted,

$$0 = \pi + C(-1)$$

so $$C = \pi$$

Then $$Y = x + \pi \cos x$$

15.7 BLOCKS, BLOCK DIAGRAMS, AND TRANSFER FUNCTIONS

The symbolic representation of system and control characteristics is typically accomplished through the use of block diagrams. These block diagrams are information-flow diagrams such as those used for system simulation in Chapters 6 and 14, except that in control diagrams the variables are usually expressed in the s domain, rather than in the time domain. The diagrams are composed of blocks which have precisely one input and one output, and summing points, as shown in Fig. 15-2. Blocks may represent either processes that are instantaneous or time dependent. The summing points are algebraic additions of two or more inputs to yield an output.

The "transfer function" of an element is the ratio of the output of the block to the input, both expressed in the s domain, as shown in Fig. 15-3a.

$$\text{transfer function} = TF = \frac{\mathcal{L}\{O(t)\}}{\mathcal{L}\{I(t)\}} = \frac{O(s)}{I(s)}$$

when initial values of input and output are zero

A property of transfer functions is that they can be cascaded, as illustrated in Fig. 15-3b. Thus, if $G(s)$ is the transfer function of the first block and $H(s)$ of the second, $O(s) = I(s)G(s)H(s)$.

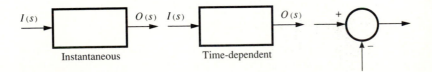

$I(s)$ ⬜ $O(s)$ $I(s)$ ⬜ $O(s)$

Instantaneous   Time-dependent

FIGURE 15-2
Symbols used in block diagrams.

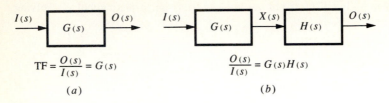

$$TF = \frac{O(s)}{I(s)} = G(s)$$

$$\frac{O(s)}{I(s)} = G(s)H(s)$$

(a) (b)

FIGURE 15-3
Transfer function and cascading of blocks.

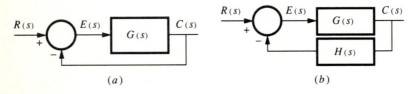

(a) (b)

FIGURE 15-4
(a) Unity feedback loop (b) nonunity feedback loop.

15.8 FEEDBACK CONTROL LOOPS

A feedback control loop is one in which the controlled variable is sensed, and this information is returned to readjust the controller to reduce the error between the sensed and the set values. Two basic forms of feedback loops are shown in Fig. 15-4. In the unity feedback loop of Fig. 15-4a the magnitude of the controlled variable $C(s)$ is compared directly to the setting, $R(s)$. The difference is the error, $E(s)$, which regulates the actuator to which the process responds. The nonunity feedback loop of Fig. 15-4b typically results when the response of the sensor is time dependent. Of special interest to a designer is the transfer function of the entire control loop, $C(s)/R(s)$, which is useful in stability analysis of the loop and in predicting the response in the time domain of disturbances to the system.

unity feedback $TF = \dfrac{G(s)}{1 + G(s)}$ (15.14)

non-unity feedback $TF = \dfrac{G(s)}{1 + G(s)H(s)}$ (15.15)

15.9 TIME-CONSTANT BLOCKS

One of the frequently occurring processes in physical (including thermal) systems is a response where the rate of change of a variable is proportional to the difference between the magnitude of a driving force and the value of the variable. Such a process can be characterized by a "time constant."

An example will illustrate a process of this class, the development of the transfer function for the process, and the inversion into the time domain.

The response of the temperature T of a sensing bulb will be sought as the fluid temperature T_f in contact with the bulb changes, as illustrated in Figs. 15-5a and 15-5b. The sensing bulb has a mass m, specific heat c, convection coefficient h, and an area A for heat transfer between the fluid and the bulb. The standard technique for developing the transfer function is as follows:

1. Write the differential equation for the process,
2. Transform the equation, and
3. Solve for the transfer function, $\mathcal{L}\{O\}/\mathcal{L}\{I\}$.

The first attempt at these three steps will give an awkward transfer function, so after seeing that result, a revised approach will be taken. An energy balance provides the differential equation:

$$mc\frac{dT}{dt} = (T_f - T)hA$$

which when transformed becomes

$$\left(\frac{mc}{hA}\right)[s\mathcal{L}\{T\} - T(0)] = \mathcal{L}\{T_f\} - \mathcal{L}\{T\}$$

Finally, divide by $\mathcal{L}\{T_f\}$ and solve for the TF,

$$\text{TF} = \frac{1 + T(0)\left[\dfrac{(mc/hA)}{\mathcal{L}\{T_f\}}\right]}{(mc/hA)s + 1} \qquad (15.16)$$

Equation (15.16) is not in a useful form because of the second term in the numerator, except for the special case where $T(0) = 0$. The next pass at expressing the transfer function will revise the differential equation,

(a)   (b)

FIGURE 15-5
(a) Response of a temperature-sensing bulb to a change in fluid temperature (b) transfer function of this time-constant block.

normalizing it with respect to the value of $T(0)$, which will be designated T_o.

$$mc\frac{d(T - T_o)}{dt} = [(T_f - T_o) - (T - T_o)]hA$$

which after transformation and division by $\mathcal{L}\{T_f - T_o\}$ yields the transfer function,

$$\text{TF} = \frac{\mathcal{L}\{T - T_o\}}{\mathcal{L}\{T_f - T_o\}} = \frac{1}{(\frac{mc}{hA})s + 1} \tag{15.17}$$

The definition of the transfer function, it may be recalled, had the stipulation that all initial conditions were zero. This is the first appearance of the normalizing process which can be accomplished by arbitrarily setting $T(0)$ to zero in Eq. (15.16) if the transfer function is never to be inverted. If the response in the time domain will ultimately be determined, the redefinition of variables (T to $T - T_o$, for example) is necessary.

The power of the transfer function, Eq. (15.17), is that the response of T to a variety of changes in T_f can be determined. If T_f experiences a step increase of Δ from T_o, as illustrated in Fig. 15-6, the transform of $(T - T_o)$ is, from Eq. (15.17),

$$\mathcal{L}\{T - T_o\} = \mathcal{L}\{T_f - T_o\}\frac{1}{(\frac{mc}{hA})s + 1} = \frac{\Delta}{s[(\frac{mc}{hA})s + 1]}$$

Inversion yields the equation for $T - T_o$ which is an exponential expression

$$T - T_o = \Delta[1 - e^{-t/(mc/hA)}] \tag{15.18}$$

The group mc/hA has the units of time and is the "time constant" τ for the process, from which the name "time-constant block" derives. In response to a step change in the input, τ has the geometric significance shown in Fig. 15-6.

The responses of a time-constant block to several other forms of inputs are explored in Problems 15.9 and 15.10.

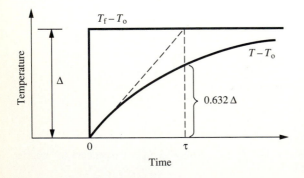

FIGURE 15-6
Step increase in fluid temperature T_f and response of the bulb temperature.

FIGURE 15-7
Time constant block with gain k.

In the heat-transfer process to the temperature-sensing bulb in Fig. 15-5, the multiplying factor in the transfer function is unity. In many processes represented by a time constant, such as the one illustrated in Fig. 15-7, a multiplying factor k, called the gain, appears. In Fig. 15-7, k has the units of

$$\frac{\text{force, N}}{\text{pressure, Pa} = \text{N/m}^2} = \text{area, m}^2$$

15.10 CASCADED TIME-CONSTANT BLOCKS

A first approximation of actual control loops can often be made by representing the elements by time-constant blocks. To gain a bit of insight into cascading time-constant blocks within a control loop, the behavior of two cascaded blocks will be examined. Suppose that the temperature-sensing bulb of Fig. 15-5 contains a liquid whose saturation pressure is transmitted to an actuator which regulates the position of a valve. As shown in Fig. 15-8a, T_a acts on the bulb temperature T_b which in turn affects T_L.

While the representation of the process in Fig. 15-8a by the block diagram of two cascaded time constants may seem reasonable, there is an important implicit assumption. This assumption becomes apparent by writing the heat balance about the bulb:

$$(T_a - T_b)h_1 A_1 = \frac{dT_b}{dt}mc + \overbrace{(T_b - T_L)h_2 A_2}^{\text{neglect}}$$

where subscript 1 refers to h and A between the air and bulb, and subscript 2 to h and A between the bulb and liquid. Notice that in order for the heat-transfer process from the air to the bulb to be represented by the time

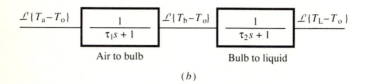

FIGURE 15-8
(a) Response of liquid temperature T_L to a change in air temperature T_a (b) cascaded time-constant blocks to represent the dynamic process.

constant, mc/h_1A_1, the rate of heat transferred to the liquid from the bulb must be negligible. When those relative rates of heat transfer do prevail, the product of the individual transfer functions is the transfer function of the cascaded process. Suppose that T_a experiences a step increase of magnitude Δ from T_o. What is the response of the liquid temperature T_L?

$$\mathcal{L}\{T_L - T_o\} = \frac{\Delta}{s}\left(\frac{1}{\tau_1 s + 1}\right)\left(\frac{1}{\tau_2 s + 1}\right) \tag{15.19}$$

The inversion of Eq. (15.19) is

$$\frac{T_L - T_o}{\Delta} = 1 - \frac{\tau_1 \tau_2}{\tau_1 \tau_2 - \tau_2^2}e^{-t/\tau_1} - \frac{\tau_1 \tau_2}{\tau_1 \tau_2 - \tau_1^2}e^{-t/\tau_2} \tag{15.20}$$

The response of T_b and T_L to a step change Δ of T_a is shown in Fig. 15-9. Several conclusions can be reached by appropriate processing of Eq. (15.20):

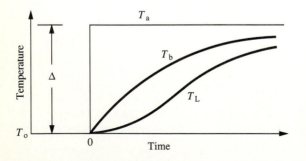

FIGURE 15-9
Response of single and cascaded time-constant blocks to a step change in the input.

1. $T_L - T_o = 0$ at $t = 0$.

2. $d(T_L - T_o)/dt = 0$ at $t = 0$.

3. As one τ becomes very short relative to the other, $\tau_2 \ll \tau_1$ for example,

$$T_L - T_o = \Delta(1 - e^{-t/\tau_1})$$

4. When $\tau_1 = \tau_2$, Eq. (15.19) must be reinverted (see Prob. 15.13) to obtain

$$\frac{T_L - T_o}{\Delta} = 1 - e^{-t/\tau} - \frac{te^{-t/\tau}}{\tau} \tag{15.21}$$

15.11 STABILITY ANALYSIS OF A CONTROL LOOP USING FREQUENCY RESPONSE AND THE PRINCIPLES OF THE BODE DIAGRAM

Two techniques will be presented for analyzing a control loop to predict whether or not the control loop will be stable. In the next section the nature of the transfer function will be analyzed to indicate stability. In this section the Bode diagram, which uses the frequency response concept, will be the tool to predict stability. While we speak of the Bode diagram, we will normally calculate and not draw a graph to make the analysis.

The Bode diagram is a practical method of determining stability, but the concept has another advantage. The logic leading up to the criteria for stability when shown on the Bode diagram offers an excellent technique for explaining the mechanics of instability to a person not well-versed in Laplace transforms.

Frequency response, which is the idea on which the Bode diagram is based, is the examination of the output of an element when subjected to a sinusoidal input, as illustrated in Fig. 15-10. The output of a time-constant block when a sinusoidal input of $\Delta \sin(2\pi f t)$ is applied (see Prob. 15.9), is

$$\frac{\Delta}{\sqrt{1 + (2\pi f \tau)^2}} \sin[2\pi f t - \tan^{-1}(2\pi f \tau)]$$

so the amplification ratio, output to input, is

$$\frac{1}{\sqrt{1 + (2\pi f \tau)^2}}$$

and the phase lag is $\tan^{-1}(2\pi f \tau)$. The amplification ratio decreases and the phase lag increases as the frequency of the input sine wave increases.

The principle that leads to the Bode criterion for stability is based on transmission of sine waves throughout the loop. The sum of the phase lags around the loop and the product of the amplification ratios are computed. Consider the example of the air heater in Fig. 15-11 in which an electric

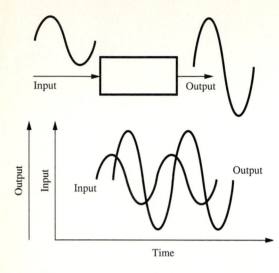

FIGURE 15-10
Frequency response in which a sinusoidal input provides an output lagging the input and also exhibiting a different amplitude than the input.

heater is controlled by a loop that senses the outlet air temperature T_a and converts this sensed temperature T_b to a control voltage V_c. The electric power provided to the air stream is proportional to the difference $V_{set} - V_c$. The block diagram is shown in Fig. 15-11b, and it is assumed that both the temperature sensor and the heater possess thermal capacity that introduce phase lags at those two elements.

The visualization suggested by the principle of the Bode diagram is that for some reason T_a experiences a disturbance that is the top half of a sine wave, as shown in Fig. 15-12. The sensed temperature T_b lags the variation in T_a by an angle ϕ_1 and also varies with a smaller amplitude. The variation in T_b translates to a half sine wave of $V_{set} - V_c$, but reversed in sign, as shown by the dashed line. This reversal is needed because as T_b increases the power input should decrease. This reversal is equivalent to a phase lag of 180°. Because of the thermal capacity in the heater a phase lag of ϕ_2 exists, providing a variation of T_a that is a continuation of the half

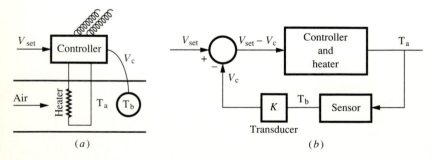

FIGURE 15-11
(a) Air heater (b) control block diagram.

FIGURE 15-12
Perpetuation of sinusoidal disturbances throughout the control loop of the air heater in Fig. 15-11.

sine wave that started the oscillation. From the position marked x a bottom-half sine wave begins another sequence, and the oscillations continue.

The Bode diagram proposes that a crucial frequency is that which causes the sum of the phase lags to be 180°, as illustrated in Fig. 15-13a. In combination with the built-in 180° reversal, a 360° displacement is provided which continues the oscillation and begins to appear unstable. Another element is involved, however, because if the oscillation continues but at progressively lower amplitude ratios, the loop would be stable. The amplitude ratios of the elements in the loop which are additive on the logarithmic scale of Fig. 15-13b. If this product is greater than 1.0, then the new amplitude is greater than the past one. The above visualization leads to the following criterion for stability from the Bode diagram:

> If at the frequency where the sum of the phase lags in the loop is 180° the product of the amplitude ratios is greater than 1.0, then the loop is unstable.

The diagram, as in Fig. 15-13, shows graphical behavior of the trends, and indeed it would be possible to solve problems by adding phase lags and the logarithms of amplitude ratios graphically, but it is usually easier to perform the operations numerically.

A few comments about the concepts on which the Bode diagram is based provide more understanding about the physical nature of instability. The first suspicion might be that it would be highly unlikely to get the sine wave disturbance of the precise frequency that would perpetuate itself. When considering the mathematical representation of arbitrary functions by a Fourier analysis, however, it can be realized that harmonics of many frequencies are embedded in disturbances. The loop can select a sine wave having a frequency close to the critical one; the loop is a form of frequency filter that identifies its critical frequency. When instability does occur the oscillations will take place at the critical frequency.

FIGURE 15-13
(*a*) Sum of the phase lags in the Bode diagram (*b*) product of the amplification ratios.

The concepts of the Bode diagram offer a practical method of predicting loop stability without the use of Laplace transforms, and furthermore suggest a means of visualizing how instability occurs and what might be done to correct or prevent instability.

15.12 STABILITY ANALYSIS USING THE LOOP TRANSFER FUNCTION

One of the strengths of the Laplace transform representation of a control loop is that stability analyses can be performed without the need of inverting back to the time domain. The reason for this ease of analysis is that the loop transfer function, for example,

$$\text{TF}_{\text{loop}} = \frac{as + b}{cs^3 + ds^2 + es + f}$$

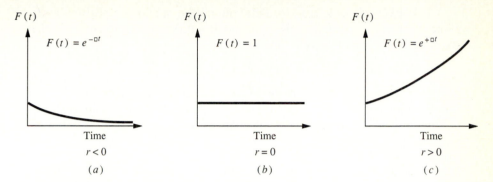

FIGURE 15-14
Inverses $F(t)$ are dependent upon the sign of the roots in $(\text{const})/(s - r)$ terms.

can be decomposed into components,

$$\frac{A}{s - r_1} + \frac{B}{s - r_2} + \frac{C}{s - r_3} \tag{15.22}$$

The values of r_1, r_2, and r_3 are the roots of what is called the characteristic equation, $cs^3 + ds^2 + es + f$. The nature of the roots indicates stability or instability, because the inverses of the terms in Eq. (15.22) are exponents as shown in Fig. 15-14. If a root is less than zero the contribution of that term decays indicating a stable response. If, on the other hand, the root is positive as shown in Fig. 15-14c, the magnitude grows with time so the response is unstable. When the root is zero, as in Fig. 15-14b, the response is a constant value, so the loop response is neutral.

Figure 15-14 implies that the roots are real numbers, but roots that are imaginary numbers appear frequently as well. If the root is $a + ib$, for example, the inverse is $e^{at} e^{ibt}$. Since $e^{ibt} = \cos(bt) + i \sin(bt)$, this contribution is oscillatory, neither growing nor decaying with time. The foregoing observations lead to the Nyquist criterion,[6] which states that a control loop is unstable if the real part of any of the roots of the characteristic equation is positive, as shown in Fig. 15-15.

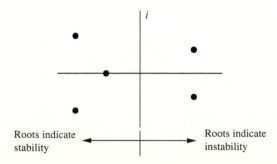

Roots indicate stability

Roots indicate instability

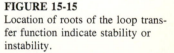

FIGURE 15-15
Location of roots of the loop transfer function indicate stability or instability.

A convenient technique for determining the number of positive roots without actually calculating them is the Routh-Hurwitz criterion.[7] If the characteristic equation has the form

$$a_n s^n + a_{n-1} s^{n-1} + a_{n-2} s^{n-2} + \ldots + a_1 s + a_0 = 0 \qquad (15.23)$$

the a, b, and c coefficients are arranged in the form shown in Fig. 15-16. The Routh-Hurwitz criterion states that the number of roots with positive real parts is equal to the number of changes in sign in the left column of the array.

Example 15.8. How many roots with positive real parts are there of the characteristic equation

$$s^4 + 7s^3 - 10s^2 - 116s - 80?$$

Solution. Using Routh-Hurwitz, $a_4 = 1$, $a_3 = 7$, $a_2 = -10$, $a_1 = -116$, and $a_0 = -80$.

$$b_3 = \frac{-1}{7} \begin{vmatrix} 1 & -10 \\ 7 & -116 \end{vmatrix} = 6.57$$

$$b_{n-3} = b_1 = \frac{-1}{7} \begin{vmatrix} 1 & -80 \\ 7 & 0 \end{vmatrix} = -80$$

$$c_3 = \frac{-1}{6.57} \begin{vmatrix} 7 & -116 \\ 6.57 & -80 \end{vmatrix} = -30.8$$

so the array is

$$
\begin{array}{ccc}
1 & -10 & -80 \\
7 & -116 & \\
6.57 & -80 & \\
-30.8 & &
\end{array}
$$

The one sign change down the left column indicates one root with a positive real part. (The roots are 4, -5, $-3 \pm 2i$).

Array

$$
\begin{array}{ccc}
a_n & a_{n-2} & a_{n-4} \\
a_{n-1} & a_{n-3} & a_{n-5} \\
b_{n-1} & b_{n-3} & b_{n-5} \\
c_{n-1} & c_{n-3} & c_{n-5}
\end{array}
$$

Where $\quad b_{n-1} = \dfrac{-1}{a_{n-1}} \begin{vmatrix} a_n & a_{n-2} \\ a_{n-1} & a_{n-3} \end{vmatrix} \qquad b_{n-3} = \dfrac{-1}{a_{n-1}} \begin{vmatrix} a_n & a_{n-4} \\ a_{n-1} & a_{n-5} \end{vmatrix}$

$$c_{n-1} = \frac{-1}{b_{n-1}} \begin{vmatrix} a_{n-1} & a_{n-3} \\ b_{n-1} & b_{n-3} \end{vmatrix}$$

FIGURE 15-16
The Routh-Hurwitz array derived from a characteristic equation of the form of Eq. (15.32).

Some techniques are available for determining the specific value of the roots should that be required. Here we will only discuss second and third degree polynomials. The roots of the characteristic equation that is a second degree polynomial can readily be found by use of the quadratic formula. If the characteristic equation is a cubic one, the roots fall into one of three classes:

1. one real root and two roots that are a conjugate pair
2. three real roots, two of which are equal
3. three real roots, all different

Subjecting the characteristic equation to the Newton-Raphson technique, perhaps with some different trial values, should lead to one real root. Thereafter the quadratic formula can be used to find the other roots. The Routh-Hurwitz criterion can determine how many roots have positive real parts, which may assist in choosing a trial value for the application of Newton-Raphson.

15.13 NORMALIZING THE VARIABLES
FOR INVERSION TO THE TIME DOMAIN

The next several sections address the challenge of translating the characteristics of the physical equipment into the symbols of the control block diagram along with the specification of the transfer functions. When the loop transfer function must be inverted to determine the response of one or more variables in the loop in the time domain, a normalization of the variables is usually necessary. Consider, for example, the controller in Fig. 15-17a that regulates the air pressure in a reservoir by admitting a stream of air from a high-pressure source. The spring attached to the piston is under tension, and the displacement of the valve stem is proportional to the pressure difference:

$$x, \text{ mm} = K(p_{set} - p, \text{ kPa})$$

The flow rate admitted to the reservoir is proportional to the position of the valve stem:

$$w_{in}, \text{ kg/s} = L\,x$$

The block diagram of the loop with p_{set} as the reference value and the reservoir pressure p as the controlled variable is shown in Fig. 15-17b.

If the only analysis planned is that of determining stability based on the loop transfer function, some short cuts can be taken. If w_s is constant, that summing point can be ignored, and the loop transfer function TF is written directly as

$$\text{TF} = \frac{KLG(s)}{1 + KLG(s)}$$

(a)

(b)

FIGURE 15-17
(a) The hardware for regulating the air pressure in a reservoir (b) the control block diagram.

If, on the other hand, the response of variables is to be determined through an inversion, the variables must be normalized. Suppose that the equation of the controlled pressure p is sought in response to a step change in p_{set} while w_s remains constant. The need for the normalization becomes apparent when developing the transfer function of the process incorporating the reservoir. The difference of flow rates in and out of the reservoir, $\Delta w = dM/dt$. Use the ideal gas equation, $M = pV/RT$, with constant values of temperature T, gas constant R, and volume V.

$$\Delta w = \left(\frac{V}{RT}\right)\frac{dp}{dt}$$

Transforming both sides of the equation,

$$\mathcal{L}\{\Delta w\} = \left(\frac{V}{RT}\right)\mathcal{L}\{\frac{dp}{dt}\} = \left(\frac{V}{RT}\right)[s\mathcal{L}\{p\} - p(0)]$$

To avoid the dangling term of the pressure at time $=$ zero, $p(0)$ or p_0, the variable must be redefined as was first done in Sec. 15.9. In this case when the pressure variable is revised to $p - p_0$,

$$G(s) = \frac{RT}{(Vs)}$$

But the consequence of revising p to $p - p_0$ is that the other variables must be restructured as well. In many instances the procedure is simply one of redefining all variables as their transient values minus their values at the initial steady state, as in Fig. 15-18a. At the initial steady state $p = p_0$, $p_{set} = p_{set,0}$, etc., so all variables are zero. The normalization of Fig. 15-18a now justifies the following treatment of the w summing point. Since the process to be examined is one where w_s remains constant, $w_s = w_{s,0}$,

FIGURE 15-18
(*a*) Normalized variables in pressure control loop (*b*) with *w* summing point omitted.

so $w_{in} - w_{in,0}$ passes directly to the $RT/(Vs)$ transfer function, as in Fig. 15-18*b*, and the summing point disappears.

> **Example 15.9.** In the pressure controller of Fig. 15-17, $V = 3$ m³, $T = 300$ K, $R = 0.287$ kJ/(kg · K), $K = 0.2$ mm/kPa, and $L = 0.3$ (kg/s)/mm. (*a*) If $p_{set,0} = 200$ kPa and $w_{s,0} = 1.2$ kg/s, what is p_0? (*b*) If at zero time p_{set} is abruptly changed to 225 kPa, what is the equation for p as a function of time?
>
> *Solution*
> (*a*) $w_{s,0} = w_{in,0} = 0.3x_0 = (0.3)(0.2)(200 - p_0)$ and since $w_{s,0} = 1.2$, $p_0 = 180$ kPa.
> (*b*) $G(s) = (0.3)(0.2)(0.287)(300)/(3s) = 1.722/s$
>
> $$p - p_0 = \mathcal{L}^{-1}\left\{\frac{25}{s}\left[\frac{1.722/s}{1 + 1.722/s}\right]\right\}$$
>
> $$p = 180 + 25(1 - e^{-t/0.5807})$$
>
> The new steady-state pressure is 205 kPa, so the same pressure difference, $p_{set} - p$, is restored because the same steady-state flow rate of 1.2 kg/s once again prevails.

Of more practical interest might be the choice of w_s as the reference variable rather than p_{set}, because p_{set} is likely to be adjusted rarely, while w_s may change frequently during operation. The first pass at the block diagram with the variables normalized is shown in Fig. 15-19*a* and an examination of this figure shows two potential problems. One problem is that the reference value is not entering the summing point as a positive value, as is standard, and the other concern is that a sign reversal at the summing point of the pressure will frustrate the attempt to drop out this summing point for the constant p_{set} case. Figure 15-19*b* shows changes in the variables that still

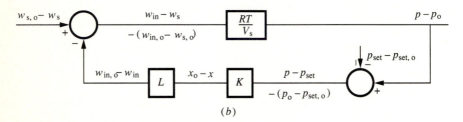

FIGURE 15-19
Pressure controller with w_s as the reference value translated to (a) the preliminary block diagram (b) with sign revisions to provide the standard form.

abide by the physical requirements, but eliminate the two above-mentioned problems. The transfer function of this loop is

$$\text{TF} = \frac{\dfrac{RT}{(Vs)}}{1 + KL\left[\dfrac{RT}{(Vs)}\right]}$$

If the initial steady-state conditions of Example 15.9 prevail and w_s increases to 1.4 from 1.2 kg/s at $t = 0$,

$$p - p_0 = \mathcal{L}^{-1}\left\{\frac{-0.2}{s}\left[\frac{28.7}{s + 1.722}\right]\right\}$$

and

$$p = 180 - 3.333(1 - e^{-t/0.5807})$$

The new steady-state tank pressure is 176.67 kPa. The pressure changes with the same time constant as when the setpoint was changed, but this may not be the case for loops having more than one dynamic element.

15.14 RESTRUCTURING THE BLOCK DIAGRAM

In the attempt to develop a transfer function of a loop, the task may be facilitated by converting a complex block diagram into a unity or nonunity feedback loop. Some of the elementary restructuring operations that may be useful in simplifying the loop are:

FIGURE 15-20
Combining two transfer functions in series.

FIGURE 15-21
Exchanging two adjacent summing points.

1. Combine two transfer functions in series, as in Fig. 15-20.
2. Exchange two adjacent summing points, as in Fig. 15-21.
3. Move a summing point upstream or downstream of a block, as in Fig. 15-22.
4. Move a takeoff point, as in Fig. 15-23.

> **Example 15.10.** Convert the loop[8] shown in Fig. 15-24 into a nonunity feedback loop.
>
> **Solution.** First move the summing point and combine it with the summing point for the reference variable, and move the takeoff of the lower reverse path, as in Fig. 15-25a. Finally, combine the upper and lower reverse paths to obtain the nonunity feedback loop shown in Fig. 15-25b.

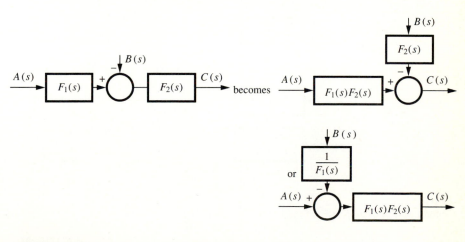

FIGURE 15-22
Moving a summing point around a block.

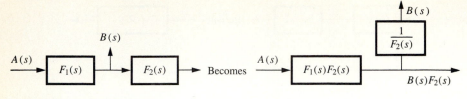

FIGURE 15-23
Moving a takeoff point.

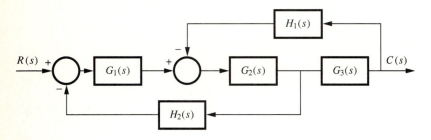

FIGURE 15-24
Control block diagram in Example 15.10.

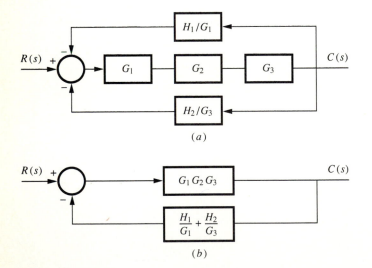

FIGURE 15-25
Modifications of loop in Example 15.10.

15.15 TRANSLATING THE PHYSICAL SITUATION INTO A BLOCK DIAGRAM

An illustration will show how to represent a physical system in block diagram symbols such that the form is ready for an inversion into the time domain. The system is an electric heater in an air duct along with its control, as shown in Fig. 15-26. The controller provides power to the heater according to the equation, $q_h = K(T_{set} - T_s)$ where T_s is the sensor temperature. The rate of heat transfer from the heater to the flowing air q_a is represented by the rate equation,

$$q_a = wc(T_h - T_i)(1 - e^{-hA/wc}) = W(T_h - T_i)\epsilon$$

where w = rate of air flow
c = the specific heat of the air
T_h = heater temperature
T_i = temperature of entering air
h = convection coefficient
A = heat-transfer area

For convenience, let $wc = W$, and the effectiveness of the heat exchanger is represented by ϵ. The initial block diagram of the control system is shown in Fig. 15-27 where the two dynamic elements are those associated with the thermal capacity of the heater M and the time constant of the sensor τ_s. The structure chooses T_{set} as the reference variable and anticipates an analysis in which the entering air temperature remains constant. Table 15.2 shows variables corresponding to the numbered positions in Fig. 15-27. Also shown in Table 15.2 is a column of variables normalized by subtracting steady-state values from each. A next, minor revision of a small section is shown in Fig. 15-28 in which, to avoid two positive values entering the summing point, the input (variable 10) is changed to $T_{i,0} - T_i$.

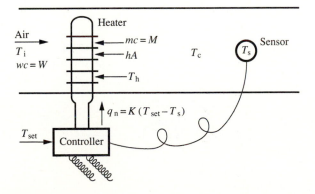

FIGURE 15-26
Air heating system and its control.

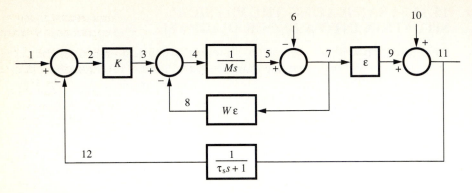

FIGURE 15-27
Air heating system and its control.

TABLE 15.2
Designations of variables in block diagram of Fig. 15-27.

Position	Nonnormalized	Normalized
1	T_{set}	$T_{set} - T_{set, 0}$
2	$T_{set} - T_s$	$(T_{set} - T_s) - (T_{set, 0} - T_{s, 0})$
3	q_h	$q_h - q_{h, 0}$
4	$q_h - q_a$	$(q_h - q_a) - (q_{h, 0} - q_{a, 0})$
5	T_h	$T_h - T_{h, 0}$
6	T_i	$T_i - T_{i, 0}$
7	$T_h - T_i$	$(T_h - T_i) - (T_{h, 0} - T_{i, 0})$
8	q_a	$q_a - q_{a, 0}$
9	$T_c - T_i$	$(T_c - T_i) - (T_{c, 0} - T_{i, 0})$
10	T_i	$T_i - T_{i, 0}$
11	T_c	$T_c - T_{c, 0}$
12	T_s	$T_s - T_{s, 0}$

If the variation of the controlled temperature is sought in response to step change in T_{set}, the next stage of consolidation is shown in Fig. 15-29. At the initial steady state $q_{a,0} = q_{h,0}$, $T_i = T_{i,0}$, and $T_{c,0} = T_{s,0}$.

Two summing points can be eliminated. The final restructuring derives from recognizing an interior feedback loop that has $q_h - q_{h,0}$ as the reference

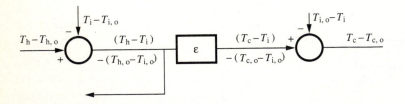

FIGURE 15-28
Change of input to summing point.

FIGURE 15-29
Diagram after elimination of two summing points.

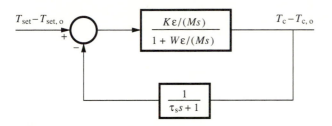

FIGURE 15-30
Simplified nonunity feedback loop for air heater controller.

value and $T_h - T_{h,0}$ as the output. When that loop transfer function is substituted, the final diagram in the form of a nonunity feedback loop is that shown in Fig. 15-30. The expression for $T_c - T_{c,0}$ in the time domain can be obtained by inverting the product of the loop transfer function and the transform of the $T_{set} - T_{set,0}$ disturbance.

15.16 PROPORTIONAL CONTROL

The pressure controller in Fig. 15-17 and the temperature controller in Fig. 15-26 employ the proportional mode of control. In both systems the actuator (the valve position x in Fig. 15-17 and the power q_h in Fig. 15-26) are proportional to the error between the set and sensed values of the controlled variables. The proportional controller is simple, which helps keep its cost low, and provides adequate control in a wide variety of applications. The proportional controller is characterized by an offset of the value of the controlled variable from the setpoint at all but one value of the load. In the pressure controller of Fig. 15-17, the offset is shown in Fig. 15-31 to be a function of the demand, w_s for a given gain. The offset experienced using a low gain in Fig. 15-31a can be reduced by increasing the gain as in Fig. 15-31b. A further influence of the high gain is that it may introduce oscillation as the system adjusts to a disturbance. Some control systems become unstable when the gain is increased beyond a certain value.

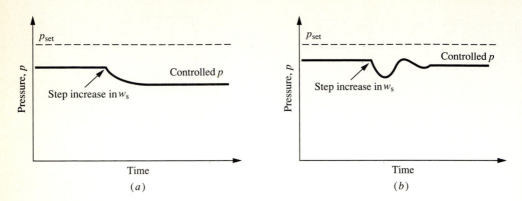

FIGURE 15-31
Pressure controller (*a*) with low gain (*b*) with high gain.

Adjustment of the proportional constant to achieve an adequate compromise between low offset and stability is called "tuning" the control.

Maintenance of the desirable qualities of the proportional controller but deemphasizing its drawbacks can be achieved by combining the proportional mode with the integral and derivative modes as explained in the next two sections.

15.17 PROPORTIONAL-INTEGRAL (PI) CONTROL

The purpose of the integral mode is to eliminate the offset associated with the proportional mode. The contribution to the control signal is

$$K_I \int (\text{error}) dt \tag{15.24}$$

so continued operation with an offset builds a contribution to the control signal that tends to eliminate the offset. The transfer function of the I-mode is K_I/s which may be illustrated as in Fig. 15-32. In the time domain, if the input experiences a step change of Δ at time zero, the integration gives the output $K_I(\Delta)t$. The transfer function is the transform of the output divided by the transform of the input

$$\text{TF} = \frac{K_I(\Delta)/s^2}{\Delta/s} = \frac{K_I}{s} \tag{15.25}$$

In the control block diagram the combination of the P- and I-modes are as shown in Fig. 15-33. Each combination of system characteristics, K_P, and K_I exhibits a unique response, but to show the behavior of a PI control, consider the example of the air heater shown in Fig. 15-26. The block diagram of this control loop, shown in Fig. 15-30, must be revised by

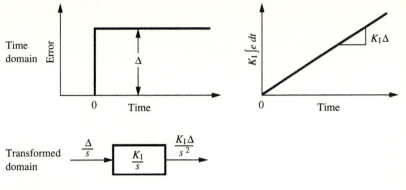

FIGURE 15-32
Transfer function of the I-mode.

changing K in the forward path to $(K_P + K_I/s)$. Introduce the following numerical values: $\epsilon = 0.7$, $M = 50$ kJ/K, $W = 0.6$ kW/K, $\tau = 4$ s, and the entering temperature of the air, $T_i = 22°C$. The loop transfer function then becomes

$$TF_{loop} = \frac{0.0035(K_p s + K_I)(4s + 1)}{s^3 + 0.258s^2 + (0.0021 + 0.0035K_P)s + 0.0035K_I} \quad (15.26)$$

The response of the controlled temperature to a step change in T_{set} from 42 to 47°C will be examined for four different combinations of K_P and K_I:

Case I. $K_P = 2.4$, $K_I = 0$

Case II. $K_P = 2.4$, $K_I = 0.02$

Case III. $K_P = 2.4$, $K_I = 0.1$

Case IV. $K_P = 2.4$, $K_I = 1.0$

The transform of the input, $\mathcal{L}\{T_{set} - T_{set,o}\} = 5/s$, so the transform of the output, $\mathcal{L}\{T_c - T_{c,o}\}$ is the product of Eq. (15.26) and $5/s$. This transform

(a)   (b)

FIGURE 15-33
Block diagram symbols of the PI control.

is the one to invert for all four cases:

$$\frac{0.0175(K_p s + K_I)(4s + 1)}{s[s^3 + 0.258s^2 + (0.0021 + 0.0035K_p)s + 0.0035K_I]} \tag{15.27}$$

Case I: $K_P = 2.4, K_I = 0.$

$$T_c - T_{c,o} = \mathcal{L}^{-1}\left\{\frac{0.168s + 0.042}{s^3 + 0.258s^2 + 0.0105s + 0}\right\} \tag{15.28}$$

From the steady-state calculation $T_{c,o} = 38°C$, and from the inversion of Eq. (15.28),

$$T_c = 42 - 4.220e^{-t/19.74} + 0.220e^{-t/4.823}$$

A graph of T_c with P-mode control is shown in Fig. 15-34. The initial offset of T_c is 4°C, and the offset after the change in setpoint is 5°C. This response will be the basis of comparison with the PI mode which will now be developed by combining the I-mode with the P-mode.

Cases II, III and IV are all PI control with varying magnitudes of K_I and showing responses as in Fig. 15-35.

Case II: $K_P = 2.4, K_I = 0.02.$

Instead of there being an offset between T_{set} and T_c at $t = 0$ as occurs in the proportional controller of Fig. 15-34, it is assumed the I-mode in Cases II to IV has brought T_c equal to T_{set} when the step change is made. With the low value of K_I in Case I the I-mode slowly corrects the offset and T_c approaches T_{set} asymptotically.

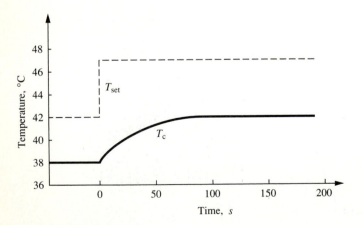

FIGURE 15-34
Response of the air heating system with the proportional-only control, $K_p = 2.4$ and $K_I = 0$ to a step change in setpoint.

FIGURE 15-35
Response of the air heating system when using PI modes with various values of K_I.

Case III: $\qquad\qquad K_P = 2.4,\ K_I = 0.1.$

With the increased value of K_I the correction of the offset takes place rapidly and T_c experiences a slight oscillation around the final steady state of 47°C with the excursions quickly dying out.

Case IV: $\qquad\qquad K_P = 2.4,\ K_I = 1.0$

This case shows that when K_I is set too high the response is unstable.

The illustrations in Fig. 15-35 apply to only one value of K_P, and it is clear that the two constants must be selected in conjunction with one another (tuning the controller) to give the desired control characteristics for the particular process being regulated. An active field of development is the application of an on-line computer program to self-tune the control.

15.18 PROPORTIONAL-INTEGRAL-DERIVATIVE (PID) CONTROL

A skilled operator controlling a process manually will instinctively change the actuator position more if the deviation of the controlled variable from the setpoint is increasing rapidly. In a similar manner, if the controlled variable is currently deviating from the setpoint but approaching the setpoint rapidly, the operator will back off on the correction in an effort to home the controlled variable in on the setpoint. In the above situations the operator is making an adjustment based on the rate of change of the controlled variable, or more precisely on the derivative of the error with respect to time. The differentiation process in the time domain and in the s domain are shown in Fig. 15-36a and 15-36b, respectively. Dividing the output of the block in Fig. 15-36b by the input shows that the transfer function of the differentiation process is s. If the block representing the PI controller in Fig. 15-33 is extended to incorporate derivative control with a constant K_D, the transfer function of the PID controller becomes

$$K_P + \frac{K_I}{s} + K_D s \tag{15.29}$$

To illustrate the influence of the D-mode of control, apply PID control to the pressure regulator of Fig. 15-17. Examine the case where $p - p_o$ is to be determined in response to a change in the outlet airflow rate $w_s - w_{s,o}$ as represented by the block diagram in Fig. 15-19. The same parameters as in Example 15.9 will be used, namely, $V = 3$ m^3, $T = 300$ K, $R = 0.287$ kJ/(kg $\cdot$ K), and $L = 0.3$ (kg/s)/mm. The pressure setting p_{set} is 200 kPa which remains unchanged. Replacing the proportional constant K by Eq. (15.29) yields the loop transfer function

$$\text{TF}_{\text{loop}} = \frac{86.1s}{(3 + 25.83 K_D)s^2 + 25.83 K_P s + 25.83 K_I} \tag{15.30}$$

If steady state has been permitted to develop, the I-mode will be assumed to drive the controller pressure to the set value of 200 kPa. For a step increase of w_s from 1.2 to 1.4 kg/s the transform to invert is

$$p - 200 = \mathcal{L}^{-1}\left\{ \frac{-17.22}{(3 + 25.83 K_D)s^2 + 25.83 K_P s + 25.83 K_I} \right\} \tag{15.31}$$

Figure 15-37 shows the response of the pressure with and without the D-mode for values of $K_P = 0.2$ and $K_I = 0.05$. Following the step increase in w_s, the pressure drops rapidly and sinks to 2.7 kPa below the set value when

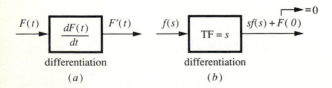

FIGURE 15-36
The differentiation process in (a) the time domain, (b) the transformed domain.

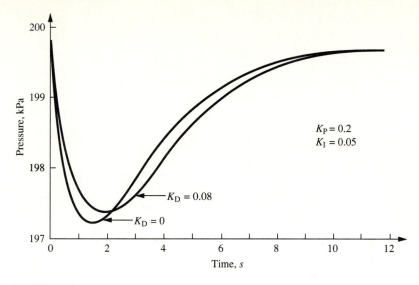

FIGURE 15-37
Influence of the D-mode on the response of pressure controller of Fig. 15-17 to a step increase in the outlet flow rate.

no D-mode is applied. Using PID control with $K_D = 0.08$ the maximum deviation is 2.46 kPa. In this case the PID control experiences a short interval where the deviation is greater than for the PI control. This difference is attributable to the I-mode, which was accumulating a greater correction during the period of high deviation occurring with the PI control.

15.19 FEEDFORWARD CONTROL

An additional mode that can be combined with PI or PID in some situations to provide superior characteristics is feedforward control. By itself, feedforward is open loop control that supplies the gross contribution, which is then supplemented by the fine tuning of PI or PID control. An example of feedforward control is the steam pressure and excess-oxygen regulator[9] for a furnace/boiler, as shown in Fig. 15-38.

The intent of the controller is to maintain a constant steam pressure by regulating the rate of fuel supplied to the burner. To maintain efficient combustion the percent of excess oxygen in the stack gas is also regulated. The excess oxygen is a function of the combination of flow rates of air and fuel, so the flow rate of combustion air must be adjusted as the flow rate of fuel changes. Two possible control concepts for the air flow rate are (1) PI control to regulate the desired percent excess oxygen, or (2) an open-loop controller that sets the rate of combustion air based on the rate of fuel

stack gas

FIGURE 15-38
Controller of fuel and air flow rate to a furnace using a combination of feedforward and PI modes.

flow—perhaps in direct proportion. Mode 1 using PI control alone is likely to suffer large excursions when an abrupt change in fuel flow occurs, or alternately it may require a long time for a very stable controller with low settings of K_P and K_I to bring the rate flow of combustion air back into range. The open-loop control of Mode 2 will react in concert with the fuel control, but the percent of excess oxygen is not likely to be precisely held because of the nonlinearities in the actuators or the combustion process. The combination of feedforward and PI control permits a rapid response of the air controller to keep the flow rate of combustion air in the proper range, and still permits the PI controller to constantly refine the setting.

Example 15.11. An electric heater, shown in Fig. 15-39, raises the temperature of a stream of water having a flow rate of 1.2 kg/s from T_i to T_{set} which is 65°C. The temperature of water in the tank, which is the same as the outlet temperature T_o, has reached a steady value of 65°C with the inlet temperature T_i of 20°C when at $t = 0$ T_i changes to 15°C. Determine responses of T_o as

FIGURE 15-39
Electric water heater and storage tank in Example 15.11.

functions of time if the modes of control are (*a*) PI control alone with $K_P = 15$ kW/°C and $K_I = 0.02$ kW/(°C · s), and (*b*) if the PI controller of part (*a*) is combined with a feedforward control keyed to T_i such that its contribution is $q_{ff} = (1.0)(T_{set} - T_i)$.

Solution. (*a*) The block diagram incorporating both the PI and feedforward control modes is shown in Fig. 15-40. For PI control alone the loop transfer function is

$$\frac{\mathcal{L}\{T_o - 65\}}{\mathcal{L}\{T_i - 65\}} = \frac{(1.2)(4.19)s}{(750)(4.19)s^2 + [K_P + (1.2)(4.19)]s + K_I} \quad (15.32)$$

which when multiplied by $-5/s$ and inverted yields the results labeled PI in Fig. 15-41.

(*b*) If the feedforward control had been set for 1.2 kg/s, its contribution would have been

$$q_{ff} = (1.2 \text{ kg/s})(4.19)(T_{set} - T_i)$$

and this open-loop control would keep the outlet temperature at precisely 65°C

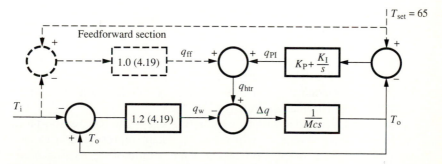

FIGURE 15-40
Block diagram of outlet water temperature in Example 15.11.

FIGURE 15-41
Response of the outlet temperature of the water heater in Example 15.11 to PI and PI-plus-feedforward control.

without the need of the PI modes. The flow rate for this particular heater is normally 1.0 kg/s which explains the setting, and at this moment the flow rate is different than expected. The loop transfer function of the complete loop shown in Fig. 15-40 that includes the feedforward control is precisely 1/6 of that expressed in Eq. (15.32). Problem 15.23 asks for the execution of this solution.

15.20 NONLINEARITIES

The analyses presented so far in this chapter have been based on ideal performance of components and systems. Actual equipment often functions with losses, friction, and delays. Most valves and other mechanical actuators suffer from hysteresis so that a reversal in the direction that the control signal is changing does not immediately reverse the direction of motion of the actuator. In thermal processes there may be some heat transfer to or from unaccounted, extraneous sources. When fluid flows through pipes there is a finite time required for a given mass of fluid to move from one location to another. The existence of nonlinearities does not repudiate the linear control analysis, because the solutions for the linear cases indicate the directions to strive for good control. Sometimes the extent of nonlinearity is small enough that the linear analysis suffices.

One nonlinearity that will be explored is dead time or transportation lag, which is often combined with a time-constant representation to adequately model certain complex processes. An example of the existence of a transportation lag is shown in Fig. 15-42 where the sensor of outlet air temperature is located some distance downstream from the coil. If the temperature at the coil is a function of time $X(t)$, the sensor receives that function

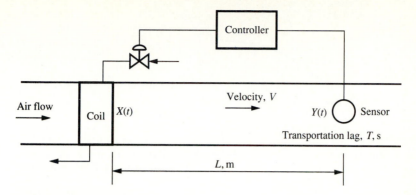

FIGURE 15-42
Transportation lag due to the distance between the coil outlet and the temperature sensor in an air-temperature controller.

$T = L/V$ seconds later. The Laplace transform[10] of the temperature seen by the sensor is

$$\mathcal{L}\{Y(t)\} = y(s) = \int_0^\infty X(t - T)e^{-st}\,dt$$

The integrand is zero for $0 < t < T$, so

$$y(s) = \int_T^\infty X(t - T)e^{-st}\,dt$$

Let $t^* = t - T$, then $dt = dt^*$ which changes the lower limit of integration to 0.

$$y(s) = \int_0^\infty X(t^*)e^{-st^*}e^{-sT}\,dt^* = e^{-sT}x(s) \qquad (15.33)$$

Multiplication of the transform of a function by e^{-sT} is equivalent to imposing a delay of T seconds on the original function.

Some physical components can be modeled, at least roughly, by combining a time delay with a time constant. In a heat exchanger where air is cooled or heated by water flowing through the tubes, there will be a transportation lag attributable to the time required for the water to pass through the water circuit.[11] Thus, the actual response of the outlet air temperature to a step change in water flow rate or water temperature can be approximated by the combination of a delay and a time constant, as shown in Fig. 15-43.

The transfer function of this coil is

$$\frac{ke^{-Ts}}{\tau_c s + 1} \qquad (15.34)$$

FIGURE 15-43

Change of outlet air temperature leaving an air heating coil in response to a step increase in water temperature or water flow rate.

The e^{-Ts} term introduces nonlinearity into the transfer function of the loop which often precludes a closed form for the inversion. Some approximations may be possible, as for example if the coil with the transfer function of Eq. (15.34) is in the forward path and the transfer function of a sensor, $1/(\tau_s s + 1)$, is in the reverse path. The transfer function of the loop is

$$\text{TF}_{\text{loop}} = \frac{k(\tau_s s + 1)}{e^{Ts}(\tau_c s + 1)(\tau_s s + 1) + k} \qquad (15.35)$$

The exponential, e^{Ts} can be expressed as a series

$$e^{Ts} = 1 + Ts + \frac{(Ts)^2}{2} + \dots$$

and inclusion of the first two or even the first three terms of the series gives a transfer function that can still be inverted conveniently. Also, if the value of Ts is small, the resulting inverse is reasonably accurate.

15.21 SELECTING CONTROL VALVES

We turn now from somewhat mathematical considerations to issues of practical hardware, although a point that will be made in this section flows naturally from analytical insight. Suppose that a control valve is to be selected to regulate the hot or chilled water flow through a coil, as shown schematically in Fig. 15-44. In selecting a valve, essentially three items are specified: pipe size, C_v value, and characteristics (quick-opening, linear, or equal-percentage).

1. The pipe size is often specified to match that of the pipe to and from the valve, although sometimes the pipe size of the valve is smaller in order to reduce the cost of the valve.

FIGURE 15-44
Valve and coil combination.

2. The C_v valve is defined by the equation

$$\text{Flow rate} = C_v \Delta p \text{ L/s} \tag{15.36}$$

where Δp is the pressure drop in kilopascals across the valve when the valve is in its wide open position.

3. The three different valve characteristics commonly available are shown in Fig. 15-45.

The valve manufacturer provides a certain characteristic by means of the design chosen for the seat and the plug. The influences of the choice of valve characteristic and C_v value are demonstrated by two different choices of C_v, designated case I and case II. Suppose that the coil has a pressure drop

$$\Delta p_{\text{coil}} = 2.5 \, Q^2 \text{ kPa}$$

where Q = flow rate in L/s and the available pressure difference across both the coil and valve is constant at 80 kPa, as shown in Fig. 15-44. A valve with linear characteristics is used in both cases, but the C_v in case I is 0.6 and in case II is 1.2.

For the valve with linear characteristics,

$$Q = \frac{\text{percent stem stroke}}{100} C_v \sqrt{\Delta p}$$

or
$$\text{percent stem stroke} = \frac{100 \, Q}{C_v \sqrt{\Delta p}} \tag{15.37}$$

For a given flow rate, say, 2 L/s, the percent stem stroke can be computed for the valve. The pressure drop through the coil would be $2.5(2.0^2)$ = 10 kPa, requiring $80 - 10 = 70$ kPa to be dissipated in the valve. In case I, with $C_v = 0.6$, Eq. (15.37) indicates the percent stem stroke to be

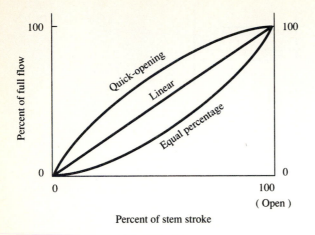

FIGURE 15-45
Three valve characteristics.

$$\text{percent stem stroke} = \frac{100(2)}{0.6\sqrt{70}} = 39.8\%$$

The relationship of the percent stem stroke to the flow rate through the coil-valve combination for the two cases is shown in Fig. 15-46, from which two observations can be made: (1) even though the flow-stem stroke

FIGURE 15-46
Flow-stem-stroke relationship of control valve and coil combinations.

relationships of the valves are linear, the flow-stem stroke relation of the combination is not linear; (2) the combination with the valve of the high C_v yields the characteristic with less linearity.

The curves in Fig. 15-46 are significant, because the slopes of the curves represent a portion of the gain of the control loop. The curvature of the $C_v = 1.2$ curve in Fig. 15-46 is less desirable than that of the valve whose $C_v = 0.6$. With the C_v value of 1.2 the gain is low when the valve is nearly fully opened, which would cause high throttling range if the controller is only proportional. When the valve is nearly closed the gain is high, and the loop may be unstable. This behavior may explain some systems being stable at moderate and heavy loads and unstable at low loads. Another means of approaching a linear flow–stem stroke relation is to choose a valve with equal-percentage characteristics as is done in Prob. 15.25.

15.22 SOLVING PARTIAL DIFFERENTIAL EQUATIONS USING LAPLACE TRANSFORMS

An important class of dynamic thermal system problems is the one where there are changes with respect to both time and some additional variable, such as distance. Transient changes in temperature along a tube or the pressure of a vapor as it flows in a pipe are examples of cases where the fundamental relationship is a partial differential equation. A short extension of the use of Laplace transforms provides a convenient means of solving this type of differential equation. An illustration will be the prediction of temperatures in a wall, shown in Fig. 15-47, in which originally the temperatures at both surfaces and the interior are T_0. At time $t = 0$ the temperature of the surface where $x = b$ is changed to T_s, while the temperature of the left surface remains constant at T_0. The governing differential equation is

$$\frac{\partial^2 T}{\partial x^2} = \frac{\rho c}{k} \frac{\partial T}{\partial t} = \frac{1}{\alpha} \frac{\partial T}{\partial t}$$

subject to the $T(x,t)$ boundary conditions:

$$T(x,0) = T_0; \, T(0,t) = T_0; \quad \text{and} \quad T(b,t) = T_s$$

Constant at T_0

Wall T

Originally T_0, then changed to T_s

b

x

FIGURE 15-47
Transient temperature distribution in a wall.

For convenience normalize the temperature by defining $U = T - T_o$, so that

$$\frac{\partial^2 U}{\partial x^2} = \frac{1}{\alpha} \frac{\partial U}{\partial t} \tag{15.38}$$

subject to

$$U(x,0) = 0; \quad U(0,t) = 0; \quad \text{and} \quad U(b,t) = T_s - T_o \tag{15.39}$$

The solution which will be attained through the use of Laplace transforms is

$$\frac{T - T_o}{T_s - T_o} = \frac{x}{b} + \frac{2}{\pi} \sum_{n=1}^{\infty} \frac{(-1)^n}{n} e^{-\alpha n^2 \pi^2 t/b^2} \sin\left(\frac{n \pi x}{b}\right) \tag{15.40}$$

The inversion technique is a general statement of one of the methods cited in Sec. 15.5:

$$\mathcal{L}^{-1}\left\{\frac{N(s)}{D(s)}\right\} = \sum_{j=1}^{n} \frac{N(a_j)}{D'(a_j)} e^{a_j t} \tag{15.41}$$

where a_j is a *simple* (nonrepeated) pole of $D(s)$. An example of the application of Eq. (15.41) is to invert $(s + 1)/(s^2 + 2s)$. $D(s) = s^2 + 2s$, so $D'(s) = 2s + 2$. The poles or roots of $D(s)$ are $s = 0$ and $s = -2$, so

$$\mathcal{L}^{-1}\left\{\frac{s + 1}{s^2 + 2s}\right\} = \frac{0 + 1}{2(0) + 2} e^{0t} + \frac{-2 + 1}{2(-2) + 2} e^{-2t} = \frac{1}{2}(1 + e^{-2t})$$

The steps in the solution of a partial differential equations are as follows:

1. Transform one of the independent variables in the differential equation; for example, transform t to s. Introduce a boundary condition, if helpful. The result is an ordinary differential equation relating $u(x,s)$ to x and s.
2. Solve the ordinary differential equation. If the form is a familiar one, simply write the solution. Otherwise, transform x to its own Laplace variable, perhaps $ȿ$, to derive a function $u(ȿ,s)$. Solve for $u(ȿ,s)$ and introduce either constants or transformed boundary conditions directly, then invert $ȿ$ to x.
3. By either of the techniques in Step 2, an ordinary differential equation for $u(x,s)$ is now available. The final step is to invert s to t —a step that usually entails the most effort of all. The individual steps will now be executed on the problem at hand.

Step 1

Transform t to s:

$$u_{xx}(x,s) = \left(\frac{1}{\alpha}\right)[s \cdot u(x,s) - U(x,0)]$$

One of the boundary conditions in Eq. (15.39) indicates that $U(x,0) = 0$, so

$$u_{xx}(x,s) - \left(\frac{s}{\alpha}\right)u(x,s) = 0 \qquad (15.42)$$

Step 2

The ordinary differential equation, Eq. (15.42), is a familiar one whose solution is

$$u(x,s) = C_1 \sinh\sqrt{\frac{s}{\alpha}}x + C_2 \cosh\sqrt{\frac{s}{\alpha}}x \qquad (15.43)$$

subject to the transformed boundary conditions $U(0,s) = 0$ and $u(b,s) = (T_s - T_o)/s$. Application of the first of these two boundary conditions yields

$$u(x,s) = C_1 \sinh(x\sqrt{s/\alpha}),$$

and from the second boundary condition

$$\frac{u(x,s)}{T_s - T_o} = \frac{\sinh\sqrt{\frac{s}{\alpha}}x}{s(\sinh\sqrt{\frac{s}{\alpha}}b)} \qquad (15.44)$$

which is the function that will be inverted using the form of Eq. (15.41).

Step 3

A series representation of the hyperbolic functions will repeatedly appear in what follows, so for future reference, recall that

$$\sinh y = y + \frac{1}{3!}y^3 + \ldots \qquad \text{and} \qquad \cosh y = 1 + \frac{1}{2!}y^2 + \ldots \qquad (15.45)$$

If $s = 0$ is substituted into $N(s)/D'(s)$, from (15.44)

$$\frac{N(s)}{D'(s)}\bigg|_{s\to 0} = \frac{\sinh\sqrt{\frac{s}{\alpha}}x}{\sinh\sqrt{\frac{s}{\alpha}}b + \frac{s}{2\sqrt{s\alpha}}\cosh\sqrt{\frac{s}{\alpha}}b}\bigg|_{s\to 0}$$

If the sinh and cosh functions are expressed in the series of Eqs. (15.45) and $\sqrt{s/\alpha}$ canceled, the result is

$$\frac{x}{b + \frac{b^2}{2}} e^{0t} = \frac{x}{b + \frac{b^2}{2}}$$

which may look reasonable, but is wrong. The reason for the error is that $s = 0$ is a compound pole, not a simple pole of $D(s)$. The problem can be circumvented by expressing both $N(s)$ and $D(s)$ of Eq. (15.44) in series form, then canceling $\sqrt{s/\alpha}$ leaving

$$\frac{x + \frac{1}{3!}\frac{s}{\alpha}x^3 + \cdots}{s(b + \frac{1}{3!}\frac{s}{\alpha}b^3 + \cdots)}$$

Now s is a simple pole and

$$\frac{N(s)}{D'(s)} e^{0t} = \frac{x}{b} \tag{15.46}$$

which is the expected steady-state value.

The other poles exist when $\sinh(b\sqrt{s/\alpha}) = 0$. Recalling that $i \sin z = \sinh iz$, and that $\sin z = 0$ when $z = 0, \pi, 2\pi, \ldots, n\pi$, then poles exist when $b\sqrt{s/\alpha} = in\pi$ for $n = 0, 1, 2, \ldots$ Thus,

$$s = -\frac{n^2\pi^2\alpha}{b^2} \quad \text{for } n = 0, 1, 2, \ldots$$

but the $n = 0$ pole has already been accounted for, so the n values of interest are 1, 2, 3, etc. Applying Eq. (15.41) to perform the inversion,

$$\sum_{j=1}^{n} \frac{N(s)}{D'(s)}\Big|_{s \to a_j} e^{a_j t} = \frac{\sinh\sqrt{\frac{s}{\alpha}}x}{\sinh\sqrt{\frac{s}{\alpha}}b + \frac{sb}{2\sqrt{s\alpha}}\cosh\sqrt{\frac{s}{\alpha}}b}\Big|_{s \to a_j} e^{a_j t}$$

Substituting $s/\alpha = in\pi/b$

$$\mathcal{L}^{-1}\{\}_{\text{transient}} = \sum_{n=1}^{\infty} \frac{\sinh(in\pi\frac{x}{b})}{\sinh(in\pi) + (\frac{in\pi}{2})\cosh(in\pi)} e^{-n^2\pi^2\alpha t/b^2}$$

$$\mathcal{L}^{-1}\{\}_{\text{transient}} = \sum_{n=1}^{\infty} \frac{i\sin(n\pi x/b)}{0 + (in\pi/2)\cos(n\pi)} e^{-n^2\pi^2\alpha t/b^2}$$

$$\mathcal{L}^{-1}\{\}_{\text{transient}} = \frac{2}{\pi}\sum_{n=1}^{\infty} \frac{(-1)^n}{n} e^{-n^2\pi^2\alpha t/b^2} \sin n\pi\frac{x}{b}$$

which when combined with the steady-state portion, Eq. (15.46), gives the complete solution, Eq. (15.40).

One of the strengths of the Laplace transform approach to solving partial differential equations is that, as was true for the solution of ordinary differential equations, the solution is a methodical process that does not demand special insight. Also, the eigenvalues in the summation term follow directly from the zeros determined from trignometric functions.

15.24 SUMMARY

This chapter addressed numerous topics related to the dynamic behavior and control of thermal systems. The intent has been to remain close to thermal hardware and to use mathematical tools of only moderate difficulty to provide more understanding of the response of actual systems. The chapter also sought to provide practice in translating physical equipment and systems to a symbolic form that can be attacked by the mathematics of control systems.

PROBLEMS

15.1. Compute the time necessary to raise the pressure of air in the storage vessel from 200 kPa to 900 kPa with the two-stage compression system shown in Fig. 15-48. Data and conditions in addition to those shown in Fig. 15-48 are as follows:

Adiabatic compression efficiency of both compressors

$$\eta_c = \frac{\text{isentropic work}}{\text{actual work}}(100) = 75\%$$

Volumetric efficiency of both compressors

$$\eta_v, \% = \frac{\text{volume rate measured at compressor suction}}{\text{displacement rate of compressor}}(100)$$

$$= 100 - 12[(p_{\text{disch}}/p_{\text{suction}}) - 1]$$

Displacement rates

low-stage compressor, 0.12 m³/s
high-stage compressor, 0.075 m³/s

Properties of air

$$R_{\text{air}} = 287 \text{ J/(kg} \cdot \text{K)}$$

$$c_p = 1.00 \text{ kJ/(kg} \cdot \text{K)}$$

$$c_v = 0.712 \text{ kJ/(kg} \cdot \text{k)}$$

FIGURE 15-48
Two-stage air compressor in Problem 15.1.

Assume no heat transfer between the environment and the compressor, intercooler, and storage vessel.

The principles of filling processes apply to air entering the tank, namely, the temperature of air after entering is greater than before entering the tank.

The air at point 1 is dry enough that no moisture condenses at either heat exchanger.

Use a time step of 4 s.

Ans.: Some values when the tank pressure has reached 900 kPa: $T_2 = 130.9°C$, $T_3 = 27.4°C$, $T_{tank} = 144.4°C$, $p_2 = 218$ kPa.

15.2. A recompression unit servicing a liquid ammonia storage vessel, Fig. 15-49, consists of a compressor, condenser, and expansion valve. The purpose of the recompression unit is to draw off vapor and thereby cool the stored liquid. The vapor is condensed and passed through a throttling valve back to the tank.

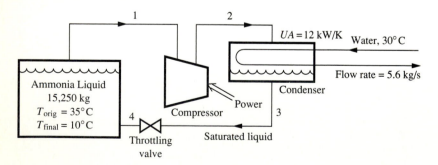

FIGURE 15-49
Recompression unit for reducing the temperature of liquid ammonia in a storage vessel.

In the process under consideration the temperature of the ammonia liquid is to be brought down from 35°C to 10°C. Using a steady state simulation plus one dynamic element, determine the time required for this reduction in temperature.

Compressor characteristics

Volume rate flow measured at the compressor suction m^3/s:

$$0.018 - 0.003\left[\left(\frac{p_2}{p_1}\right)^{1.2} - 1\right]$$

Work required by the compressor which is absorbed by the vapor during compression:

$$\Delta h, \text{kJ/kg} = 2010\left[\left(\frac{p_2}{p_1}\right)^{0.07} - 1\right]$$

Condenser heat transfer

Compute the log-mean temperature difference in the condenser as though the ammonia were at a constant temperature (the condensing temperature).

Ammonia properties as a function of temperature, °C

h of saturated liquid, kJ/kg $= 200. + 4.599t + 0.004177t^2$

h of saturated vapor, kJ/kg $= 1461.5 + 1.125t - 0.0103t^2$

specific volume of saturated vapor,
$$\text{m}^3/\text{kg} = 0.2854 - 0.00826t + 0.0000776t^2$$

saturation pressure-temperature relationship, where p is in kPa:

$$\ln p = 16.1 - \frac{2740}{(t + 273.15)}$$

Suggested time step $= 200$ s.

Ans.: Two variables for checking at final state: $T_3 = 37.7$°C, mass flow rate through the compressor, 0.0596 kg/s.

15.3. Substitute the following functions of time into the definition of the Laplace transform, Eq. (15.8), and integrate to verify the transforms shown for these functions on Table 15.1: (a) e^{at}, (b) $\cosh(at)$, and (c) $\cos(at)$, making use of $\cosh(jat)$.

15.4. Use trigonometric identities and transforms from Table 15.1 where possible to verify the following transforms:

(a) $\mathcal{L}\{\sin t + 2\cos t\} = \dfrac{1 + 2s}{s^2 + 1}$

(b) $\mathcal{L}\{\sin(t)\cos(t)\} = \dfrac{1}{s^2 + 4}$

(c) $\mathcal{L}\{\cos^2(at)\} = \dfrac{s^2 + 2a^2}{s^3 + 4sa^2}$

15.5. Use partial fractions and Table 15.1 where possible to show that

(a) $\mathcal{L}^{-1}\left\{\dfrac{7s^3 - 4s^2 + 20s - 16}{s^4 - 16}\right\} = 2e^{2t} + 4e^{-2t} + \cos(2t)$

(b) $\mathcal{L}^{-1}\left\{\dfrac{6s^2 + 10s + 20}{s^3 + 2s^2 + 5s}\right\} = 4 + 2e^{-t}\cos(2t)$

15.6. Solve the following differential equations using Laplace transformations:

(a) $Y''(t) + 2kY'(t) + k^2 Y(t) = 0$

   Ans.: $Y = e^{-kt}[C_1 + (C_1 k + C_2)t]$
   $= e^{-kt}(C_1 + C_3 t)$

(b) $Y''(t) + 4Y(t) = \sin t$ with boundary conditions: $Y(0) = Y'(0) = 0$

   Ans.: $Y = (1/3)\sin t - (1/6)\sin(2t)$

(c) $Y''(t) - 2Y'(t) + Y(t) = 0$ with boundary conditions: $Y(0) = 0$, $Y(1) = 2$

   Ans.: $Y(t) = 2te^{(t-1)}$

15.7. Solve the following differential equations using Laplace transformations:

(a) $Y'(t) - Y(t) = e^{-t}$ with boundary conditions: $Y(0) = 2$

   Ans.: $Y(t) = 2e^t + \sinh t$

(b) $Y''(x) + Y'(x) = 2$ with boundary conditions: $Y(0) = 0$ and $Y'(0) = 4$

   Ans.: $Y(x) = 2x + 2 - 2e^{-x}$

(c) $Y''(t) + Y(t) = t$ with boundary conditions: $Y(0) = 0$ and $Y(\pi/2) = 2$

   Ans.: $Y(t) = t + (2 - \pi/2)\sin t$

15.8. Electric current passes through a small cylinder of semiconductor material, shown in Fig. 15-50. The following conditions prevail:

$T = 20°C$
$K = 1.5$ W/m·K
$\rho = 10^{-5}\,\Omega\cdot$m
$h = 12$ W/m^2·K
Ambient temperature $= 20°C$
$T = 20°C$

$I = 12$ A

40 mm

10 mm

FIGURE 15-50
Semiconductor element in Prob. 15.8.

diameter of cylinder, 10 mm

length of cylinder, 40 mm

thermal conductivity, $k = 1.5$ W/m $\cdot$ k

current, $I = 12$ A

electrical resistivity, $\rho = 10^{-5}$ Ω-m

convection coefficient for heat transfer from external surface of the cylinder to air, $h = 12$ W/m$^2 \cdot$ K

The temperatures of both ends of the cylinder are 20°C, and the ambient air is also at 20°C

The electrical resistance of a length of the material is ρ (length in direction of flow)/(cross-sectional area).

(a) What is the differential equation (including numerical coefficients) and boundary conditions relating the temperature T to the distance x along the element?

(b) Using Laplace transforms, solve the differential equation to develop the expression for T as a function of x.

Ans.: At the midlength, $T = 40.2$°C.

15.9. The temperature-sensing bulb in Fig. 15-5a is subjected to a fluid temperature T_f that varies sinusoidally according to the equation

$$T_f - T_m = \Delta \sin(2\pi f t),$$

where T_m is constant

Δ = amplitude

f = frequency

(a) Determine the transfer function $\mathcal{L}\{T - T_m\}/\mathcal{L}\{T_f - T_m\}$ if at $t = 0$, $T = T_m$.

(b) What is the equation for the temperature T of the block after the process has been in operation for a long time?

(c) Develop two graphs—one of the phase shift in degrees and the other of $(T - T_m)/\Delta$—against $2\pi f \tau$ where $\tau = mc/hA$. Graph $2\pi f \tau$ as the abscissa on a log scale over the range of 0.1 to 100.

Ans.: to (b):

$$T - T_m = \frac{\Delta}{\sqrt{1 + (2\pi f \tau)^2}} \sin[2\pi f t - \tan^{-1}(2\pi f \tau)]$$

15.10. The temperature-sensing bulb illustrated in Fig. 15-5a is subjected to a fluid temperature T_f that varies linearly according to the equation, $T_f = T_o + gt$, where g is a constant.

(a) Determine the expression for T if $T = T_o$ when $t = 0$.

(b) Sketch the variation of T_f and T with respect to time.

Ans.: $T - T_o = g\left\{t - \left(\frac{mc}{hA}\right)[1 - e^{-t/(mc/hA)}]\right\}$

15.11. A room heated by warm air as shown in Fig. 15-51 has a volume of 225 m^3 and is supplied with 0.36 kg/s of air. The air in the system is at

FIGURE 15-51
Heating system in Prob. 15.11.

a pressure of 101 kPa, its c_p = 1.0 kJ/(kg · K), and R = 0.287 kJ/(kg · K). The heat is supplied in an on/off manner in that a thermostat switches on an electric heater when the room temperature drops to 24°C whereupon 38°C air immediately begins entering the room. When the room temperature rises to 26°C the thermostat turns off the heater and the temperature of the supply air drops immediately to that of the return air. The heat loss from the room is constant at 3.8 kW.

(*a*) How long a period is the heater on?

(*b*) How long a period is the heater off?

15.12. For purposes of dynamic analysis, a steam coil that heats air is considered to be a mass of metal that is at a uniform temperature throughout at any given time, as shown in Fig. 15-52. The magnitudes of some of the parameters are as follows:

hA = product of heat-transfer coefficient and area between steam and coil, 5.7 kW/K

C = product of mass and specific heat of coil, 95 kJ/K D = product of mass rate of flow and specific heat of air, 1.9 kW/K

The effectiveness of the coil in transferring heat to the air is 0.5, thus $(T_o - T_i)/(T_c - T_i) = 0.5$

FIGURE 15-52
Steam coil in Prob. 15.12.

If at time $t = 0$ the steam temperature T_s experiences a step change from 65°C to 95°C, what is the equation for the outlet air temperature T_o as a function of time?

Ans.: $T_o = 46.43 - 12.86e^{-0.07t}$

15.13. Invert the transfer function for the cascaded time constant blocks, Eq. (15.19), subject to the step input to develop Eq. (15.21) for the case where $\tau_1 = \tau_2$.

15.14. The humidity is controlled in a laboratory that uses 100% outdoor air by adding water vapor as needed. In the system shown in Fig. 15-53, the humidity sensor in the exhaust air measures the humidity ratio H, which is the kg of water vapor associated with a kg of dry air. The humidity sensor has a time constant of 30 s. The controller regulates the power supplied to the electric heater in the humidifier and thus the rate of water vaporized into the air stream, w_{wv}, according to the equation

$$w_{wv}, \text{kg/s} = 300(H_{set} - H_{sensed})$$

Due to the thermal capacity of the heater and the water reservoir, the response of the humidifier is not instantaneous, but has a time constant of 4 min. Assume that the entering outdoor air contains negligible water vapor, and that the mixing of water vapor throughout the laboratory is uniform.

Use frequency response and the principles of the Bode diagram to determine (*a*) the frequency resulting in the loop phase lag of 180°, and (*b*) the combined loop gain when the loop phase lag is 180°. (*c*) Is the control stable or unstable?

Ans.: (*b*) 1.2

15.15. The expansion valve regulating the flow rate of refrigerant to an evaporator in a refrigeration system is the type that controls the number of degrees of superheat at the outlet of the evaporator. The control loop is represented by a combination of time-constant blocks, as shown in Fig. 15-54. The

FIGURE 15-53
Humidity controller in Prob. 15.14.

FIGURE 15-54
Block diagram of evaporator-expansion valve control loop in Prob. 15.15.

characteristics of the evaporator and certain characteristics of the valve are fixed and are designated in Fig. 15-54.

Use frequency response and the principles of the Bode diagram to determine the maximum permissible gain of the valve, k_2, in order to make this loop stable. This gain in kg/s per °C is predominantly dependent on the size of the valve port.

Ans.: 0.028 kg/s per °C.

15.16. The transfer function of a feedback control loop is

$$\frac{14s^2 - 2s + 3}{s^3 + 6s^2 - 59s + 156}$$

(a) Use the Routh-Hurwitz criterion to determine the number of roots of the characteristic equation that have positive real parts.

(b) What are the roots of the characteristic equation?

Ans.: (a) 2

15.17. The control loop shown in Fig. 15-55 consists of three time-constant blocks.

(a) If $\tau_1 = \tau_2 = \tau_3$, use the Routh-Hurwitz criterion to determine the product $k_1k_2k_3$ resulting in the loop being on the borderline between stability and instability.

(b) Choose any arbitrary combination of nonequal values of the τ's and compute the $k_1k_2k_3$ for borderline stability. *Comment*: the least stable

FIGURE 15-55
Control loop of three time-constant blocks in Prob. 15.17.

combination is when all τ's are equal. Sometimes a loop can be made more stable by spreading apart the values of the τ's.

15.18. The temperature T_c of an environmental chamber and its interior mass, shown schematically in Fig. 15-56, is regulated by a controller that adjusts the rate of electric power input, q_h. The thermal capacity of the interior mass of the chamber is M kJ/K and its temperature T_c is assumed to be uniform throughout the interior of the chamber, but varies with time. The instantaneous rate of heat transfer from the chamber to the ambient, q_a, is represented by the equation

$$q_a = H(T_c - T_a),$$

where T_a = ambient temperature. The temperature of the sensor is T_s and its time constant is τ_s. The characteristic of the controller is $q_h = B(T_{set} - T_s)$

(a) Construct the block diagram of the control loop, inserting appropriate transfer functions and using the ambient temperature T_a as the input and the chamber temperature T_c as the output.

(b) Normalize the variables associated with the block diagram in anticipation of determining the response of T_c to a change in T_a while T_{set} remains constant.

(c) Simplify the loop to a unity or nonunity feedback loop.

(d) Some numerical values are: M = 300 kJ/K, B = 5 kW/K, H = 0.5 kW/K, and τ_s = 10 s. If T_a = 30°C and T_{set} = 60°C, what are steady-state values of T_s and T_c?

(e) If the ambient temperature T_a changes abruptly to 20°C, what is the equation for the response of T_c with respect to time?
   Ans.: $T_c = 56.36 - 0.085e^{-0.07824t} + 0.99e^{-0.02344t}$

15.19. The pressure in the evaporator of Fig. 15-57 is regulated by controlling the rate of vapor pumped by the compressor, w_c. The purpose of the evaporator is to cool a fluid that flows through tubes immersed in the liquid. The control is of the proportional type following the equation

$$w_c = K(p - p_{set})$$

FIGURE 15-56
Environmental chamber in Prob. 15.18.

FIGURE 15-57
Evaporator with controlled pressure in Prob. 15.19.

Some other applicable relationships are as follows:

> mass of vapor in vessel, $m_v = Bp$
> latent heat of evaporating fluid $= h_{fg}$
> evaporation rate, $w_{ev} = C(p_f - p)$
> vapor pressure of liquid, $p_f = DT_f$
> thermal capacity of liquid in evaporator $= M$
> rate of heat transfer from tube to liquid $= hA(T_t - T_f)$
> the state of liquid entering the evaporator is essentially saturated liquid

at temperature T_f

(a) Construct the block diagram of the control loop with the tube temperature T_t as the reference variable and the pressure p as the controlled variable. Indicate the variables (nonnormalized) and the transfer functions of the components.
(b) Normalize the variables.
(c) Restructure the block diagram into a unity or nonunity feedback loop, and specify the transfer function of the loop.

Ans.: (c)

$$\frac{DC\,hA}{BM\,s^2 + (BhA + BCDh_{fg} + KM + CM)s + KhA + KCDh_{fg} + ChA}$$

15.20. The air-heating system of Fig. 15-26 is subjected to a step change in T_{set} from 42 to 47°C which yields the loop transfer function shown in Eq. (15.27). The control chosen is integral only, with $K_P = 0$ and $K_I = 0.02$.
(a) Develop the equation for the response of the controlled temperature T_c as a function of time.
(b) Plot T_c in the 0–400 s time interval.

Ans.: (b) to check results, T_c first reaches 47°C in 113.2 s.

15.21. Figure 15-35 shows the response of the controlled temperature T_c to a step change in T_{set} from 42 to 47°C for several different combinations of K_P and K_I. Invert Eq. (15.27) for the case where $K_P = 2.4$ and $K_I = 0.1$ and compare with Fig. 15-35.

    Ans.: to check results, peak of overshoot is 48.377°C at 62.1 s.

15.22. A higher value of the derivative constant K_D than used in Fig. 15-37 for the pressure controller is to be explored. Compute and plot the pressure for 20 s following the 0.2 kg/s step increase in outlet flow rate when $K_P = 0.2$, $K_I = 0.05$, and $K_D = 0.4$.

    Ans.: to check results, maximum drop in pressure is 2.03 kPa at 3.7 s.

15.23. (a) Derive the loop transfer function, Eq. (15.32), for the PI-only control of the water heater in Example 15.11.

    (b) Multiply the transform of Eq. (15.32) by the transform of the input, $-5/s$, and then invert to verify the response of the PI-only control shown on Fig. 15-41.

    (c) Restructure the complete block diagram of Fig. 15-40 (both PI and feedforward control) into a unity or nonunity feedback loop to verify that the loop transfer function is 1/6 that given in Eq. (15.32).

    Ans.: a check value in part (b) is 64.02°C at 300 s.

15.24. A heating coil is modeled in Fig. 15-43 as a combination of a time constant and a transportation lag. In the Bode diagram procedure for assessing stability of a loop, as described in Sec. 15.11, a time-constant block is shown to have a phase lag that only approaches 90°. When combined with dead time, however, the phase lag can exceed 90°. If the heating coil has a time constant of 20 s and a dead time of 3 s, above what frequency does the phase lag exceed 90°?

15.25. Section 15.21 described the flow characteristics of a coil regulated by a valve with linear characteristics. The equation of the flow–stem position for another type of valve mentioned in Sec. 15.21, the equal-percentage valve, is

$$\frac{Q}{C_v \sqrt{\Delta p}} = A^x \quad \text{where } x = \frac{\text{percent of stem stroke}}{100} - 1$$

If such a valve with $A = 20$ and $C_v = 1.2$ is applied to controlling the coil in Fig. 15-44 with $\Delta p_{coil} = 2.5Q^2$ and the total pressure drop across the valve and coil of 80 kPa, what is the flow rate when the valve stem stroke is at the halfway position? (Compare with a linear-characteristic valve in Fig. 15-46.)

    Ans.: 2.21 L/s.

15.26. An underground electric cable, as in Fig. 15-58, that is 5,000 m long is cooled by pumping oil through a pipe that surrounds the cable. The oil has a density of 800 kg/m³, a specific heat of 2.0 kJ/kg · K, its flow rate is 8 kg/s or 0.01 m³/s, and it enters the pipe with a temperature of 15°C. The cross-sectional area for oil flow is 0.02 m². All the oil in the pipe is initially at 15°C when at time zero an electrical load of 50 W/m is imposed on the cable which must be absorbed by the oil.

FIGURE 15-58
Underground cable in Prob. 15.26.

(a) Develop the differential equation for the oil temperature as a function of time and distance.

(b) Solve the partial differential equation of part (a) by Laplace transforms using, if needed, the transform

$$\mathcal{L}^{-1}\left\{\frac{e^{-ks}}{s^2}\right\} = \begin{bmatrix} 0 & \text{when} & 0 < t < k \\ t - k & \text{when} & t > k \end{bmatrix}$$

Sketch the result for several different values of t on a $T - x$ diagram.

Ans.: (a) $2\dfrac{\partial T}{\partial t} + \dfrac{\partial T}{\partial x} - 0.00313 = 0$

15.27. Water flows through a tube that rejects heat to cooler air, as in Fig. 15-59. The air is at a uniform temperature of 20°C. The water flows at a rate of 0.5 kg/s in the tube that has a cross-sectional area of 0.001 m² and a length of 2 m. The specific heat of water is 4.19 kJ/kg K and its density is 1,000 kg/m³. The heat transfer coefficient on a *length* basis is 70 W/K · m. The tube has negligible specific heat. Steady-state conditions had been permitted to develop with the water entering the tube at 30°C when at time $t = 0$ the temperature of the entering water suddenly changes to 40°C.

(a) Develop the differential equation that relates T to x and t.

(b) Specify the boundary and initial conditions.

(c) Develop the complete transform with $x \rightarrow s$ and $t \rightarrow w$.

Ans.: (a) $\dfrac{\partial U}{\partial x} + 2\dfrac{\partial U}{\partial t} + 0.0334U = 0$ where $U = T - 20$

(c) $q = \dfrac{20}{w(2w + s + 0.0334)} + \dfrac{20}{(s + 0.0334)(2w + s + 0.0334)}$

20°C air

Water
0.5 kg/s

2 m

Area
= 0.001 m²

FIGURE 15-59
Transfer of heat from tube in Prob. 15.27.

REFERENCES

1. J. A. Lock, *Transient Responses in a Large Steam System*, Petroleum Mechanical Engineering Workshop and Conference, ASME, 1982.
2. C. L. Wilson, "Fan-Motor Time-Torque Relations," *Heating, Piping, and Air Conditioning*, vol. 43, no. 5, pp. 75–77, May 1971.
3. D. F. Baker, "Surge Control for Multistate Centrifugal Compressors," *Chemical Engineering*, vol. 89, no. 11, pp. 117–122, May 31, 1982.
4. F. G. Shinskey, *Process-Control Systems*, 2d. ed., McGraw-Hill, New York, 1979.
5. P. O. Danig and B. Teislev, "A Generalized Program for Dynamic Analysis of Non-linear Systems," Report F29-77.01, Refrigeration Laboratory, Danish Technical University, 1977.
6. H. Nyquist, "Regeneration Theory," *Bell System Technical Journal*, January 1932, pp. 126–147; also in *Automatic Control: Classical Linear Theory*, G.J. Thaler, ed., pp. 105–126, Dowden, Hutchinson and Ross, Stroudsburg, PA, 1974.
7. E. J. Routh, *Dynamics of a System of Rigid Bodies*, Macmillan, New York, 1892.
8. H. L. Harrison and J.G. Bollinger, *Introduction to Automatic Controls*, International Textbook Co., Scranton, Penn., 1969.
9. G. R. Sapienza, "Using Fixed and Dynamic Performance Curves to Optimize HVAC Equipment Operation," Paper from seminar: "Using Your DDC for More Than PID Control," American Society of Heating, Refrigerating, and Air-Conditioning Engineers, New York, January 1987.
10. D. L. Auslander, Y. Takahashi, and M.J. Rabins, *Introducing Systems and Control*, McGraw-Hill, New York, 1974.
11. C. G. Nesler, "Direct Digital Control of Discharge Air Temperature Using a Proportional-Integral Controller," M.S. Thesis in Mechanical Engineering, University of Illinois at Urbana-Champaign, 1983.

CHAPTER
16

CALCULUS
METHODS
OF
OPTIMIZATION

16.1 CONTINUED EXPLORATION
OF CALCULUS METHODS

The introduction to calculus methods of optimization was provided in Chapter 8's coverage of Lagrange multipliers. The function to which Lagrange multiplier optimization directly applies is

$$y = y(x_1, x_2, \ldots x_n) \tag{16.1}$$

subject to the equality constraints,

$$\phi_1(x_1, x_2, \ldots x_n) = 0 \tag{16.2}$$

$$\vdots$$

$$\phi_m(x_1, x_2, \ldots x_n) = 0 \tag{16.3}$$

The optimal values of the x's are found by the simultaneous solution of the $n + m$ scalar equations, some of which may be non-linear:

$$\nabla y - \sum_{i=1}^{m} \lambda_i \nabla \phi_i = 0 \tag{16.4}$$

$$\phi_1(x_1, x_2, \ldots x_n) = 0 \tag{16.5}$$

$$\vdots$$

$$\phi_m(x_1, x_2, \ldots x_n) = 0 \tag{16.6}$$

The method of Lagrange multipliers is useful and powerful and could very well be the best choice for many practical optimization problems. For large problems the explicit solution of Eqs. (16.4) to (16.6) may be prohibitive, but since the task is one of solving a set of nonlinear simultaneous equations, the methods of Chapter 14 apply. This chapter also suggests the structure of a generalized computer program using an iterative technique to solve the equations when provided the statement of the problem—Eqs. (16.1) to (16.3)—and trial values of the x's. This chapter also provides more insight into the topic touched in Sec. 8.11 of using Lagrange multipliers for optimization where one or more of the constraints are inequalities.

A major portion of this chapter deals with principles—some of which facilitate solutions using calculus methods and some others helpful in search methods. Substantiation for the Lagrange multiplier equations, the physical interpretation of λ, and tests for maxima-minima are also provided.

16.2 THE NATURE OF THE GRADIENT VECTOR

The condition for optimality expressed by the Lagrange-multiplier equation, Eq. (16.4), calls for gradient vectors of the obejctive function y and the ϕ equations. Three characteristics of the gradient vector are (1) that it is normal to the surface of constant y, (2) that it indicates the direction of maximum rate of change of y with respect to the x distance, and (3) that it points in the direction of increasing y.

Section 8.7 showed for two-dimensional space that ∇y is perpendicular to the line of constant y passing through the point. A more general proof is available from vector algebra using the fact that the dot product of two normal vectors is zero. An arbitrary vector of small length with dx's as the components and the $\mathbf{i}$'s as the unit vectors is

$$dx_1 \mathbf{i}_1 + dx_2 \mathbf{i}_2 + dx_3 \mathbf{i}_3 \tag{16.7}$$

For the vector to be tangent to the surface, the change in y from the start to the end of the vector, dy, must be zero. Furthermore, from calculus

$$dy = \frac{\partial y}{\partial x_1}dx_1 + \frac{\partial y}{\partial x_2}dx_2 + \frac{\partial y}{\partial x_3}dx_3 = 0 \qquad (16.8)$$

From Eq. (16.8)

$$dx_1 = -\frac{(\partial y/\partial x_2)dx_2 + (\partial y/\partial x_3)dx_3}{\partial y/\partial x_1}$$

which can be substituted into Eq. (16.7) to develop the tangent vector,

$$\mathbf{T} = -\left[\frac{\left(\frac{\partial y}{\partial x_2}\right)dx_2 + \left(\frac{\partial y}{\partial x_3}\right)dx_3}{\frac{\partial y}{\partial x_1}}\right]i_1 + dx_2 i_2 + dx_3 i_3$$

The dot product is a scalar formed by multiplying the components of one vector by the corresponding components of the other vector, and then summing the products. The dot product of ∇y and the tangent vector is

$$\mathbf{T}\cdot\nabla y = 0$$

When the dot product of two vectors is zero the vectors are normal to one another, and ∇y is normal to all tangent vectors at the point in question, so it is normal to the surface.

The second characteristic of the gradient vector is that it indicates the direction of the maximum rate of change of the function y with respect to distance in the $x_1 - x_n$ space. Perhaps the property that ∇y is normal to a surface of constant y strongly suggests a direction in which the rate of change is most rapid, but the characteristic can be proven more convincingly. The components $dx_1, dx_2, \ldots$ of a very short vector will be sought to give the maximum rate or change in y. From calculus

$$dy = \frac{\partial y}{\partial x_i}dx_1 + \frac{\partial y}{\partial x_2}dx_2 + \cdots + \frac{\partial y}{\partial x_n}dx_n \qquad (16.9)$$

The relative values of the dx's are sought that give the maximum dy, but abiding by the limitation of moving a specific distance r, where r is given by the equation

$$(dx_1)^2 + (dx_2)^2 + \cdots + (dx_n)^2 = r^2 \qquad (16.10)$$

In two dimensions, Eq. (16.10) describes a circle of radius r, and in three dimensions a sphere, as shown in Fig. 16-1a and 16-1b, respectively.

The task at hand is a constrained optimization problem to determine the optimal values of $dx_1, \ldots dx_n$ to give a maximum value of dy in Eq. (16.9) subject to the constraint of Eq. (16.10). The method of Lagrange

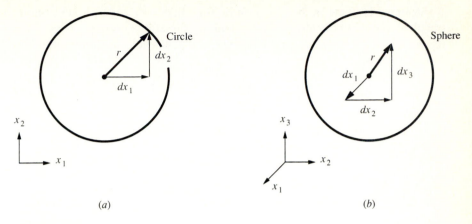

FIGURE 16-1
Maximum change in y when moving a distance r in (a) two dimensions (b) three dimensions.

multipliers will be chosen to perform the optimization—the same method that will be proven in this chapter.

The multiplier equation in the i direction is

$$\frac{\partial y}{\partial x_1} - 2\lambda dx_i = 0$$

so

$$dx_i = \frac{1}{2\lambda} \frac{\partial y}{\partial x_i}$$

which when expressed in vector form is

$$dx_1 \mathbf{i}_1 + dx_2 \mathbf{i}_2 + \cdots + dx_n \mathbf{i}_n$$

or

$$\frac{1}{2\lambda} \left[\frac{\partial y}{\partial x_1} \mathbf{i}_1 + \frac{\partial y}{\partial x_2} \mathbf{i}_2 + \cdots + \frac{\partial y}{\partial x_n} \mathbf{i}_n \right] = \frac{1}{2\lambda} \nabla y$$

Of interest here is that ∇y indicates the direction of maximum change for a given distance in the space.

The third characteristic answers a question that often arises in the execution of an optimization problem: Does ∇y point in the direction of increasing or decreasing y, and indeed is there any predictability? In a small move in the $x_1 - x_2$ space the change in y is represented by

$$dy = \frac{\partial y}{\partial x_1} dx_1 + \frac{\partial y}{\partial x_2} dx_2 \tag{16.11}$$

If the move is made in the direction of ∇y, then dx_1 and dx_2 are proportional to the components in the x_1 and x_2 directions, respectively. Thus $dx_1 = c(\partial y/\partial x_1)$ and $dx_2 = c(\partial y/\partial x_2)$, which when substituted into Eq. (16.11) yields

$$dy = c\left[\left(\frac{\partial y}{\partial x_1}\right)^2 + \left(\frac{\partial y}{\partial x_2}\right)^2\right] \tag{16.12}$$

so dy is equal or greater than zero, showing that y always increases in the direction of ∇y.

16.3 TEST FOR MAXIMUM OR MINIMUM OF AN UNCONSTRAINED OPTIMUM

Section 8.9 showed that the second derivatives at the optimal point in an unconstrained optimization are the indicators of whether the optimum is a maximum or minimum. In a single variable optimization, for example, if d^2y/dx^2 is positive, the optimum is a minimum. The role of the first and second derivatives can be visualized by imagining that the optimum has been found at the point $x_1^*, \ldots x_n^*$. That this point is a minimum, for example, can be demonstrated by writing a Taylor series expansion about that point, as was done in Eq. (8.24), arbitrarily move one or more of the x's, and verify that no lower value of y can be found. In matrix notation the Taylor series is

$$y = y^* + [\nabla y]^T[\Delta x] + [\Delta x]^T[Q][\Delta x] + \cdots \tag{16.13}$$

where $\quad \Delta x_i = x_i - x_i^*$

Q = matrix of second derivatives, called the "Hessian matrix"

y^* = optimal value being tested.

If $[\nabla y]^T[\Delta x]$ can be made to go negative, y^* is not truly the minimum, and the only way to avoid this condition for arbitrarily chosen signs of the Δx's is the ∇y to be zero. The second-degree terms in Eq. (16.13)

$$(\Delta x)^T(Q)(\Delta x) = (\Delta x_1 \ldots \Delta x_n)\begin{bmatrix} \dfrac{\partial^2 y}{\partial x_1^2} & \cdots & \dfrac{\partial^2 y}{\partial x_1 \partial x_n} \\ \dfrac{\partial^2 y}{\partial x_1 \partial x_2} & & \\ \vdots & & \\ & & \\ & & \dfrac{\partial^2 y}{\partial x_n^2} \end{bmatrix}\begin{bmatrix} \Delta x_1 \\ \vdots \\ \vdots \\ \vdots \\ \Delta x_n \end{bmatrix}$$

$$\tag{16.14}$$

and must always be positive if y^* is a minimum, or always negative if y^* is a maximum. Section 8.9 showed the conditions necessary for a 2 x 2 Hessian matrix to provide a positive sum in Eq. (16.14). A more general description of the condition is that the Hessian matrix must be *positive definite* for Eq. (16.14) to be uniformly positive or *negative definite* for the sum to be negative.

The tests for the positive-definite condition of a symmetric matrix, such as $[Q]$, are (1) that all the diagonal elements are positive, and (2) that all *leading principal diagonal* determinants are positive. The leading principal diagonal determinants of the matrix

$$\begin{bmatrix} 2 & 1 & -1 \\ 1 & 3 & 2 \\ -1 & 2 & 4 \end{bmatrix}$$

are

$$|2|$$

$$\begin{vmatrix} 2 & 1 \\ 1 & 3 \end{vmatrix}$$

and

$$\begin{vmatrix} 2 & 1 & -1 \\ 1 & 3 & 2 \\ -1 & 2 & 4 \end{vmatrix}$$

and since all of the determinants are positive, the matrix is positive definite. The test for a negative-definite matrix is to reverse the signs of all the elements and test whether the revised matrix is positive definite.

Example 16.1. For positive values of the x's, determine whether the optimum of the function

$$y = x_1 x_2 + \frac{1}{x_1 x_3} + \frac{4}{x_2} + x_3$$

is a maximum, minimum, or not definite.

Solution. By setting $\nabla y = 0$, the optimal values are found to be: $x_1^* = 0.5$, $x_2^* = 2.828$, and $x_3^* = 1.414$. The Hessian matrix is

$$\begin{bmatrix} 11.31 & 1 & 2 \\ 1 & 0.354 & 0 \\ 2 & 0 & 1.414 \end{bmatrix}$$

which is a positive definite matrix, since all of the diagonal elements are positive, and the leading principal diagonal determinants are 11.31, 3.0, and 2.831 — all positive. Because the matrix of the second derivatives is positive definite, the optimum is a minimum.

16.4   CONSTRAINED OPTIMIZATION

Most important engineering optimization problems are of the constrained type. While it is possible in principle to solve for one variable in each constraint equation and reduce the problem to an unconstrained optimization, execution of this strategy may not always be possible if the variables appear in an implicit form in the constraints. Furthermore, the algebra may become formidable, bringing with it the possibility of errors. The Lagrange multiplier equations, Eqs. 16.4 to 16.6, thus represent the basis for the solution of practical constrained optimization problems.

Proofs of the validity of the Lagrange multiplier equations appear only rarely in the literature. The next several sections present not a proof in the mathematical sense but a demonstration of the legitimacy of the equations. The target proof is that called the Jacobian method developed by Wilde and Beightler.[1] Rather than presenting the verification in full generality which risks bogging down in details, the next several sections proceed step-by-step to progressively more complex situations—each time showing the soundness of the Lagrange multiplier equations. There are at least two reasons for establishing the truth of the Lagrange multiplier equations. First, they are such classical relations that understanding their background is a broadening experience. Second, these demonstrations offer the means of solving certain practical problems related to optimization.

The sequence will be to show that the Lagrange multiplier equations apply to the following cases: two variables with one constraint, three variables with one constraint, and three variables with two constraints. By that point the direction toward full generality will have been established.

16.5   TWO VARIABLES AND ONE CONSTRAINT

The procedure in the next several sections will be to analyze the first-degree terms in the Taylor series expansion of the y function, shown in Eq. (16.13). When optimizing $y(x_1, x_2)$ subject to $\phi(x_1, x_2) = 0$, the first-degree terms are

$$\left(\frac{\partial y}{\partial x_1}\right)\Delta x_1 + \left(\frac{\partial y}{\partial x_2}\right)\Delta x_2 \tag{16.15}$$

The base point about which the expansion takes place is on the constraint and is the presumed optimal point. In this constrained problem Δx_1 and Δx_2 are not free to change independently, but must change in such a way that the new position continues to satisfy the constraint. Thus,

$$d\phi = \left(\frac{\partial \phi}{\partial x_1}\right)\Delta x_1 + \left(\frac{\partial \phi}{\partial x_2}\right)\Delta x_2 = 0$$

so
$$\Delta x_1 = -\frac{\dfrac{\partial \phi}{\partial x_2}}{\dfrac{\partial \phi}{\partial x_1}}\Delta x_2 \tag{16.16}$$

which when substituted into Eq. (16.15) yields

$$\left[-\frac{\partial y}{\partial x_1}\frac{\dfrac{\partial \phi}{\partial x_2}}{\dfrac{\partial \phi}{\partial x_1}} + \frac{\partial y}{\partial x_2}\right]\Delta x_2 \tag{16.17}$$

In the search for a position that improves the objective function in comparison to the presumed optimum, Δx_2 is permitted to shift either positively or negatively. In order for no improvement of y to be possible, the term in the brackets of Eq. (16.17) must be zero.

Define a term

$$\lambda = \frac{\dfrac{\partial y}{\partial x_1}}{\dfrac{\partial \phi}{\partial x_1}} \tag{16.18}$$

which when substituted into the zero term in the brackets of Eq. (16.17) gives

$$\frac{\partial y}{\partial x_2} - \lambda\frac{\partial \phi}{\partial x_2} = 0 \tag{16.19}$$

Revise the form of Eq. (16.18) to

$$\frac{\partial y}{\partial x_2} - \lambda\frac{\partial \phi}{\partial x_1} = 0 \tag{16.20}$$

Equations (16.19) and (16.20), along with the constraint equation which appeared by requiring the base point to be on the constraint, are the Lagrange multiplier equations for the two-variable, one-constraint case.

Example 16.2. A best fit of the equation $\ln p = -A/T + B$ is being developed to express the saturation pressure-temperature relation of methane. The equation is to represent the critical point (190.6 K, 4595 kPa) precisely and provide a minimum deviation D to tabular data elsewhere. In the search process the current values of A and B are 1029.18 and 13.8328, respectively, which satisfy the critical point. By numerical differentiation, the partial derivatives are $\partial D/\partial A = 0.0023$ and $\partial D/\partial B = -0.65$. The step size to the new value of B is 0.01. Should B be increased or decreased, and what should be the change in A?

Solution. Were the deviation $D(A,B)$ unconstrained, A would be decreased and B increased based on the signs of $\partial D/\partial A$ and $\partial D/\partial B$. The function is constrained, however, by

$$\phi = \ln p + A/T - B = 0$$

The derivatives at the critical point are $\partial \phi / \partial A = 0.0052$ and $\partial \phi / \partial B = -1$. The expression in the brackets of Eq. (16.17) is

$$-(0.0023)(-1)/(0.0052466) - 0.65 = -0.2116$$

In order to decrease D, ΔB must be positive, thus B should be increased. To hold the critical-point constraint, from Eq. (16.16)

$$\Delta A = -(-1/0.0052466)(0.01) = 1.906.$$

So even though $\partial D / \partial A$ is positive, suggesting a decrease in A to reduce D, the constraint requires A to increase.

16.6 THREE VARIABLES AND ONE CONSTRAINT

Here the function to be optimized is $y(x_1, x_2, x_3)$ subject to the constraint $\phi(x_1, x_2, x_3) = 0$. The first-degree terms in the Taylor series are

$$\left(\frac{\partial y}{\partial x_1}\right)\Delta x_1 + \left(\frac{\partial y}{\partial x_2}\right)\Delta x_2 + \left(\frac{\partial y}{\partial x_3}\right)\Delta x_3 \qquad (16.21)$$

The presumed optimum satisfies the constraint, and any motion of the x's must be such that the constraint remains satisfied, so

$$\left(\frac{\partial \phi}{\partial x_1}\right)\Delta x_1 + \left(\frac{\partial \phi}{\partial x_2}\right)\Delta x_2 + \left(\frac{\partial \phi}{\partial x_3}\right)\Delta x_3 = 0 \qquad (16.22)$$

Any two of the three variables can be moved independently, but the motion of the third variable must abide by Eq. (16.22). Arbitrarily choose x_1 and x_3 as the independent ones; these will be called *decision variables*. The other variable, x_2, (which will be called the *state variable*), can be expressed in terms of the others in Eq. (16.22) and substituted into Eq. (16.21) to obtain

$$\frac{\partial y}{\partial x_1}\Delta x_1 - \frac{\partial y}{\partial x_2}\left[\frac{\frac{\partial \phi}{\partial x_1}}{\frac{\partial \phi}{\partial x_2}}\Delta x_1 + \frac{\frac{\partial \phi}{\partial x_3}}{\frac{\partial \phi}{\partial x_2}}\Delta x_3\right] + \frac{\partial y}{\partial x_3}\Delta x_3 \qquad (16.23)$$

In order for no improvement to be possible in the function in comparison to the presumed optimum, the coefficients of Δx_1 and Δx_3 in Eq. (16.23) must be zero:

$$\frac{\partial y}{\partial x_1} - \frac{\partial y}{\partial x_2}\frac{\frac{\partial \phi}{\partial x_1}}{\frac{\partial \phi}{\partial x_2}} = 0 \qquad (16.24)$$

and

$$\frac{\partial y}{\partial x_3} - \frac{\partial y}{\partial x_2} \frac{\dfrac{\partial \phi}{\partial x_3}}{\dfrac{\partial \phi}{\partial x_2}} = 0 \qquad (16.25)$$

Define

$$\lambda = \frac{\dfrac{\partial y}{\partial x_1}}{\dfrac{\partial \phi}{\partial x_1}}$$

or

$$\frac{\partial y}{\partial x_1} - \lambda \frac{\partial \phi}{\partial x_1} = 0$$

Then from Eq. (16.24)

$$\frac{\partial y}{\partial x_2} - \lambda \frac{\partial \phi}{\partial x_2} = 0$$

and from Eq. (16.25)

$$\frac{\partial y}{\partial x_3} - \lambda \frac{\partial \phi}{\partial x_3} = 0$$

The above three equations are the scalar form of the Lagrange multiplier equation

$$\nabla y - \lambda \nabla \phi = 0$$

The Lagrange multiplier relation, therefore, holds for the case of three variables and one constraint.

16.7 THREE VARIABLES AND TWO CONSTRAINTS

This case continues the patterns set by the previous two cases, but introduces a new situation as well. The assignment is to optimize y, where

$$y = y(x_1, x_2, x_3) \qquad (16.26)$$

subject to

$$\phi_1(x_1, x_2, x_3) = 0 \qquad (16.27)$$

$$\phi_2(x_1, x_2, x_3) = 0 \qquad (16.28)$$

If a point (x_1^*, x_2^*, x_3^*) has been located which lies on the two constraints and is to be tested at the optimum, the first-degree terms in the expansion of y around that point are

$$\sum_{i=1}^{3} \frac{\partial y}{\partial x_i} \Delta x_i \qquad (16.29)$$

where $\Delta x_i = x_i = x_i^*$

and in order for movements of the x's to hold the point on the constraints

$$\sum_{i=1}^{3} \frac{\partial \phi_1}{\partial x_i} \Delta x_i = 0 \qquad (16.30)$$

and

$$\sum_{i=1}^{3} \frac{\partial \phi_2}{\partial x_i} \Delta x_i = 0 \qquad (16.31)$$

A matrix equation can be formed from Eqs. (16.30) and (16.31) to solve two Δx's in terms of the third. Suppose that Δx_1 and Δx_2 are expressed in terms of Δx_3, then

$$\begin{bmatrix} \dfrac{\partial \phi_1}{\partial x_1} & \dfrac{\partial \phi_1}{\partial x_2} \\ \dfrac{\partial \phi_2}{\partial x_1} & \dfrac{\partial \phi_2}{\partial x_2} \end{bmatrix} \begin{bmatrix} \Delta x_1 \\ \Delta x_2 \end{bmatrix} = - \begin{bmatrix} \dfrac{\partial \phi_1}{\partial x_3} \\ \dfrac{\partial \phi_2}{\partial x_3} \end{bmatrix} \Delta x_3$$

In the foregoing matrix equation x_1 and x_2 are called *state variables*, and x_3 the *decision variable*. Applying Cramer's rule,

$$\Delta x_1 = - \frac{\dfrac{\partial(\phi_1, \phi_2)}{\partial(x_3, x_2)}}{\dfrac{\partial(\phi_1, \phi_2)}{\partial(x_1, x_2)}} \Delta x_3 \qquad (16.32)$$

and

$$\Delta x_2 = - \frac{\dfrac{\partial(\phi_1, \phi_2)}{\partial(x_1, x_3)}}{\dfrac{\partial(\phi_1, \phi_2)}{\partial(x_1, x_2)}} \Delta x_3 \qquad (16.33)$$

where the designation of the Jacobian determinant is

$$\begin{vmatrix} \dfrac{\partial F}{\partial x_1} & \dfrac{\partial F}{\partial x_2} \\ \dfrac{\partial G}{\partial x_1} & \dfrac{\partial G}{\partial x_2} \end{vmatrix} = \frac{\partial(F, G)}{\partial(x_1, x_2)} \qquad (16.34)$$

The expressions for Δx_1 and Δx_2 can be substituted into the summation of first-degree terms, Eq. (16.29), and regrouped. Then, Δx_3 is factored out:

$$\left[\frac{\partial y}{\partial x_3} - \frac{\frac{\partial(y,\phi_2)}{\partial(x_1,x_2)}}{\frac{\partial(\phi_1,\phi_2)}{\partial(x_1,x_2)}} \frac{\partial \phi_1}{\partial x_3} - \frac{\frac{\partial(y,\phi_1)}{\partial(x_2,x_1)}}{\frac{\partial(\phi_1,\phi_2)}{\partial(x_1,x_2)}} \frac{\partial \phi_2}{\partial x_3} \right] \Delta x_3 \qquad (16.35)$$

Define the coefficients of $\partial \phi_1/\partial x_3$ and $\partial \phi_2/\partial x_3$ as λ_1 and λ_2, respectively. The x_3 variable is free to move, so the only way in which Δx_3 can change either negatively or positively and yet prevent the entire expression from changing y in a favorable manner is for the terms enclosed in the brackets to be zero, thus

$$\frac{\partial y}{\partial x_3} - \lambda_1 \frac{\partial \phi_1}{\partial x_3} - \lambda_2 \frac{\partial \phi_2}{\partial x_3} = 0 \qquad (16.36)$$

Equation (16.36) was the result when x_3 was declared the decision variable. If x_1 were chosen as the decision variable the resulting equation is

$$\frac{\partial y}{\partial x_1} = \lambda_1' \frac{\partial \phi_1}{\partial x_1} - \lambda_2' \frac{\partial \phi_2}{\partial x_1} = 0 \qquad (16.37)$$

and, when x_2 is the decision variable,

$$\frac{\partial y}{\partial x_2} - \lambda_1'' \frac{\partial \phi_1}{\partial x_2} - \lambda_2'' \frac{\partial \phi_2}{\partial x_2} - 0 \qquad (16.38)$$

Equations (16.36) to (16.38) are the Lagrange multiplier equations, except that it still must be shown that

$$\lambda_1 = \lambda_1' = \lambda_1'' \qquad \text{and} \qquad \lambda_2 = \lambda_2' = \lambda_2'' \qquad (16.39)$$

Concentrate on showing just one equality, namely

$$\lambda_1 = \lambda_1'$$

Equation (16.35), with x_3 as the decision variable, resulted in λ_1; by analogy, with x_1 as the decision variable it results in λ_1':

$$\lambda_1 = \frac{\frac{\partial(y,\phi_2)}{\partial(x_1,x_2)}}{\frac{\partial(\phi_1,\phi_2)}{\partial(x_1,x_2)}} \qquad \lambda_1' = \frac{\frac{\partial(y,\phi_2)}{\partial(x_2,x_3)}}{\frac{\partial(\phi_1,\phi_2)}{\partial(x_2,x_3)}}$$

For two functions $F(x,y,z)$ and $G(x,y,z)$, a theorem from advanced calculus states that

$$\frac{dz}{dx} = - \frac{\frac{\partial(F,G)}{\partial(x,y)}}{\frac{\partial(F,G)}{\partial(y,z)}}$$

and

$$\frac{dy}{dx} = -\frac{\dfrac{\partial(F,G)}{\partial(x,z)}}{\dfrac{\partial(F,G)}{\partial(y,z)}}$$

For the problem at hand

$$y\,(x_1, x_2, x_3) \qquad [1]$$

$$\phi_1\,(x_1, x_2, x_3) \qquad [2]$$

and

$$\phi_2\,(x_1, x_2, x_3) \qquad [3]$$

From [1] and [3] and from [2] and [3]

$$\frac{dx_3}{dx_1} = -\frac{\lambda_1 \text{ numerator}}{\lambda_1' \text{ numerator}} \tag{16.40a}$$

$$\frac{dx_3}{dx_1} = -\frac{\lambda_1 \text{ denominator}}{\lambda_1' \text{ denominator}} \tag{16.40b}$$

Equating the two expressions from (16.40*a*) and (16.40*b*) shows that

$$\lambda_1 = \lambda_1'$$

All the other equalities in Eq. (16.39) can be verified in an analogous manner.

The preceding development was still a special case (three variables and two constraints), but all of the principles that would be needed in larger problems have now been demonstrated.

16.8 ALTERNATE EXPRESSION OF CONSTRAINED OPTIMIZATION PROBLEM

The optimization of a constrained function,

optimize $\qquad\qquad y\,(x_1, x_2) \qquad\qquad$ (16.41)

subject to $\qquad\qquad \phi(x_1, x_2) = b \qquad\qquad$ (16.42)

can be restructured as an unconstrained function,

$$L(x_1, x_2) = y\,(x_1, x_2) - \lambda[\phi(x_1, x_2) - b] \tag{16.43}$$

The optimum occurs where $\nabla L = 0$,

$$\frac{\partial L}{\partial x_1} = \frac{\partial y}{\partial x_1} - \lambda\frac{\partial \phi}{\partial x_1} = 0 \tag{16.44}$$

and

$$\frac{\partial L}{\partial x_2} = \frac{\partial y}{\partial x_2} - \lambda \frac{\partial \phi}{\partial x_2} = 0 \tag{16.45}$$

along with conformity to the constraint

$$\phi(x_1, x_2) - b = 0 \tag{16.46}$$

Equations (16.44) to (16.46) are the Lagrange multiplier equations, but the form of the problem as expressed in Eq. (16.43) will be convenient in developing tests for maxima and minima as well as in developing the Kuhn-Tucker conditions in Chapter 17.

16.9 INTERPRETATION OF λ AS THE SENSITIVITY COEFFICIENT

Section 8.10 described the sensitivity coefficient as the rate of change of the optimal value of the function, y^*, with respect to the change in the constraint constant b as it appears in Eq. (16.42). For a special case Sec. 8.10 showed that the sensitivity coefficient SC or $\partial y^*/\partial b$ equals λ. A more general verification, at least for two variables and one constraint, as in Eqs. (16.41) and (16.42), is provided through the chain rule:

$$SC = \frac{\partial y^*}{\partial b} = \frac{\partial y^*}{\partial x_1} \frac{\partial x_1}{\partial b} + \frac{\partial y^*}{\partial x_2} \frac{\partial x_2}{\partial b} \tag{16.47}$$

Also,

$$\frac{\partial \phi}{\partial b} = \frac{\partial \phi}{\partial x_1} \frac{\partial x_1}{\partial b} + \frac{\partial \phi}{\partial x_2} \frac{\partial x_2}{\partial b} - 1 = 0 \tag{16.48}$$

Multiply Eq. (16.48) by λ where in one case

$$\lambda = \frac{(\partial y^*/\partial x_1^*)}{(\partial \phi/\partial x_1^*)}$$

and in the other

$$\frac{(\partial y^*/\partial x_2^*)}{(\partial \phi/\partial x_2^*)}$$

Then

$$\frac{\partial y}{\partial x_1} \frac{\partial x_1}{\partial b} + \frac{\partial y}{\partial x_2} \frac{\partial x_2}{\partial b} - \lambda = 0$$

which when compared to Eq. (16.47) shows that

$$SC = \lambda$$

The above procedure can be extended to a larger number of variables and constraints (see Prob. 16.5).

16.10   TEST FOR MAXIMUM-MINIMUM OF A CONSTRAINED FUNCTION

Section 16.3 showed that the nature of the Hessian matrix indicates for an unconstrained optimization whether the optimum is a maximum or a minimum. Also, for a constrained optimum, it is the second-degree terms in the Taylor series expansion that indicate the nature of the optimum. The possible motion of the variables is restricted such that the constraint(s) continue to be satisfied. The maximum-minimum test for the simplest of the constrained problems (two variables and one constraint) resolves to a check of the determinant:

$$K = \begin{vmatrix} F_{x_1 x_1} & F_{x_1 x_2} & \dfrac{\partial \phi}{\partial x_1} \\[2mm] F_{x_2 x_1} & F_{x_2 x_2} & \dfrac{\partial \phi}{\partial x_2} \\[2mm] \dfrac{\partial \phi}{\partial x_1} & \dfrac{\partial \phi}{\partial x_2} & 0 \end{vmatrix} \qquad (16.49)$$

where

$$F_{x_i} = \frac{\partial y}{\partial x_i} - \lambda \frac{\partial \phi}{\partial x_i}$$

and

$$F_{x_i x_j} = \frac{\partial (F_{x_i})}{\partial x_j}$$

If $K < 0$ the optimum is a minimum, and if $K > 0$ the optimum is a maximum.

Tests for three-variable problems are presented in advanced calculus texts, but it may be more practical just to investigate several positions in the neighborhood of the optimum, abiding, of course by the constraints.

Example 16.3. The UA product of a heat exchanger is to be 24 kW/K. The cost is proportional to the area A m^3 and increases as the 3/2 power of the U value according to the equation $Cost = 12A + 4U^{1.5}$. The optimum is $A^* = 5.102$ and $U^* = 4.704$. The optimum cost is 102.03. Check points in the neighborhood of this optimum to determine whether the point is a maximum or minimum.

Solution Try A values of 5.002 and 5.202 for which the constraint requires U values of 4.798 and 4.614, respectively. The costs at these two points are 102.06 and 102.07 which when compared to 102.03 indicate that the optimum is a minimum.

16.11 OPTIMAL MULTIBRANCH DUCT SYSTEM

An optimization problem that has been addressed by several authors[2,3,4] for which Lagrange multipliers seems to be a natural choice is the selection of duct diameters for a multibranch air supply system. In a system such as shown in Fig. 16-2, the geometric layout, the required flow rates at each outlet, and the static pressure at the fan outlet are specified. The task is to select the duct diameters of each section (0–1, 1–2, 1–3, 3–4, and 3–5) such that the total first cost is a minimum. A first approximation is that the cost of a section of duct is proportional to the length and the diameter. Thus, the total cost C of the duct network is

$$C = k_{0-1} D_{0-1} + k_{1-2} D_{1-2}$$
$$+ k_{1-3} D_{1-3} + k_{3-4} D_{3-4} + k_{3-5} D_{3-5} \quad (16.50)$$

where the k's embody the length of the given section. The constraints must somehow insure that the required flow rates, Q_A, Q_B, and Q_C are supplied. The underlying philosophy of the set of constraint equations listed as Eqs. (16.51) is that all of the available static pressure, SP, is used to overcome friction in the straight duct and fittings:

$$\Delta p_{0-1} + \Delta p_{1-2} \qquad = SP$$
$$\Delta p_{0-1} + \Delta p_{1-3} + \Delta p_{3-4} = SP \qquad (16.51)$$
$$\Delta p_{0-1} + \Delta p_{1-3} + \Delta p_{3-5} = SP$$

The physical implication of Eqs. (16.51) is that even though dampers would certainly be installed in the section leading to each outlet, the intent is that these dampers could be left wide open. A collateral advantage of this optimized duct network is, then, that it should be perfectly balanced and not need final damper adjustment.

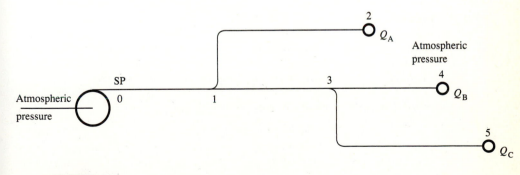

FIGURE 16-2
Optimization of a multibranch air supply system.

The pressure drop in each section, Δp_{ij} is

$$\Delta p_{ij} = f_{ij} \frac{L_{ij}}{D_{ij}} \frac{V_{ij}^2}{2} \rho$$

where f_{ij} = friction factor
L_{ij} = length of straight duct plus the equivalent
length of such fittings as elbows, branches, and tees

Since

$$V = \frac{4Q}{\pi D^2}$$

an alternate expression for Δp_{ij} is

$$\Delta p_{ij} = \frac{8}{\pi^2} f_{ij} \frac{L_{ij}}{D_{ij}^5} Q^2 \rho \qquad (16.52)$$

Equation (16.52) can be solved for the diameter of a section

$$D_{ij} = \left(\frac{8}{\pi^2} f_{ij} L_{ij} Q^2 \rho \right)^{1/5} (\Delta p_{ij})^{-1/5} = a_{ij} (\Delta p_{ij})^{-1/5} \qquad (16.53)$$

and this expression for the diameter can be substituted into the objective function, Eq. (16.50), to yield

$$C = b_{0-1}(\Delta p_{0-1})^{-1/5} + b_{1-2}(\Delta p_{1-2})^{-1/5} + b_{1-3}(\Delta p_{1-3})^{-1/5}$$

$$+ b_{3-4}(\Delta p_{3-4})^{-1/5} + b_{3-5}(\Delta p_{3-5})^{-1/5} \qquad (16.54)$$

where $b_{ij} = k_{ij} a_{ij}$

which is the objective function for cost that along with the constraints, Eqs. (16.51), provide the mathematical statement of the problem.

Example 16.4. The following specifications apply to the multibranch duct system in Fig. 16-2: $Q_A = 0.9$, $Q_B = 0.7$, and $Q_C = 1.5 \ m^3/s$; the lengths of the sections of duct are $L_{0-1} = 30$, $L_{1-2} = 40$, $L_{1-3} = 25$, $L_{3-4} = 25$, and $L_{3-5} = 35$ m; assume a friction factor of 0.02; the cost of the duct is \$63 per m² of surface area. The static pressure at the fan outlet is 500 Pa above the atmospheric pressure. Determine the duct diameters that result in a minimum-first-cost system while supplying the required flow rates.

Solution The coefficients in the objective function are

$$k_{i-j} = 63 L_{i-j} \pi$$

and

$$a_{i-j} = \left[\frac{8}{\pi^2} (0.02) L_{i-j} Q_{i-j}^2 \rho \right]^{1/5}$$

The air density $\rho = 1.2$ kg/m³. $Q_{0-1} = 3.1$, for example. When the appropriate values are substituted, the statement of the optimization problem is as follows:

$$\text{Cost} = 8383(\Delta p_{0-1})^{-1/5} + 7218(\Delta p_{1-2})^{-1/5} + 5868(\Delta p_{1-3})^{-1/5}$$
$$+ 3713(\Delta p_{3-4})^{-1/5} + 7545(\Delta p_{3-5})^{-1/5}$$

subject to the constraints

$$\Delta p_{0-1} + \Delta p_{1-2} \qquad = 500$$
$$\Delta p_{0-1} + \Delta p_{1-3} + \Delta p_{3-4} = 500$$
$$\Delta p_{0-1} + \Delta p_{1-3} + \Delta p_{3-5} = 500$$

Application of the Lagrange multiplier equations to the above problem statement and the subsequent solution yields the following optimal values: $\Delta p_{0-1} = 137.8$, $\Delta p_{1-2} = 362.2$, $\Delta p_{1-3} = 133.1$, $\Delta p_{3-4} = 229.1$, and $\Delta p_{3-5} = 229.1$ Pa. The diameters of the duct sections corresponding to these pressure drops are, from Eq. (16.53), $D_{0-1} = 0.527$, $D_{1-2} = 0.281$, $D_{1-3} = 0.446$, $D_{3-4} = 0.253$, and $D_{3-5} = 0.367$m. The total first cost of the optimal system is $11,356.

Example 16.4 is an illustration of the general class of problems directed toward the optimization of flow networks. The task illustrated in Ex. 16.4 could be considered an optimization within an optimization. The static pressure was specified as 500 Pa, but a more global problem would be to choose the static pressure and the optimal duct system that resulted in minimum present cost of the combination of the first cost of the duct system with the lifetime energy costs for the fan.[4] As Fig. 16-3 illustrates, the

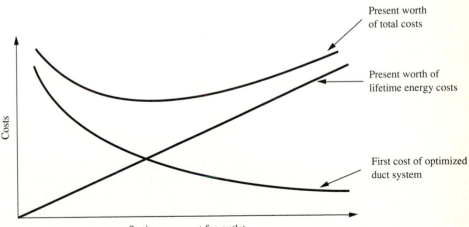

FIGURE 16-3
Total owning costs as a function of static pressure with optimized duct system.

choice of a higher static pressure reduces the duct-system cost, but increases the present worth of the lifetime energy costs. An approach to the solution of this extended problem is to choose a series of static pressures, compute the energy costs, and the cost of the optimized duct system for each static pressure.

16.12   INEQUALITY CONSTRAINTS

The classic calculus method of optimization requires that the constraints be of the equality type. It may be possible, however, by repeated application of equality constraints to optimize a problem where the constraints are actually inequalities. The usual process is one of first assuming that the inequality constraint does not apply. If the preliminary solution violates the inequality constraint, it may then be inserted as an equality constraint. There could be situations where the initial omission of an inequality constraint results in a physically impossible situation. In that case the constraint must be included, but reruns should be made with the inequality more restrictive than actual, in order to determine whether the optimum occurs without the inequality constraint pushed to its limit. An illustration of the appearance of inequality constraints is the modification of the duct optimization of Example 16.4. Problem 16.8 places an inequality constraint on the diameter of one of the sections. A possible approach is to first assume that the inequality constraint will not be violated. If the diameter exceeds the constraint, then the problem can be reworked with the condition expressed as an equality constraint.

16.13   GENERALIZED LAGRANGE MULTIPLIER COMPUTER PROGRAM

The wide applicability and the systematic nature of the Lagrange multiplier method suggest the possibility of a generalized computer program for its solution. The overall operation of the computer program should permit the user to introduce the equations for the objective function, the constraints, and also trial values of the optimal x's and λ's. The computer program should apply a Newton-Raphson technique that numerically extracts the partial derivatives and solves the appropriate linear algebraic equations in the interactive solutions.

The statement of the optimization problem is described in Eqs. (16.1) to (16.3), and it should be the role of the generalized program to adjust trial values of the x's and λ's until Eqs. (16.4) to (16.6) are satisfied. With the trial values, $x_1, \ldots x_n$ and $\lambda_1 \ldots \lambda_m$, nonzero values of the f's result:

$$y_i' - \lambda_1 \phi_{1,i}' - \cdots - \lambda_m \phi_{m,i}' = f_i \tag{16.55a}$$

$$\phi_j(x_1, \cdots x_n) = f_j \tag{16.55b}$$

where the shorthand notation is

$$\frac{\partial y}{\partial x_k} = y'_k$$

$$\frac{\partial \phi_i}{\partial x_j} = \phi'_{i,j}$$

and

$$\frac{\partial^2 \phi_i}{\partial x^2_{jk}} = \phi'_{i,jk}$$

When any of Eqs. (16.55a) is expressed as a Taylor series the result is

$$f_i = f_i(x_{1c}, \ldots x_{nc}, \lambda_{1c} \ldots \lambda_{mc})$$

$$+ \frac{\partial}{\partial x_1}(y'_i - \lambda_1 \phi'_{1,i} - \cdots - \lambda_m \phi'_{m,i})\Delta x_1 + \cdots$$

$$+ \frac{\partial}{\partial x_n}(y'_i - \lambda_1 \phi'_{1,i} - \cdots - \lambda_m \phi'_{m,i})\Delta x_n \qquad (16.56)$$

$$+ \frac{\partial}{\partial \lambda_1}(\qquad)\Delta \lambda_1$$

$$+ \frac{\partial}{\partial \lambda_m}(\qquad)\Delta \lambda_m$$

where the x_c's and λ_c's are the correct values,

$$\Delta x_j = x_j - x_{jc}$$

and

$$\Delta \lambda_k = \lambda_{kc}$$

Since both the x's and the λ's are variables the differentiation of a $\lambda_i \phi'_{j,k}$ term must be treated as differentiation of a product. The Taylor series expansion of any of Eqs. (16.55b) is

$$f_j = f_j(x_{1c}, \ldots x_{nc}) + \phi_{j,1}\Delta x_1 + \ldots + \phi_{j,n}\Delta x_n \qquad (16.57)$$

Since

$$f_i(x_{ic} \ldots x_{nc}, \lambda_{ic} \ldots \lambda_{mc}) = 0$$

and

$$f_{n+1}(x_{1c} \ldots x_{nc}) = 0$$

the equations become

$$(y_{11}'' - \lambda_1 \phi_{1,11}'' - \lambda_2 \phi_{2,11}'' - \cdots - \lambda_m \phi_{m,11}'')\Delta x_1 + \cdots$$

$$(y_{1n}'' - \lambda_1 \phi_{1,1n}'' - \lambda_2 \phi_{2,1n}'' - \cdots - \lambda_m \phi_{m,1n}'')\Delta x_n \qquad (16.58)$$

$$-\phi_{1,1}' \Delta \lambda_1 - \cdots - \phi_{m,1}' \Delta \lambda_m = f_1$$

The $(n + 1)$th equation is

$$\phi_{1,1}' \Delta x_1 + \cdots + \phi_{1,n}' \Delta x_n = f_{n+1}$$

and the $(n + m)$th equation is

$$\phi_{m,1}' \Delta x_1 + \cdots + \phi_{m,n}' \Delta x_n = f_{n+m}$$

So in matrix form

$$
\begin{bmatrix}
A & \vdots & B \\
n \times n & \vdots & n \times m \\
\cdots & \vdots & \cdots \\
C & \vdots & D \\
m \times n & \vdots & m \times m
\end{bmatrix}
\begin{bmatrix}
\Delta x_1 \\
\Delta x_n \\
\Delta \lambda_1 \\
\Delta \lambda_m
\end{bmatrix}
=
\begin{bmatrix}
f_1 \\
f_n \\
f_{n+1} \\
f_{n+m}
\end{bmatrix}
$$

Where the A matrix is

$$\left[y_{ij}'' - \sum_{k=1}^{m} \lambda_k \phi_{k,ij}'' \right]$$

where i = row 1 to n

j = column 1 to n

The B matrix is

$$
\begin{bmatrix}
-\phi_{1,1}' & -\phi_{2,1}' \cdots & -\phi_{m,1}' \\
-\phi_{1,n}' & -\phi_{2,n}' \cdots & -\phi_{m,n}'
\end{bmatrix}
$$

C matrix is

$$
\begin{bmatrix}
\phi_{1,1}' & \phi_{1,n}' \\
\phi_{m,1}' & \phi_{m,n}'
\end{bmatrix}
$$

and the D matrix is $[D] = [0]$

In the solution process the second partial derivatives are computed numerically according to the procedure:

$$y_{ij}'' = \frac{y_i'(x_1, x_2, \ldots x_j + \delta, \ldots x_n) - y_i'(x_1, x_2, \ldots x_j, \ldots x_n)}{\delta}$$

where

$$y_i' = \frac{y(x_1, x_2, \ldots x_i + \Delta, \ldots x_n) - y(x_1, x_2, \ldots x_i, \ldots x_n)}{\Delta}$$

As is typical of the Newton-Raphson procedure, trial values must be entered. Hopefully the user of the generalized program will be familiar enough with the physical system to arrive at reasonable trial values of the x's. Usually, however, the magnitude of the λ's is difficult to predict from the physical understanding of the problem. Trial values of the λ's may be estimated by introducing the trial x's in Eqs. (16.55a) with the f's = 0, and solving for the λ's using m of these linear equations.

16.14 THE UTILITY OF CALCULUS METHODS

Calculus methods of optimization are both instructive and practical. They are instructive in providing insight into certain other methods of optimization, such as search methods. Calculus optimization will often turn out to be the most workable method, particularly when the constraints are of the equality type, but even when one or more of the constraints are inequalities.

PROBLEMS

16.1. The equation for the two-dimensional temperature distribution in the metal plate shown in Fig. 16-4 is

$$T, °C = 4x_1^2 - 2x_1 + 6x_1x_2 + x_2^2 - 4x_2 + 25$$

The magnitude of thermal stress is influenced by the change of temperature

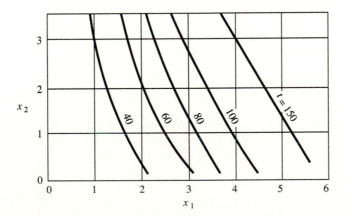

FIGURE 16-4
Two-dimensional temperature distribution.

in unit distance. What is the maximum rate of change of temperature in °C/m in the plate at the point (3,2)?

Ans.: 38.5 °C/m.

16.2. Two tangent vectors at a point on a surface are

$$\mathbf{T}_1 = 4\mathbf{i}_1 + 2\mathbf{i}_2 + 3\mathbf{i}_3$$

and

$$\mathbf{T}_2 = -2\mathbf{i}_1 - 3\mathbf{i}_2 + \mathbf{i}_3$$

Using the property of the dot product of two normal vectors, determine the unit vector normal to the surface at the point.

Ans. $-0.652\mathbf{i}_1 + 0.592\mathbf{i}_2 + 0.474\mathbf{i}_3$

16.3. (a) The primal problem:

minimize $\qquad\qquad y = 16x_1^2 + x_2^2 + 4x_3^2$

subject to $\qquad\qquad x_1x_2x_3 = b = 8$

(b) The dual problem:

maximize $\qquad y' = x_1x_2x_3$

subject to $\qquad 16x_1^2 + x_2^2 + 4x_3^2 = b' = 48$

(c) What is the relation of $y*$ and b'?
What is the relation of $y'*$ and b?
What is the relation of λ and λ'?

16.4. A function $y = f(x_1, x_2, x_3)$ is subject to the constraint

$$\frac{354 \ln(x_1/x_2)}{x_1 - x_3} - (x_3 - x_2)(x_3 - x_1) = 0$$

The function y is messy, but the partial derivatives are $y_1' = 2.10, y_2' = 3.20$, and $y_3' = 6.70$ at the point $x_1 = 3.1, x_2 = 5.8$, and $x_3 = 10.2$. (This point satisfies the constraint.) At this point the derivatives of the constraint are: $\partial\phi/\partial x_1 = -7.32, \partial\phi/\partial x_2 = 15.72$, and $\partial\phi/\partial x_3 = -15.91$.

Variables x_1 and x_2 are both to be shifted by a magnitude of 0.1 in such a way that the function y is reduced, but the constraint must continue to be satisfied. (a) In which direction should x_1 and x_2 be changed, and (b) what is the new value of x_3?

Ans.: (a) increase, decrease.

16.5. Show by an extension of the technique in Sec. 16.9 that $SC_1 = \lambda_1$ for a three-variable, two-constraint problem.

16.6. Develop the Lagrange multiplier equations from the mathematical statement of the optimization problem given in the solution to Example 16.4. Verify that the Δp's given in the solution satisfy these Lagrange multiplier equations.

16.7. Use the concept of λ as the sensitivity coefficient in Example 16.4 to determine the reduction in cost if a static pressure of 510 Pa is available rather than 500 as stated in the problem.

Ans.: $45.40.

16.8. Space is limited in the region through which duct section 1–3 in Ex. 16.4 passes, such that D_{1-3} can be no greater than 0.4 m. Determine the new optimal diameters for a minimum-cost duct system.

Ans.: for checking results, total cost $= \$11,521$.

REFERENCES

1. D. J. Wilde and C. S. Beightler, *Foundations of Optimization*, Prentice-Hall, Englewood Cliffs, N.J., 1967.
2. M. Kovarik, "Automatic Design of Optimal Duct Systems," *Use of Computers for Environmental Engineering Related to Buildings*, Building Science Series 39, National Bureau of Standards, 1971.
3. W. F. Stoecker, R. C. Winn, and C. O. Pedersen, "Optimization of an Air-Supply Duct System," *Use of Computers for Environmental Engineering Related to Buildings*, Building Science Series 39, National Bureau of Standards, 1971.
4. H. Arkin and A. Schitzer, "Study of Engineering and Economic Parameters Related to the Cost of an Optimal Air-Supply Duct System," *ASHRAE Transactions*, vol. 85, part 2, pp. 363–374, 1981.

CHAPTER
17

VECTOR
AND
REDUCED
GRADIENT
SEARCHES

17.1 SEARCH METHODS AND AN
EVALUATION OF THEIR USEFULNESS

Chapter 9 introduced search methods of optimization and presented several techniques for both unconstrained and constrained problems. Several classical techniques for single-variable searches were explained, but because most of the efficient single-variable search methods apply only to unimodal functions, the use of them is risky. The intent of this chapter is to push further into practical search methods that apply to constrained optimization problems. A challenge is to select one, two, or three methods from the extensive variety that is available. Optimization problems fitting the mold of linear or dynamic programming are usually identifiable, and are usually best attacked by those methods. The method of Lagrange multipliers and search methods compete especially on constrained problems. When the objective function and constraints are capable of expression in equation form, Lagrange multipliers may be preferred, particularly if a generalized program for the solution is available. When the objective function and/or constraints are not in explicit form, a search method may be the most practical choice. The most difficult class of problems is likely to be that of constrained optimizations with inequality constraints. As discussed in Sec. 16.12 this problem may sometimes be attacked by repeated application of the method of Lagrange multipliers. With either Lagrange multipliers or a search method, a computer program with some human interaction is often required.

This chapter limits its efforts in solving constrained optimization problems to three methods: (1) penalty functions, (2) vector algebra methods, and (3) the generalized reduced gradient method. Converting a constrained optimization process to an unconstrained one through the use of penalty functions was introduced in Sec. 9.16. The concept is simple, but for the method to be effective the magnitude of the penalty multiplying constants must be adjusted throughout the optimization process. Section 17.2 addresses this task. Vector algebra relations provide extremely useful means of determining search directions, at least for the limited class of two-, three- and four-variable problems. No surveys are available to indicate which method is most widely used for constrained optimization problems, but one of the front runners surely is the generalized reduced gradient technique, which will be explained. The Kuhn-Tucker conditions will be presented because of their classical importance and also their occasional usefulness in the solution of problems.

17.2 PENALTY FUNCTIONS

The straightforward manner of introducing penalty functions, as outlined in Sec. 9.16, is to convert the constrained problem, illustrated here for maximization,

$$y = y(x_1, x_2, \ldots x_n) \rightarrow \text{maximum}$$

subject to
$$\phi_1(x_1, x_2, \ldots x_n) = 0$$
$$\ldots \ldots \ldots \ldots$$
$$\phi_m(x_1, x_2, \ldots x_n) = 0$$

to the unconstrained problem

$$y = y - P_1(\phi_1)^2 - P_2(\phi_2)^2 - \ldots - P_m(\phi_m)^2 \rightarrow \text{maximum}$$

Squaring the ϕ functions and choosing negative values of the P's penalizes the function for either positive or negative deviations from zero. Squaring the ϕ's introduces a parabolic penalty[1] in that large deviations of ϕ from zero are magnified. This form of a penalty has the advantage of being continuous and also having continuous derivatives. It is usual to adjust the magnitudes of the P's during the course of the search in order to balance the influences of the objective function and the constraints. Near the end of the search process the P's must be large in order to draw the final position onto the constraints as the ϕ's approach zero.

Zahradnik[2] proposed a systematic procedure of repeated solutions using progressively larger values of P. The starting point of each search is the final result of the previous search with a lower value of P.

Example 17.1. Using a search method with penalty functions, determine x_1^* and x_2^* that minimize

$$y = 2x_1^2x_2 + x_1x_2^3$$

subject to the constraint

$$\phi = x_1x_2^2 - 6 = 0$$

Solution. Establish a new function to minimize

$$Y = y + P\ \phi^2 \tag{17.1}$$

This simple problem could be solved first analytically to provide a basis for comparison of the penalty function method. The minimum $y^* = 19.60$ occurs where $x_1^* = 1.0$ and $x_2^* = 2.45$. Also for this simple problem the penalized objective function in Eq. (17.1) can be solved analytically. Table 17.1 shows how the magnitude of P influences the degree of adherence to the constraint. As P increases, ϕ drives toward zero and the x's drive toward the constrained minimum. Next, perform a steepest-descent search, choosing a starting point of $x_1 = 0.25$ and $x_2 = 5$. Select the step size of $x_1 = 0.0001$ and move in the direction indicated by the gradient vector until a new minimum is reached. Some results are as follows:

With $P = 1$, progress stops after 40 searches with $x_1 = 0.8303$, $x_2 = 2.2329$, and $Y = 15.783$

With $P = 100$, progress stops after 413 searches with $x_1 = 0.4304$, $x_2 = 3.7272$, and $Y = 23.7098$

Neither of these results are satisfactory. With $P = 1$ the result deviates too much from the constraint. With $P = 100$ the search hangs up short of the minimum because the distance between the contours is so small. Figure 17-1 shows several contours with these two values of P suggesting that the tight intervals for $P = 100$ require a smaller step size and greater computation time.

TABLE 17.1
Analytical solution of the penalized function, Eq. (17.1), with different values of P

P	x_1^*	x_2^*	Y^*	y^*	ϕ
1	0.81319	2.25347	15.785	12.2860	−1.8705
10	0.96889	2.44600	19.184	18.7722	−0.2033
100	0.99623	2.44995	19.554	19.5127	−0.0204
1000	0.99924	2.45000	19.592	19.5876	−0.00204

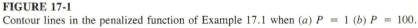

FIGURE 17-1
Contour lines in the penalized function of Example 17.1 when (*a*) $P = 1$ (*b*) $P = 100$.

If Zahradnik's strategy were applied, the search with $P = 1$ would first be performed and the starting values for the $P = 100$ search would be $x_1 = 0.8303$ and $x_2 = 2.2329$. That second series stops after one search direction at $x_1 = 1.0447$, $x_2 = 2.3923$, and $Y = 19.5697$. This result is much closer to the expected optimum for $P = 100$ listed in Table 17.1, but still may be unsatisfactory. But now since there are likely to be a modest number of search directions to execute, the step size could be reduced. With the step size reduced to 0.00001, seven search directions would bring the result to $x_1 = 1.0246$, $x_2 = 2.4157$, and $Y = 19.559$.

17.3 SEARCH DIRECTIONS FROM VECTOR ALGEBRA

For two- and three-variable constrained problems vector algebra using dot products and cross products provides straightforward means of specifying search directions. The basic operations are alternately moving toward the constraint(s) in a favorable direction with respect to y. Thus the three classes of moves for a three-variable constrained problem are as follows:

1. Moving freely in the direction of greatest rate of change of ϕ. This operation occurs when moving from the trial point to a constraint plane, or when moving back to a constraint after a move tangent to a ϕ-surface.
2. Moving tangent to one ϕ-surface in the most favorable direction with respect to another ϕ-surface or with respect to y.
3. Moving tangent to two ϕ-surfaces simultaneously in the direction of improving y.

The first class of moves is simply in the direction of $\nabla \phi$ or $-\nabla \phi$, depending on whether ϕ is negative or positive, respectively.

The second class of moves covers the general situation of moving tangent to one ϕ-surface in the most favorable direction with respect to the objective function y or with respect to the other constraint ϕ_2. Propose that the current point is on the ϕ-surface, having reached that surface through a Class 1 move. Next determine $\nabla \phi$ and ∇y at the point on the constraint, as in Fig. 17-2. The cross product of $\nabla \phi$ and ∇y, called A, is

$$A = \nabla \phi \times \nabla y = \begin{vmatrix} \mathbf{i}_1 & \mathbf{i}_2 & \mathbf{i}_3 \\ \partial \phi / \partial x_1 & \partial \phi / \partial x_2 & \partial \phi / \partial x_3 \\ \partial y / \partial x_1 & \partial y / \partial x_2 & \partial y / \partial x_3 \end{vmatrix} \tag{17.2}$$

The A vector is normal to both $\nabla \phi$ and ∇y, and because $\nabla \phi$ is normal to the ϕ-surface A is tangent to this surface. The next task is to find a vector that is the rotation of A back to the $\nabla \phi - \nabla y$ plane, but still tangent to ϕ. That vector B is the cross product,

$$B = A \times \nabla \phi \tag{17.3}$$

So B is the desired vector that points in the direction tangent to the ϕ-surface, yet also moving in the direction of maximum rate of change of y.

The other necessary information is whether B points in the direction of an increase or a decrease of y. Because ∇y points in the direction of increasing y, as was shown in Sec. 16.2, B when computed from the sequences in Eqs. (17.2) and (17.3) will also point in the direction of increasing y, and should be used for a maximization process. If y is being minimized, the direction should be opposite to B.

The third class of moves for which vector algebra gives a simple indication of direction is that where the current point lies on both the ϕ_1- and ϕ_2- surfaces and a move tangent to both constraints is desired. As

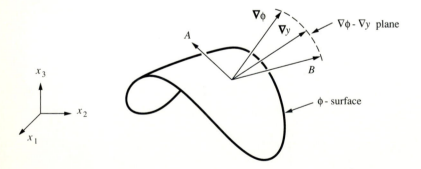

FIGURE 17-2
Vector that is tangent to a constraint and points in the most favorable direction with respect to y.

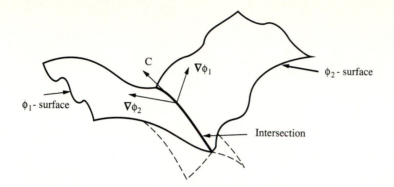

FIGURE 17-3
The cross product of $\nabla \phi_1$ and $\nabla \phi_2$ is tangent to both constraints.

shown in Fig. 17-3, the cross product

$$c = \nabla \phi_1 \times \nabla \phi_2 \qquad (17.4)$$

is normal to both $\nabla \phi_1$ and $\nabla \phi_2$ so is tangent to both planes. Whether to move in the c or $-c$ direction must be determined from the trend of the y values in the c direction. Since ∇y points toward an increase of y, a positive dot product $c \cdot \nabla y$ indicates an increase of y in the c direction; conversely if $c \cdot \nabla y$ is negative, c points toward a decrease in y.

The cross and dot products are applicable to two- and three-variable spaces, but an intriguing question is whether any of the concepts extend to higher-dimensional space. One special situation[3] where similar formulations hold is the four-variable optimization when there are two equality constraints, ϕ_1 and ϕ_2. From a point that satisfies the two constraints, the vector is sought that indicates the most favorable change of y while also remaining tangent to the two constraints. The procedure is first to compute a vector $\mathcal{D}$ by crossing ∇y with $\nabla \phi_1$ and $\nabla \phi_2$ as follows:

$$\mathcal{D} = \begin{vmatrix} \mathbf{i}_1 & \mathbf{i}_2 & \mathbf{i}_3 & \mathbf{i}_4 \\ \dfrac{\partial y}{\partial x_1} & \dfrac{\partial y}{\partial x_2} & \dfrac{\partial y}{\partial x_3} & \dfrac{\partial y}{\partial x_4} \\ \dfrac{\partial \phi_1}{\partial x_1} & \dfrac{\partial \phi_1}{\partial x_2} & \dfrac{\partial \phi_1}{\partial x_3} & \dfrac{\partial \phi_1}{\partial x_4} \\ \dfrac{\partial \phi_2}{\partial x_1} & \dfrac{\partial \phi_2}{\partial x_2} & \dfrac{\partial \phi_2}{\partial x_3} & \dfrac{\partial \phi_2}{\partial x_4} \end{vmatrix} \qquad (17.5)$$

This vector $\mathcal{D}$ which has components of d_1, d_2, d_3, and d_4 is crossed again with ϕ_1 and ϕ_2 to determine the vector $\mathcal{E}$ which indicates the direction sought.

$$E = \begin{vmatrix} i_1 & i_2 & i_3 & i_4 \\ d_1 & d_2 & d_3 & d_4 \\ \dfrac{\partial \phi_1}{\partial x_1} & \dfrac{\partial \phi_1}{\partial x_2} & \dfrac{\partial \phi_1}{\partial x_3} & \dfrac{\partial \phi_1}{\partial x_4} \\ \dfrac{\partial \phi_2}{\partial x_1} & \dfrac{\partial \phi_2}{\partial x_2} & \dfrac{\partial \phi_2}{\partial x_3} & \dfrac{\partial \phi_2}{\partial x_4} \end{vmatrix} \qquad (17.6)$$

The unit vector is $E/|E|$, and the validity of the direction (Problem 17.7) can be demonstrated by comparison with the result of maximizing the constrained form of Eqs. (16.9) and (16.10):

$$dy = \frac{\partial y}{\partial x_1} dx_1 + \frac{\partial y}{\partial x_2} dx_2 + \frac{\partial y}{\partial x_3} dx_3 + \frac{\partial y}{\partial x_4} dx_4 \rightarrow \text{maximum} \qquad (17.7)$$

Subject to

$$\frac{\partial \phi_1}{\partial x_1} dx_1 + \frac{\partial \phi_1}{\partial x_2} dx_2 + \frac{\partial \phi_1}{\partial x_3} dx_3 + \frac{\partial \phi_1}{\partial x_4} dx_4 = 0 \qquad (17.8)$$

$$\frac{\partial \phi_2}{\partial x_1} dx_1 + \frac{\partial \phi_2}{\partial x_2} dx_2 + \frac{\partial \phi_2}{\partial x_3} dx_3 + \frac{\partial \phi_2}{\partial x_4} dx_4 = 0 \qquad (17.9)$$

and

$$(dx_1)^2 + (dx_2)^2 + (dx_3)^2 + (dx_4)^2 = 1. \qquad (17.10)$$

17.4   GENERALIZED REDUCED GRADIENT

One of the popular search methods for optimizing constrained functions is the generalized reduced gradient (GRG) technique.[4] The GRG method uses a repeated sequence of steps that (a) starts on the constraint(s), (b) moves tangent to the constraint(s) in a favorable direction with respect to y, and (c) returns to the constraint(s). The individual steps for a problem of n variables and m constraints are as follows:

1. Start with a trial point on the constraints. Such a position may be located by (a) arbitrarily assigning values to $n\text{-}m$ variables, (b) substituting these variables into the m constraints, and (c) solving for the m variables, perhaps by the Newton-Raphson technique.
2. At the point on the constraints compute components of the gradient vectors ∇y, $\nabla \phi_1$, $\ldots \nabla \phi_m$.
3. Apply the reduced gradient concept to move tangent to the constraints in a favorable direction with respect to y. If the step size chosen results in a less favorable value of y than existed at the starting point, reduce the step size.

4. Return to the constraints by adjusting m of the variables through application of Newton-Raphson to the m simultaneous equations provided by the constraints.

5. If in the process of returning to the constraints the value of y is degraded relative to its value at the starting point, return to Step 3 with a smaller step size.

6. If no improvement is possible with a sufficiently-small step size, terminate, otherwise, return to Step 2.

The central feature of the GRG method is Step 3, which determines the direction that is tangent to the constraints and at the same time indicates a favorable direction for y. Consider an example of optimizing a three-variable function subject to two constraints, a case also considered in Sec. 16.7 using calculus methods. The GRG development parallels the calculus method of Chapter 16. A differential change in y, Δy, is

$$\Delta y = (\partial y/\partial x_1)\Delta x_1 + (\partial y/\partial x_2)\Delta x_2 + (\partial y/\partial x_3)\Delta x_3 \qquad (17.11)$$

When the constraints are applied to Eq. (17.11) Δy equals the grouping shown in Eq. (16.35). That grouping reappears in the GRG method with two modifications. The first difference is that in the calculus method Δy is equated to zero to give the conditions for the optimum, while in the GRG search the process is one of continually improving y. The second difference is in the form that the grouping appears. The typical conventions used in the GRG method for the example in question are

$$y_{\text{new}} = y_{\text{old}} + y'_1 \Delta x_1 + y'_2 \Delta x_2 + y'_3 \Delta x_3 \qquad (17.12)$$

$$\Delta \phi_1 = \phi'_{11} \Delta x_1 + \phi'_{12} \Delta x_2 + \phi'_{13} \Delta x_3 = 0 \qquad (17.13)$$

$$\Delta \phi_2 = \phi'_{21} \Delta x_1 + \phi'_{22} \Delta x_2 + \phi'_{23} \Delta x_3 = 0 \qquad (17.14)$$

where $\phi'_{ij} = \partial \phi_i/\partial x_j$

Arbitrarily choose x_3 as the decision variable and denote

$$J = \begin{bmatrix} \phi'_{11} & \phi'_{12} \\ \phi'_{21} & \phi'_{22} \end{bmatrix} \quad \text{and} \quad K = \begin{bmatrix} \phi'_{13} \\ \phi'_{23} \end{bmatrix}$$

The solution for Δx_1 and Δx_2 from Eqs. (17.13) and (17.14) is

$$\begin{bmatrix} \Delta x_1 \\ \Delta x_2 \end{bmatrix} = -J^{-1} K \, \Delta x_3 \qquad (17.15)$$

so
$$y_{\text{new}} = y_{\text{old}} + \{-[y'_1 \quad y'_2] J^{-1} K + y'_3\} \Delta x_3 \qquad (17.16)$$

or, in the GRG symbols

$$y_{\text{new}} = y_{\text{old}} + \{-[\nabla \hat{y}]^T J^{-1} K + \nabla \bar{y}\} \Delta x_3 \qquad (17.17)$$

where $\nabla\hat{y}$, J and κ are in terms of the state variables, and $\nabla\overline{y}$ and $\nabla\tilde{y}$ are in terms of the decision variable. The expression in the braces of Eq. (17.17) is the GRG or $\nabla\hat{y}$.

The importance of Eq. (17.17) is that the sign of the GRG indicates whether x_3 should be increased or decreased. After step size Δx_3 is chosen, Δx_1 and Δx_2 can be computed from Eq. (17.15), such that new values for all the variables are then available. As the final portion of Step 3, compute y_{new} to assure improvement in y. If y is not improved by the step, reduce Δx_3 and try again.

Example 17.2. Use the GRG method to minimize

$$y = x_1^2 x_2 + 2x_2 x_3^{0.5}$$

subject to

$$x_1 + 3x_2^2 + x_3 = 8$$

$$x_1 x_2 x_3 = 6$$

Solution. To find the starting point arbitrarily choose $x_2 = 1$ which when substituted into the constraints yields two feasible points, $(2,1,3)$ with $y = 7.46$, and $(3,1,2)$ with $y = 11.83$. Since the problem is one of minimization, choose the former point. The partial derivatives are

$$\partial y/\partial x_i = 4.0, \ 7.464, \ 0.577$$

$$\partial\phi_1/\partial x_i = 1, \ 6, \ 1$$

and

$$\partial\phi_2/\partial x_i = 3, \ 6, \ 2$$

Arbitrarily select x_3 as the decision variable and choose an initial step size of 0.5.

$$J = \begin{bmatrix} 1 & 6 \\ 3 & 6 \end{bmatrix}$$

$$J^{-1} = \begin{bmatrix} -0.5 & 0.5 \\ 0.25 & -0.0833 \end{bmatrix}$$

and

$$\kappa = \begin{bmatrix} 1 \\ 2 \end{bmatrix}$$

The GRG, $\nabla\tilde{y} = y_3' - [4.0 \quad 7.464] \, J^{-1}\kappa = -2.045$, so for minimization x_3 should be increased.

$$\begin{bmatrix} \Delta x_1 \\ \Delta x_2 \end{bmatrix} = -J^{-1}\kappa \ \Delta x_3 = \begin{bmatrix} -0.25 \\ -0.0417 \end{bmatrix}$$

The x values after the tangential move are $(1.75, 0.9583, 3.5)$, which provide a y value of 6.5206. In the process of making the move, however, the position shifts off the constraints, so using Newton-Raphson to find the new x_1 and

TABLE 17.2
GRG Optimization of Example 17.2

Step	Δx_3	After tangential move				After return to constraints			
		x_1	x_2	x_3	y	x_1	x_2	x_3	y
1	0.5	1.750	0.9583	3.5	6.5206	1.811	0.9468	3.5	6.6465
2	0.5	1.674	0.8828	4.0	6.0060	1.721	0.8716	4.0	6.0678
3	0.5	1.679	0.7841	4.5	5.5360	1.742	0.7656	4.5	5.5702
4	0.2	1.803	0.7069	4.7	5.3645	1.814	0.7039	4.7	5.3672
5	0.1	1.871	0.6666	4.8	5.2547	1.896	0.6593	4.8	5.2589
6	0.02	1.922	0.6475	4.82	5.2360	1.923	0.6472	4.82	5.2361
7	0.02	1.955	0.6340	4.84	5.2119	1.960	0.6323	4.84	5.2126
8	0.02	2.006	0.6152	4.86	5.1868	2.026	0.6094	4.86	5.1880

x_2 while holding x_3 constant at 3.5 yields the point $(1.8106, 0.9468, 3.5)$ for which $y = 6.64565$. Table 17.2 shows the progression.

Equations (17.15) to (17.17) and Example 17.2 address a three-variable, two-constraint problem in which the reduced gradient indicates a specific vector, and the only decision is which of the two directions offered by the vector should be chosen. In problems where $n - m$ is greater than 1, there are more degrees of freedom of the resulting reduced gradient. For example, in a four-variable problem with two constraints, the equation comparable to Eq. (17.17) with x_3 and x_4 arbitrarily chosen as the decision variables would be

$$y_{\text{new}} = y_{\text{old}} + \text{GRG}_3\,\Delta x_3 + \text{GRG}_4\,\Delta x_4 \qquad (17.18)$$

The gradient vector principle now applies such that in determining the relative values of Δx_3 and Δx_4

$$\frac{\Delta x_3}{\Delta x_4} = \frac{\text{GRG}_3}{\text{GRG}_4}$$

17.5   KUHN-TUCKER CONDITIONS

Most of the work in Chapter 16 and this chapter has been devoted to optimization with equality constraints, although Chapter 16 did illustrate how to attack problems with inequality constraints by repeated application of calculus methods. The Kuhn-Tucker Conditions (KTC) apply to problems where there may be both equality and inequality constraints; they also express some fundamental insight into optimization problems.

If a function

$$y = y(x_1, \ldots, x_n)$$

is to be optimized subject to equality constraints

$$h_k(x_1, \ldots, x_n) = 0$$

where $k = 1, 2, \ldots K$

and also inequality constraints,

$$g_j(x_1, \ldots, x_n) > 0$$

where $j = 1, 2, \ldots J$

certain of the g inequality constraints may be active or binding, and others inactive or nonbinding. If it were known in advance which were binding, those would be treated as equality constraints, and the nonbinding ones would be ignored. The KTC may aid in identifying the binding constraints during the optimization process.

The KTC applicable to a minimization problem are

$$\nabla y(\mathbf{x}) - \sum_{k=1}^{K} \lambda_k \nabla h_k(\mathbf{x}) - \sum_{j=1}^{J} u_j \nabla g_j(\mathbf{x}) = 0 \qquad (17.19)$$

$$g_j(\mathbf{x}) \geq 0 \qquad (17.20)$$

$$h_k(\mathbf{x}) = 0 \qquad (17.21)$$

$$u_j g_j(\mathbf{x}) = 0 \qquad (17.22)$$

$$u_j \geq 0 \qquad (17.23)$$

If the problem is one of maximization, the negative of the objective function should be minimized.

The u terms are similar to the λs except that they apply to the inequality constraints. Equation (17.22) indicates that either or both u_j and g_j must equal zero. Imagine a three-variable problem with one equality constraint, h, and one inequality constraint, g. If g is inactive the constraint should have no influence in Eq. (17.19), so u must equal zero. If g is active, on the other hand, Equation (17.20) becomes an equality, $g(\mathbf{x}) = 0$, and u may take on a positive value.

Example 17.3. Establish the KTC for the minimization of the function subject to one of two different constraints,

$$y = (x_1 - 1)^2 + (x_2 - 2)^2$$

(a) subject to $\qquad g = x_1 + x_2 - 2 > 0$

(b) subject to $\qquad g = x_1 + x_2 - 4 > 0$

Solution. For this simple problem we can essentially start with the answer and explore the nature of u, which is a new element introduced by the KTC.

(*a*) From the diagram in Fig. 17-4 it can be seen that the minimum is the unconstrained minimum occurring at (1,2). The scalar equations corresponding to Eq. (17.19) are

$$2x_1 - 2 - u = 0$$

and
$$2x_2 - 4 - u = 0,$$

which may be combined with the second, fourth and fifth KTC,

$$x_1 + x_2 - 2 \geq 0$$

$$u(x_1 + x_2 - 2) = 0$$

and
$$u \geq 0$$

Knowing that the (x_1, x_2) solution is (1,2) leads to $g = 1$ and $u = 0$.

(*b*) With the $2x_2 - 4 - u = 0$ constraint, the constraint is active so $g = x_1 + x_2 - 4 = 0$. The other variables at the optimum are $x_1 = 1.5$, $x_2 = 2.5$, and $u = 1$. So in this case $u > 0$ and $g = 0$.

Suppose that in part (*a*) the constraint had been assumed to be active. Then $g = 0$ and the optimum would occur at $x_1 = 0.5$, $x_2 = 1.5$, and $u = -1$. The negative value of u violates Eq. (17.23), indicating that the assumption of the constraint being active was invalid.

A possible approach to using Lagrange multipliers to solve an optimization problem where one or more constraints are inequalities was to assume that the inequalities were not active. After determining the solution, substitute into the inequalities to see if any are violated. Those that are violated are then reinserted into the problem as equality constraints. An alternative approach using KTC is to assume that the inequalities are active. If the u value associated with the inequality is negative, that constraint is inactive and can be omitted in a recalculation.

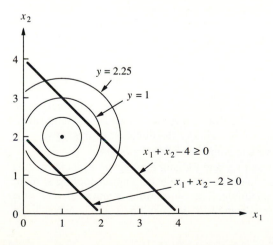

FIGURE 17-4
Constraints and contours of y in Example 17.3.

PROBLEMS

17.1. The maximum is sought for an unconstrained function $y(x_1, x_2)$. At the point $x_1 = 4$ and $x_2 = 2$ the following first and second derivatives are known:

$$\frac{\partial y}{\partial x_1} = -1.5 \qquad \frac{\partial y}{\partial x_2} = 5.2 \qquad \frac{\partial^2 y}{\partial x_1^2} = 6.4$$

$$\frac{\partial^2 y}{\partial x_1 \partial x_2} = 7.2 \qquad \frac{\partial^2 y}{\partial x_2^2} = -12.4$$

Using the above information, estimate x_1^* and x_2^*.

Ans.: $x_1^* = 3.856$.

17.2. The function

$$y = 3x_1^2 + 2x_1 x_2 + x_2^2$$

is being minimized subject to the constraint

$$x_1 x_2^{1.5} = 16$$

This constrained problem has been converted to an unconstrained problem by treating the constraint as a penalty function

$$Y = y + P(x_1 x_2^{1.5} - 16)^2$$

As a strategy the magnitude of P is to be chosen such that each move in the steepest descent of Y is equally influenced by the original objective function and by the constraint. Thus

$$|\nabla y| = P \; \nabla[(\phi)^2]$$

If the current point is $(4, 4)$, what should be the value of P?

17.3. Two constraints prevail in a three-variable optimization using a search method,

$$\phi_1 = \frac{x_1 x_2}{x_3} + \frac{x_1}{x_2 x_3} - 3.5 = 0$$

$$\phi_2 = (1 - e^{-x_1 x_2/10})(x_1 - x_3) - 0.5 = 0$$

The current point is $(3.5, 1, 2)$ which lies on ϕ_1. The assignment is to drive tangent to ϕ_1 in the most favorable direction toward ϕ_2. Using vector methods determine the unit vector defining the direction of this move.

Ans.: for checking result, component of $\mathbf{i}_1 = 0.414$

17.4. The flow rates of cooling water w_w and radiator cooling air w_a, both in kg/s, are to be chosen such that the total power in the engine cooling system, shown in Fig. 17-5, is a minimum. The outlet water temperature from the engine is 100°C and the temperature of the inlet air is 35°C. The radiator

FIGURE 17-5
Engine cooling system in Prob. 17.4.

has a *UA* value of 12 kW/K. The equations for power in kW are

$$\text{pump power} = 0.02\, w_w^3$$

and

$$\text{fan power} = 0.01\, w_a^3$$

For simplification, assume that the arithmetic-mean-temperature difference is sufficiently accurate in the heat-transfer rate equation, thus

$$12[(100 + t_w)/2 - (35 + t_a)/2] = 300 \text{ kW}$$

(a) Develop the objective function and constraint in terms of w_w and w_a.
(b) Convert this constrained problem into an unconstrained one using penalty functions, and solve by a steepest descent search. (Suggest first $P = 10$ and the final $P = 1000$).
      Ans.: constraint in (a), $35.8/w_w + 150/w_a = 40$.

17.5. An objective function to be maximized is

$$y = 3x_1 + x_2 + x_3^2$$

subject to the constraints

$$20x_1^2 + 5x_2^2 + 3x_3^2 = 127$$

and

$$x_1 x_2 x_3 = 12$$

The current point is (1,4,3) which satisfies both constraints. The next move in the search is to be tangent to the two constraints with a change of x_1 of 0.1 such that the objective function increases. Use vector methods to determine the values of the x's at the new point.
      Ans.: new value of $y = 17.8$.

17.6. The function

$$y = (1/3)x_1^3 + 2x_2^2 + 2x_3$$

is being minimized subject to the constraint

$$x_1 x_2 x_3 = 4$$

At the point $(2,1,2)$, which satisfies the constraint, use vector methods to find the unit vector pointing in the direction of steepest descent, yet also tangent to the constraint.

Ans.: for checking result, component of $i_1 = -0.913$

17.7. Choose four arbitrary values of each of the following: $\partial y/\partial x_i$, $\partial \phi_1/\partial x_i$, and $\partial \phi_2/\partial x_i$, and using Eqs. (17.5) and (17.6), determine a unit vector from a point satisfying the two constraints that points in the most favorable direction with respect to y while moving tangent to the constraints. Verify that the components of this vector agree with the Lagrange multiplier solution of Eqs. (17.7) to (17.10).

17.8. A pasteurizer in a dairy receives 1.8 kg/s of raw milk at 15°C, preheats it on one side of a regenerative heat exchanger, and then brings the temperature up to 70°C in a steam heater for pasteurization, as shown in Fig. 17-6. The hot milk is then cooled on the other side of the heat exchanger and then cooled further in a refrigeration plant to 3°C. The U value of the regenerator is 0.4 kW/(m² · K), and the specific heat of milk is 3.9 kJ/(kg · K). The sizes of the regenerator, steam heater, and refrigeration unit are to be chosen such that the total lifetime cost of the facility is a minimum. The expressions for the present worth of these lifetime costs of the three components are

> steam heater (including steam costs), $\$ = 400(70 - t_1)^{1.4}$
> regenerator, $\$ = 180(A, \text{m}^2)$
> refrigerating unit (including electric power), $\$ = 160(t_2 - 3)^{1.5}$

(a) Develop the expressions for the objective function and constraint(s) in terms of t_1, t_2, and A to minimize the total lifetime costs.

(b) Verify from the Lagrange multiplier equations that the optimum occurs when $t_1 = 61.5°C$, $t_2 = 23.5°C$, $A = 96$ m², and the total present worth of the costs = $\$40,135$.

FIGURE 17-6
Chain of heat exchangers pasteurizing milk in Prob. 17.8.

(c) If the problem were instead to be solved using the GRG search method, make one complete GRG operation, starting on the constraints where $A = 120$ m^2. Choose a step size of $A = 5$ m^2, and return to the constraints after the step, holding A constant. What are the new values of A, t_1, t_2, and y?

17.9. Section 17.4 and Example 17.2 illustrated the application of the generalized reduced gradient (GRG) method to a three-variable problem with two constraints. Develop the expressions for the GRGs when optimizing

$$y = y(x_1, x_2, x_3, x_4)$$

subject to

$$\phi_1(x_1, x_2, x_3, x_4) = 0$$

and

$$\phi_2(x_1, x_2, x_3, x_4) = 0$$

Choose x_3 and x_4 as decision variables.

17.10. A building is being constructed on a 30 × 50 m lot and requires a setback as shown in Fig. 17-7. The zoning rules restrict the height of the building h

$$h \le 5(a + b)$$

and another restriction is that the building can occupy no more than 50% of the lot area. The owner wishes to construct a building with maximum volume.

(a) Determine the optimum dimensions of the building by assuming that both constraints are active, (b) use the Kuhn-Tucker conditions to determine whether either of the constraints is nonbinding, and (c) guided by the u values in part (b) reoptimize using the active constraint(s) as equality constraints.

   Ans.: optimum volume = 94,836 m^3.

FIGURE 17-7
Setback requirements for construction of the building in Prob. 17.10.

REFERENCES

1. G. V. Reklaitis, A. Ravindran, and K. M. Ragsdell, *Engineering Optimization—Methods and Applications*, John Wiley, New York, 1983.
2. R. L. Zahradnik, *Theory and Techniques of Optimization*, Barnes and Noble, New York, 1971.
3. N. L. Baker, Graduate Student Project, University of Illinois, 1986.
4. G. A. Gabriele and K. M. Ragsdell, "The Generalized Reduced Gradient Method: A Reliable Tool for Optimal Design," *ASME Journal for Industry*, vol. 99, Series B, no. 2, pp. 394–400, May 1977.

CHAPTER 18

CALCULUS OF VARIATIONS AND DYNAMIC PROGRAMMING

18.1 THE RELATIONSHIP BETWEEN CALCULUS OF VARIATIONS AND DYNAMIC PROGRAMMING

Chapter 10 introduced dynamic programming and emphasized that the method applies to problems seeking an optimal *function*, rather than an optimal *point*. In Fig. 18-1 the function $y = y(x)$ between point 1 and 2 is to be found that gives a minimum (or maximum) value of the integral of $F(x) \, dx$. Calculus of variations (COV) and dynamic programming (DP) are companion methods since both are techniques to determine the optimal $y(x)$. A difference between COV and DP is that DP breaks the function into discrete stages, while COV treats it as a continuous function.

Which method is precise and which is an approximation depends on the problem. If the velocity of a vehicle is continuously adjusted during a trip to minimize the total energy, COV is a precise representation and DP is an approximation. Dynamic programming in that problem would represent the varying speed as a series of steps, while COV inherently assumes a continuous function with continuous derivatives. In a different problem of

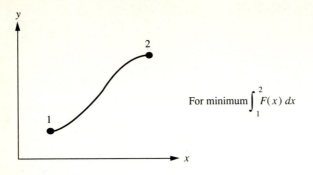

FIGURE 18-1
Calculus of variations and dynamic programming seek an optimal function.

optimizing the sizes of a series of heat exchangers, DP is the precise method and COV the approximation, because the process truly consists of a chain of steps in passing from one heat exchanger to another.

Chapter 10 explained the application of dynamic programming to a single-variable, unconstrained optimization. The chapter also showed that certain "apparently constrained" problems could be restructured to be dealt with as unconstrained problems. This current chapter begins with COV and presents the basic Euler-Lagrange equation. Then the discussion proceeds to the application of COV to unconstrained and constrained situations where there is one independent variable.

Calculus of variations has a merit of its own since some industrial problems can be solved using the method, but COV provides insight into the DP method as well. A characteristic of COV is that a differential equation must always be solved. After moving beyond the situations where the differential equation can be solved analytically, numerical solutions of the differential equation are necessary. The mathematics developed for COV suggests some approaches that are transferable to complex dynamic programming problems. The study of dynamic programming with a single independent variable will be extended to (1) staged processes with recycle, and (2) constrained optimization.

18.2 THE EULER-LAGRANGE EQUATION

The next several pages will present the development of the Euler-Lagrange (E-L) equation—a basic one in the calculus of variations. A function y is sought, where $y = y(x)$, such that the integral I is optimized:

where
$$I = \int_{x_1}^{x_2} F(x, y, y')\, dx \tag{18.1}$$

The function F is known and y is to be determined. Not all of the terms— x, y, and y'—need appear in the function F, but the E-L equation can accommodate them when they are present. The E-L equation to be developed is

$$\frac{\partial F}{\partial y} - \frac{d}{dx}\left(\frac{\partial F}{\partial y'}\right) = 0 \tag{18.2}$$

which is a differential equation that eventually must be solved to obtain the function y.

The development of the E-L equation parallels the strategy used in Eq. (16.13) to verify the existence of an optimal point. In that case the function was expanded in a Taylor series about the point being tested. To prevent the possibility of moving one or more of the x's and finding a more-favorable point, it was necessary for ∇y to equal zero.

In the development of the E-L equation a function $y(x)$ is shown in Fig. 18-2 that is expected to yield the minimum value of I (called I_0). Define another function in Fig. 18-2, $Y(x)$, that deviates only slightly from $y(x)$ and is expressed by the equation,

$$Y(x) = y(x) + \epsilon\eta(x) \tag{18.3}$$

where ϵ is a numerical value, and $\eta(x)$ is a function of x.

Substitute Y for y, and $y' = dY/dx$ for y' in the integral, Eq. (18.1),

$$I(\epsilon) = \int_{x_1}^{x_2} F(x, Y, Y') \, dx \tag{18.4}$$

The value of the integral in Eq. (18.4) is represented as a function of ϵ, namely $I(\epsilon)$, while the value of I in Eq. (18.1) is simply a numerical value. In Eq. (18.4) the value of the integral incorporates the optimal value of the integral, I_0, as well as the ϵ term. In fact, $I(\epsilon)$ can be expressed as

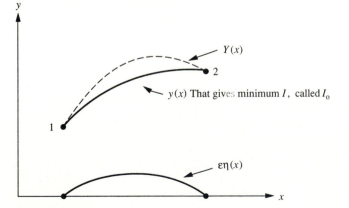

FIGURE 18-2
Function that gives the optimal value of the integral and a function Y that deviates slightly from y.

a Taylor series expansion of ϵ,

$$I(\epsilon) = I_0 + \left[\frac{\partial I(\epsilon)}{\partial \epsilon}\right]_{\epsilon=0} \epsilon + \frac{1}{2}\left[\frac{\partial^2 I(\epsilon)}{\partial \epsilon^2}\right]_{\epsilon=0} \epsilon^2 + \cdots \quad (18.5)$$

The first term is the value of the integral where $\epsilon = 0$ which is where Y has shrunk to y, and Y' to y'. By our definition of y as the function resulting in the optimal I, the first term becomes I_0. The situation represented by Eq. (18.5) is parallel to the basic notion of optimization of a function with one independent variable. If the first term in the Taylor expansion is truly the optimum, the coefficient of ϵ in the second term must be zero, otherwise either a small positive or a small negative value could be assigned to ϵ which would improve $I(\epsilon)$, making it more favorable than the supposed optimum I_0. The condition for y to be the optimum, then, is that

$$\frac{\partial I}{\partial \epsilon}\bigg|_{\epsilon=0} = \left[\frac{\partial}{\partial \epsilon}\int_{x_1}^{x_2} F(x, Y, Y')\, dx\right]_{\epsilon=0} = 0 \quad (18.6)$$

The remainder of the development will now be devoted to restructuring Eq. (18.6).

The differentiation of the integral with respect to ϵ in Eq. (18.6) can be moved within the integral sign, since the limits of integration are constants,

$$\left[\int_{x_1}^{x_2} \frac{\partial}{\partial \epsilon} F(x, Y, Y')\, dx\right]_{\epsilon=0} = 0 \quad (18.7)$$

Applying the chain rule,

$$\left[\int_{x_1}^{x_2}\left(\frac{\partial F}{\partial x}\frac{\partial x}{\partial \epsilon} + \frac{\partial F}{\partial Y}\frac{\partial Y}{\partial \epsilon} + \frac{\partial F}{\partial Y'}\frac{\partial Y'}{\partial \epsilon}\right)\right]_{\epsilon=0} dx = 0 \quad (18.8)$$

The right bracket was moved inside the dx, since x is not a function of ϵ. Equation (18.3) and its derivative with respect to x,

$$Y' = y'(x) + \epsilon\eta'(x) \quad (18.9)$$

can be utilized to express several of the derivatives in Eq. (18.8):

$\partial x/\partial \epsilon = 0$ since x is not a function of ϵ
$\partial Y/\partial \epsilon = \eta$ from Eq. (18.3)
$\partial Y'/\partial \epsilon = \eta'$ from Eq. (18.9)

So Eq. (18.8) becomes

$$\left[\int_{x_1}^{x_2}\left(\frac{\partial F}{\partial Y}\eta + \frac{\partial F}{\partial Y'}\eta'\right)\right]_{\epsilon=0} dx = 0 \quad (18.10)$$

When $\epsilon = 0$, then $Y = y$, and $Y' = y'$, so this substitution can be made in Eq. 18.10, and the two terms in the integrand can be divided into separate integrals,

$$\int_{x_1}^{x_2} \frac{\partial F}{\partial y} \eta \, dx + \int_{x_1}^{x_2} \frac{\partial F}{\partial y'} \eta' \, dx = 0 \qquad (18.11)$$

The second term in Eq. (18.11) can be integrated by parts,

$$\int_{x_1}^{x_2} \frac{\partial F}{\partial y'} \eta' \, dx = \frac{\partial F}{\partial y'} \eta \, \Big|_{\substack{x = x_1 \\ x = x_2}} - \int_{x_1}^{x_2} \eta \, \frac{d(\partial F / \partial y')}{dx} \, dx$$

$$= \frac{\partial F}{\partial y'} \big[\eta(x_2) - \eta(x_1) \big] - \int_{x_1}^{x_2} \eta \frac{d(\partial F / \partial y')}{dx} \, dx \qquad (18.12)$$

Both $\eta(x_2)$ and $\eta(x_1) = 0$, so substituting Eq. (18.12) back into Eq. (18.11) results in

$$\int_{x_1}^{x_2} \left[\frac{\partial F}{\partial y} - \frac{d(\partial F / \partial y')}{dx} \right] \eta(x) \, dx = 0 \qquad (18.13)$$

The line of reasoning used now is that since $\eta(x)$ can be arbitrary, the only way to assure that the integral is always zero is for the term in the brackets to be zero,

$$\frac{\partial F}{\partial y} - \frac{d}{dx} \left(\frac{\partial F}{\partial y'} \right) = 0 \qquad (18.2)$$

This is the Euler-Lagrange equation which is a basic tool in the calculus of variations. It is applicable where no constraint relates x, y, and y'. The function F will be known, and the task will be to determine y, which is a function, not a specific value. Since the E-L equation is a differential equation, it will be necessary to solve this differential equation to determine y.

The E-L equation can be restructured into an alternate general form, and also into simplified forms for special cases. The chain rule may be applied to the second term in Eq. (18.2) to yield

$$\frac{\partial F}{\partial y} - \frac{\partial(\partial F / \partial y')}{\partial x} \frac{dx}{dx} - \frac{\partial(\partial F / \partial y')}{\partial y'} \frac{dy'}{dx} - \frac{\partial(\partial F / \partial y')}{\partial y} \frac{dy}{dx} = 0$$

For brevity let

$$F_y = \partial F / \partial y$$

$$F_{y'} = \partial F / \partial y'$$

$$F_{y'y'} = \partial^2 F / (\partial y')^2$$

Then the alternate form of the Euler-Lagrange equation is the second-order equation

$$F_y - F_{xy'} - F_{y'y'} y'' - F_{yy'} y' = 0 \qquad (18.14)$$

Equation (18.14) can be simplified for three special cases:

1. F is a function of y' only. The remaining term in Eq. (18.14) is

$$F_{y'y'}y'' = 0$$

If $F_{y'y'}$ is not equal to zero, then

$$y'' = 0 \quad \text{and} \quad y = c_1x + c_2$$

2. F does not contain y explicitly.

$$\partial F / \partial y = 0$$

so from Eq. (18.2)

$$\frac{d}{dx}\left(\frac{\partial F}{\partial y'}\right) = 0.$$

Integrating,

$$\partial F / \partial y' = \text{constant}$$

3. F does not contain x explicitly.
   F is a function of y and y' only, so using the chain rule

$$\frac{dF}{dx} = \frac{\partial F}{\partial y} y' + \frac{\partial F}{\partial y'} y'' \tag{18.15}$$

From Eq. (18.2)

$$\frac{\partial F}{\partial y} = \frac{d}{dx}\left(\frac{\partial F}{\partial y'}\right)$$

which can be substituted into Eq. (18.15) to obtain

$$\frac{dF}{dx} = \left[\frac{d}{dx}\left(\frac{\partial F}{\partial y'}\right)\right]y' + \frac{\partial F}{\partial y'} y'' \tag{18.16}$$

The two terms on the right side of Eq. (18.16) are the derivative with respect to x of the product $[y'(\partial F / \partial y')]$, so

$$\frac{dF}{dx} = \frac{d}{dx}\left(y'\frac{\partial F}{\partial y'}\right)$$

Integrating with respect to x,

$$F = y'(\partial F / \partial y') + \text{constant} \tag{18.17}$$

Example 18.1. A pump delivers hydraulic fluid through a long pipe to a vessel that has air trapped at the top, as shown in Fig. 18-3. The process is one of building up pressure in the vessel from 100 kPa to 190 kPa in 120 s by increasing the mass m in the vessel from 0 kg when $t = 0$ s to $m = 180$ kg when $t = 120$ s. There is a pressure drop in the pipe that is attributable

FIGURE 18-3
Pump delivering hydraulic fluid through a long pipe to build up pressure in a vessel.

to laminar flow of the hydraulic fluid, $p_1 - p_2 = 15\,m'$, where $m' = $ flow rate, kg/s. The vessel pressure $p_2 = 100 + 0.5m$. The pumping power is $m_p'(p_1 - 100)$, where m_p' is the ideal pump flow rate. There is back leakage in the pump such that

$$m' = m_p' - 0.2(p_1 - 100)$$

Determine the equation for m' as a function of time t such that minimum pumping energy is required during the process.

Solution. The integral to minimize is

$$I = \int_0^{120} m_p'(p_1 - 100)\,dt = \int_0^{120} [m' + 0.2(p_1 - 100)](p_1 - 100)\,dt$$

$$p_1 - 100 = 15m' + p_2 - 100 = 15m' + 0.5m$$

So the integral becomes

$$I = \int_0^{120} [3.5mm' + 60(m')^2 + 0.05m^2]\,dt$$

Subjecting F to the E-L equation of the form in Eq. (18.14),

$$m'' - 0.0008333m = 0$$

The solution of this differential equation is

$$m = C_1 \sinh \sqrt{0.0008333}\, t + C_2 \cosh(0.02887t)$$

The boundary condition $m(0) = 0$ requires that $c_2 = 0$, so

$$m = C_1 \sinh(0.02887t)$$

The second boundary condition, $m(120) = 180$ gives

$$C_1 = 180/(\sinh 3.464) = 11.28$$

so $$m = 11.28 \sinh(0.02887t)$$

and $$m' = 0.3257 \cosh(0.02887t)$$

FIGURE 18-4
Optimal plan for flow rate to build up pressure in Example 18.1.

which results in minimum pumping energy. The profile of the flow rate is shown in Fig. 18-4 in which reduced time is allowed at high p_1 to inhibit back leakage through the pump.

18.3 THE EULER-LAGRANGE EQUATION TO OPTIMIZE A CONSTRAINED FUNCTION

The mathematical statement of the constrained problem is to find y that minimizes I

$$I = \int_{x_1}^{x_2} F(x, y, y') dx \qquad (18.18)$$

subject to

$$\int_{x_1}^{x_2} G(x, y, y') dx = J \qquad (18.19)$$

where J is a constant.

The E-L equation for the constrained case will be shown to be

$$\frac{\partial(F - \lambda G)}{\partial y} - \frac{d}{dx}\left[\frac{\partial(F - \lambda G)}{\partial y'}\right] = 0 \qquad (18.20)$$

which is of the same form as Eq. (18.2) for the unconstrained case. The procedure in determining y is to solve Eq. (18.20), but in addition use Eq. (18.19) to evaluate λ.

The development of the E-L equation for a constrained function is a short extension of that for the unconstrained function that was described in Sec. 18.2. The concept to be used is similar to that used in Chapter 16 in optimizing a function by calculus methods to find an optimal point subject to constraints.

Propose a test function $y(x)$ as the one that minimizes I in Eq. (18.18) but also satisfies J in Eq. (18.19). Let Y be a nearby solution, where

$$Y = y(x) + \alpha\eta(x) + \beta\gamma(x) \qquad (18.21)$$

where α and β are constants. These functions are shown in Fig. 18-5. In order to keep the terminal points in fixed positions

$$\eta(x_1) = \eta(x_2) = 0$$

and

$$\gamma(x_1) = \gamma(x_2) = 0$$

The deviation functions $\eta(x)$ and $\gamma(x)$ are arbitrary, but if η is selected, γ must be chosen such that the integral in Eq. (18.19) continues to equal J. The revised function Y and its derivative will replace y and y' in the integrals to determine if an improvement of I is possible.

$$I = \int_{x_1}^{x_2} F(x, Y, Y')dx \qquad\qquad K = \int_{x_1}^{x_2} G(x, Y, Y')dx$$

Next express I and K in Taylor series of α and β, which are two independent variables.

$$I(\alpha, \beta) = I(0,0) + \left.\frac{\partial I}{\partial \alpha}\right|_{\alpha, \beta = 0} \alpha + \left.\frac{\partial I}{\partial \beta}\right|_{\alpha, \beta = 0} \beta + \cdots \qquad (18.22)$$

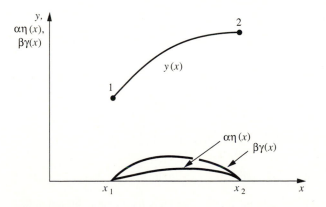

FIGURE 18-5
Graph of the function y that gives minimum value of I subject to constraint J, and functions $\alpha\eta(x)$ and $\beta\gamma(x)$ that are deviations from y.

and

$$K(\alpha, \beta) = K(0,0) + \left.\frac{\partial K}{\partial \alpha}\right|_{\alpha,\beta=0} \alpha + \left.\frac{\partial K}{\partial \beta}\right|_{\alpha,\beta=0} \beta + \cdots \qquad (18.23)$$

$I(0,0)$ in Eq. (18.22) is the integral of $y\,dx$, and $K(0,0)$ in Eq. (18.23) is J. In order for $I(\alpha, \beta)$ to be a minimum the first-degree terms in Eq. (18.22) must be zero.

$$\left.\frac{\partial I}{\partial \alpha}\right|_{0,0} \alpha + \left.\frac{\partial I}{\partial \beta}\right|_{0,0} \beta = 0 \qquad (18.24)$$

And for no deviation of K from J,

$$\left.\frac{\partial K}{\partial \alpha}\right|_{0,0} \alpha + \left.\frac{\partial K}{\partial \beta}\right|_{0,0} \beta = 0 \qquad (18.25)$$

Combining Eqs. (18.24) and (18.25),

$$-\frac{\beta}{\alpha} = \frac{\partial I / \partial \alpha}{\partial I / \partial \beta} = \frac{\partial K / \partial \alpha}{\partial K / \partial \beta} \qquad (18.26)$$

Rearrange the right two members of Eq. (18.26) and equate to a constant which will be called λ.

$$\frac{\partial I / \partial \alpha}{\partial K / \partial \alpha} = \lambda = \frac{\partial I / \partial \beta}{\partial K / \partial \beta}$$

which can be rewritten

$$\frac{\partial}{\partial \alpha} [I - \lambda K] = 0 \qquad \text{and} \qquad \frac{\partial}{\partial \beta} [I - \lambda K] = 0 \qquad (18.27)$$

Substituting the expressions for I and K into Eq. (18.27) gives

$$\frac{\partial}{\partial \alpha}\left[\int_{x_1}^{x_2} F(x, Y, Y')dx - \lambda \int_{x_1}^{x_2} G(x, Y, Y')dx \right] = 0$$

as well as a similar partial derivative with respect to β. These expressions are similar to that in Eq. (18.6) and comparable steps can then be taken to arrive at the E-L equation for a constrained function,

$$\frac{\partial(F - \lambda G)}{\partial y} - \frac{d}{dx}\left[\frac{\partial(F - \lambda G)}{\partial y'} \right] = 0 \qquad (18.20)$$

which must be solved with Eq. (18.19) to evaluate λ.

Example 18.2. One of the steps in processing a certain edible oil is to pump it from a 40 m^3 tank through a filtration unit, as shown in Fig. 18-6. The volume in m^3 pumped is designated by V, and the 40 m^3 is to be completely pumped in 20 min. The volume flow rate is designated V' m^3/s, and the

FIGURE 18-6
Pumping edible oil from a tank through a filter in Example 18.2.

pressure difference developed by the pump, Δp kPa, increases with time due to the progressive increase in resistance in the filter and the thickening of the oil. The pressure difference is represented by the equation

$$\Delta p = (3600 + 6t)V'$$

where $t =$ time from start, s.

    The flow rate is to be programmed so that minimum pumping energy is required during the 20 min pumpout operation. A restriction on the program because of operating characteristics of the filter is that the mean Δp must equal 210 kPa. Thus,

$$\frac{1}{1200} \int_0^{1200} \Delta p \, dt = 210$$

Solution. Since pump power, kW $= \Delta p \, V'$, the integral representing the pumping energy that is to be minimized is

$$\int_0^{1200} (3600 + 6t)(V')^2 \, dt \qquad\qquad (18.28)$$

subject to

$$\int_0^{1200} (3600 + 6t)V' \, dt = 252,000 \qquad\qquad (18.29)$$

Therefore

$$F = (3600 + 6t)(V')^2$$

and

$$G = (3600 + 6t)V'$$

Before solving the constrained problem, it would be interesting to solve for the minimum energy function without the constraint. Application of the E-L equation to F yields

$$V, \text{ m}_3 = 36.41 \ln(6t + 3600) - 298.15$$

$$V', \text{ m}_3/\text{s} = 218.46/(6t + 3600)$$

$$\Delta p = 218.46 \text{ kPa}$$

and energy for the operation $= 8738$ kJ.

We observe that the optimal pumping plan adjusts the flow rate so that Δp remains constant.

For the constrained problem, apply Eq. (18.20), where

$$F - \lambda G = (3600 + 6t)[(V')^2 - \lambda V')],$$

and since V, corresponding to y in Eq. (18.30), does not appear in the function

$$\frac{\partial(F - \lambda G)}{\partial V'} = (3600 + 6t)(2V' - \lambda) = \text{constant}$$

so

$$\frac{dV}{dt} = \frac{C_1}{3600 + 6t} + \frac{\lambda}{2}$$

Integrating

$$V - C_2 \ln(6t + 3600) + \lambda t/2 + C_3 \qquad (18.30)$$

Three constants are yet to be determined, C_2, C_3, and λ. The three relations are the initial and final values of (V,t), namely $(0,0)$ and $(40,1200)$, and the constraint, Eq. (18.22).

$$C_2 = 36.41 - 546.1\lambda$$

$$C_3 = 4472\lambda - 298.2$$

and

$$\lambda = -0.02616$$

The expressions for V, V', and Δp resulting in minimum energy while abiding by the constraint are

$$V = 50.7 \ln(6t + 3600) - 0.01308t - 415.2$$

$$V' = 304.2/(6t + 3600) - 0.01308$$

$$\Delta p = 257.1 - 0.07848t$$

Because of the influence of the constraint, the Δp profile is tilted from a horizontal line to a curve sloping downward. When the expressions for V' and Δp are substituted into Eq. (18.21) and integrated, the total pumping energy is found to be 8872 kJ, which is less favorable than resulted in unconstrained operation.

A final question is whether the value of λ represents the sensitivity coefficient, as it does in the method of Lagrange multipliers. What, for example, is the effect of increasing the average pressure in the constraint, Eq. (18.29)? The expression for λ in this problem is $(p_{ave} - 218.46)/323.4$, so at $p_{ave} = 210$, 214.23, and 218.46, the values of λ are -0.02616, -0.01308, and 0, respectively. Between $p_{ave} = 210$ and the unconstrained p_{ave} of 218.46, the energy decreases from 8872 to 8738. This change could have been predicted by multiplying the mean value of λ by the difference in p_{ave}, $(-0.01308)(218.46 - 210) = 133$. Indeed, also in COV the λ-term indicates the sensitivity coefficient.

18.4 DYNAMIC PROGRAMMING AND CALCULUS OF VARIATIONS

Dynamic programming and COV are members of the same family. Dynamic programming yields a series of points that when connected describe a function that optimizes a summation, as illustrated in Fig. 18-7. Calculus of variations develops an equation for the function that is continuous with continuous derivatives. This function optimizes an integral based on the function.

Another common characteristic of both DP and COV is that the terminal points, points 1 and 2 in Fig. 18-7, are known. The expectation of knowing the terminal points often helps resolve a difficulty that was discussed in Sec. 10.6 surrounding apparently constrained problems. In Example 18.1 the expression for m' is sought, and the first inclination may be to denote m' as the variable being sought (the y function in the general development of the Euler-Lagrange equation). Had m' been chosen as the function, an integral of ($m' \, dt$) would be expected in order to assure that the pump delivers 180 kg in the specified time. Also, the terminal points are not specified, because m' at the start and end of the process are not initially known. Choosing m as the function solves both difficulties, because the terminal points are then known and the problem can be treated as an unconstrained function.

Whether COV or DP is the more accurate tool for determining the optimal function depends on the nature of the problem. For continuous functions, COV is naturally more representative of the physical problem. It

FIGURE 18-7
(a) Dynamic programming (b) calculus of variations.

has no doubt been observed that the solution of a differential equation is a necessary step in finding the optimal function using COV. In some cases the analytical solution of this differential equation may be extremely difficult, and perhaps impossible. When confronting this situation a numerical solution of the differential equation is a possibility, but another option is to subdivide the continuous process into steps that are then solved by DP. On the other hand, if the physical situation is one of discrete stages, DP is more accurate and COV is an approximation.

The calculation procedures for basic DP programs have already been illustrated in Chapter 10, and the remaining topics addressed here are extensions of DP to two situations of greater complexity.

18.5 STAGED OPERATIONS WITH RECYCLE

Some chemical operations are characterized by recycled streams, as shown in Fig. 18-8. The symbols in Fig. 18-8 are similar to those first used in Fig. 10-1 in the introduction to DP, where the objective is to maximize the summation

$$\sum_{n=1}^{N} r_n$$

when the inlet concentration S, the entering flow rate Q, and the recycle rate R are known. A complication occurring in Fig. 18-8 is that S_N and S'_1 are not initially known. The values of S_n and S'_1 can be assumed, however, and DP applied to the series of stages. With the values of S_N and S'_1 determined from the first DP solution, more accurate values of S_N and S'_1 can be used in the next iteration.

18.6 DYNAMIC PROGRAMMING FOR CONSTRAINED OPTIMIZATIONS

All the problems to which DP has been applied so far in Chapter 10 and in this chapter have been truly unconstrained. Constrained problems can be

FIGURE 18-8
Recycle stream in a staged process.

encountered, however, just as has been the case in COV. A dilemma arises in making the decision of which is the optimal choice within one block of a table. No longer is it possible to select the minimum (or maximum) of the summation being optimized, but a sacrifice must be made in order to keep under control the constraint, which in DP is a summation.

The question is how much to trade off between the summation being optimized and the constraint summation. It is here that insight from COV provides the answer. The value of λ has proven to be the sensitivity coefficient, and the DP process will be conducted with internal choices made to hold a constant value of λ through the successive tables.

Example 18.3. Solve Example 18.2 by treating it as a four-stage DP problem, each stage consisting of 300 s of pumping operation.

Solution. Some understanding of the problem and results are already available from the COV solution of Example 18.2; that information will be applied here. The first need is a set of tables that express pump energy and mean pressure for each stage between a given inlet and outlet pressure. To compute such a table an assumption must be made about how the pumping is to take place—constant flow rate, constant Δp, etc. Since a constant Δp process was the optimal one in Example 18.2, it will be chosen for a given stage. No doubt the Δp, while constant for a stage, will be different for each stage. Table 18.1 provides pumping energy data, and Table 18.2 the contribution to the Δp summation for various combinations of inlet and outlet volumes.

Start at the final pumping stage, 900 to 1200 s, for which the energy and Δp values come directly from Tables 18.1 and 18.2 for the available pumping plans of 33.8, 34.2, 34.6, 35.0, and 35.4 to 40 m³. The next table, Table 18.3, applies to the 600 to 1200-second time interval and shows the cumulative energy and Δp contributions.

The extra dimension of λ is required in Table 18.3 to make the tradeoff in expending more energy in order to reduce the Δp contribution. In Example 18.3 the sensitivity coefficient is the difference between two energy quantities divided by the difference between two Δp contributions on the line just above and below within a given inlet volume block. Thus, for the inlet volume of 26.4 m³, the λ between the first two lines is -0.008 and between the second and third lines it is -0.034. From Example 18.2 the value of λ was found to be -0.02616 which will be the value chosen as the target in Table 18.3. Because of the steps in the DP solution, precise values of λ cannot be reproduced, and instead the choice of volumes pumped is the one closest to the desired value of λ. The best situation is where three lines give two values of λ that just straddle -0.02616, because the middle line then nearly reflects the desired λ.

Two more calculation tables that are omitted here would follow Table 18.3 giving the optimal result shown in Table 18.4.

The results are close to those obtained by COV, with the difference explained by the discrete steps used in the DP solution. In the COV solution the value of λ was inherently constant throughout the entire process, and the

TABLE 18.1
Pumping Energy in a Stage, Example 18.3

Volume at start	Volume at end															
	15.8	16.2	16.6	17.0	17.4	26.4	26.8	27.2	27.6	28.0	33.8	34.2	34.6	35.0	35.4	40.0
0.0	3694.1	3883.5	4077.7	4276.6	4480.2											
15.8						2343.4	2523.6	2710.5	2904.0	3104.3						
16.2						2169.9	2343.4	2523.6	2710.5	2904.0						
16.6						2003.0	2169.9	2343.4	2523.6	2710.5						
17.0						1842.9	2003.0	2169.9	2343.4	2523.6						
17.4						1689.4	1842.9	2003.0	2169.9	2343.4						
26.4											1472.4	1635.9	1808.0	1988.7	2178.0	
26.8											1317.5	1472.4	1635.9	1808.0	1988.7	
27.2											1171.3	1317.5	1472.4	1635.9	1808.0	
27.6											1033.6	1171.3	1317.5	1472.4	1635.9	
28.0											904.5	1033.6	1171.3	1317.5	1472.4	
33.8																1265.0
34.2																1107.0
34.6																959.6
35.0																822.7
35.4																696.3

TABLE 18.2
Δp **Contribution in a Stage, Example 18.3**

Volume at start	Volume at end															
	15.8	16.2	16.6	17.0	17.4	26.4	26.8	27.2	27.6	28.0	33.8	34.2	34.6	35.0	35.4	40.0
0.0	70142	71917	73693	75469	77245											
15.8						66323	68826	71329	73831	76334						
16.2						63821	66323	68826	71329	73831						
16.6						61318	63821	66323	68826	71329						
17.0						58815	61318	63821	66323	68826						
17.4						56312	58815	61318	63821	66323						
26.4											59693	62919	66146	69372	72599	
26.8											56466	59693	62919	66146	69372	
27.2											53239	56466	59693	62919	66146	
27.6											50013	53239	56466	59693	62919	
28.0											46786	50013	53239	56466	59693	
33.8																61211
34.2																57261
34.6																53312
35.0																49363
35.4																45414

TABLE 18.3
Cumulative energy and Δp from 600 to 1200 s

V at 600 s	V at 900 s	Cumulative energy	Cumulative Δp	λ
	33.8	1265.0 + 1472.4 = 2737.4	61211 + 59693 = 120,904	−0.008
	34.2	1107.0 + 1635.9 = 2742.9	57261 + 62919 = 120,180	−0.034
26.4	34.6	959.6 + 1808.0 = 2767.6	53312 + 66146 = 119,458	−0.061
	35.0	822.7 + 1988.7 = 2811.4	49363 + 69372 = 118,735	−0.087
	35.4	696.3 + 2178.0 = 2874.3	45414 + 72599 = 118,013	
	33.8	1265.0 + 1317.5 = 2582.5	61211 + 56466 = 117,677	0.0043
	34.2	1107.0 + 1472.4 = 2579.4	57261 + 59692 = 116,953	−0.022
26.8	34.6	959.6 + 1635.9 = 2595.5	53312 + 62919 = 116,231	−0.049
	35.0	822.7 + 1808.0 = 2630.7	49363 + 66146 = 115,509	−0.075
	35.4	696.3 + 1988.7 = 2685.0	45414 + 69372 = 114,786	
	33.8	1265.0 + 1171.3 = 2436.3	61211 + 53239 = 114,450	0.0163
	34.2	1107.0 + 1317.5 = 2424.5	57261 + 56466 = 113,727	−0.010
27.2	34.6	959.6 + 1472.4 = 2432.0	53312 + 59693 = 113,005	−0.037
	35.0	822.7 + 1635.9 = 2458.6	49363 + 62919 = 112,282	−0.063
	35.4	693.3 + 1808.0 = 2504.3	45414 + 66146 = 111,560	
	33.8	1265.0 + 1033.6 = 2298.6	61211 + 50013 = 111,224	0.0281
	34.2	1107.0 + 1171.3 = 2278.3	57261 + 53239 = 110,500	0.0017
27.6	34.6	959.6 + 1317.5 = 2277.1	53312 + 56466 = 109,778	−0.025
	35.0	822.7 + 1472.4 = 2295.1	49363 + 59693 = 109,056	−0.051
	35.4	693.3 + 1635.9 = 2332.2	45414 + 62919 = 108,333	
	33.8	1265.0 + 904.0 = 2169.5	61211 + 46786 = 107,997	0.040
	34.2	1107.0 + 1033.6 = 2140.8	57261 + 50013 = 107,274	0.0134
28.0	34.6	959.6 + 1171.3 = 2130.9	53312 + 53239 = 106,551	−0.013
	35.0	822.7 + 1317.5 = 2140.2	49363 + 56466 = 105,829	−0.039
	35.4	693.6 + 1472.4 = 2168.7	45414 + 59693 = 105,107	

TABLE 18.4
Optimal pumping plan for Example 18.3

Time stage, s	Volumes in/leaving	Contribution to energy, kJ	Contribution to Δ pressure, kPa
0 to 300	0/17.0	4276.6	75,469
300 to 600	17.0/27.2	2169.9	63,820
600 to 900	27.2/34.6	1472.4	59,692
900 to 1200	34.6/40.0	959.6	53,312
	Total:	8878.5 kJ	252,293 kPa

DP solution also strived to use the same value of λ in making the decisions at each stage. A less-than-optimal result would have occurred had a high value of λ been chosen in the early stages and a low one later as necessary to meet the constraint.

In solving Example 18.3 we had advance information that λ should be -0.02616, but if the problem had not been solved first by COV, this information would not have been available. When starting fresh on a constrained DP problem, it is necessary to estimate a value of λ and calculate a preliminary solution. The arbitrarily chosen value of λ would probably result in the constraint either being exceeded or underused. Based on the results with the trial λ, a new λ would be chosen.

PROBLEMS

18.1. The equation for the path through air is to be found such that minimum work results in traversing between points 1 and 2 in Fig. 18-9. The work is the product of the resistance and the distance, $R\,ds$, where R is dependent on the air density. The equation for R is

$$R = 1500/(y + 10)$$

(a) Apply the Euler-Lagrange equation to develop the differential equation for y.

(b) Integrate the differential equation to determine the path between 1 and 2.

Ans.: (a) $(y + 10)y' = \sqrt{2,250,000/c_1 - (y + 10)^2}$

18.2. A refrigerated warehouse is to be cooled down from a temperature of 25°C to 0°C in a time interval of 12,000 s. The refrigeration equipment is to be operated in such a manner that it requires minimum compressor energy for the cooldown. Specifically, we seek a function $T = y(t)$, as in Fig. 18-10, where T is the room temperature in °C and t is the time in seconds from the start of cooldown. The expression for the ratio of compressor energy to

FIGURE 18-9
Path through air resulting in minimum work.

FIGURE 18-10
Optimal temperature program for minimum energy.

refrigeration energy is

$$\frac{\text{compressor power, kW}}{\text{refrigeration rate, kW}} = 0.01(50 + 0.012t - \theta)$$

where θ = air-side temperature of the evaporator, °C. The equation reflects an increased compressor-to-refrigeration power ratio at low evaporator temperature, and also an increase of this ratio with time due to frost formation on the evaporator coil.

Consider the warehouse to be a mass of uniform temperature that changes with time, and that the thermal capacity is 400,000 kJ/K. The product of the heat-transfer coefficient and the heat-transfer area between the warehouse and air-side of the coil (between T and θ) is 20 kW/K.

Ans.: $T = (1.5 \times 10^{-7})t^2 - 0.003883t + 25$.

18.3. The speed of a large turbine-generator unit is to be brought up from zero to its operating value of 60 rev/s in 1,000 s. During that 1,000 s the unit is to complete 35,000 revolutions to achieve temperature uniformity. If the energy cost is represented by the equation,

$$\text{cost, \$} = \int_0^{1000} 120\,(\omega')^2\,dt$$

where ω' = angular acceleration, rev/s^2
   t = time, s

what is the equation for the rotative speed ω rev/s as a function of time that results in the minimum energy cost?

Ans.: minimum cost = \$468.

18.4. A rocket starts with zero velocity and travels on a horizontal path against negligible resistance. The mass of the metal of the rocket is w which is very large relative to the mass of the fuel.

Determine the velocity program, $v(t)$, that results in the maximum distance traversed in time θ while using up all the fuel of mass M. The thrust is represented by the equation

$$\text{thrust} = w\,\frac{d^2x}{dt^2} = A\sqrt{\frac{dM}{dt}}$$

where $\dfrac{dM}{dt}$ = fuel burning rate

Also determine the distance traveled in time θ when following the optimum plan.

Ans.: distance $= \dfrac{A}{w}\sqrt{\dfrac{\theta^3 M}{3}}$

18.5. The distance between two pumping stations of a natural gas (methane) pipeline is 200 km. The pipeline carries 30 kg/s and the pressure entering the pipeline is 4200 kPa and the leaving pressure is 2000 kPa.

Instead of choosing a pipe of constant diameter, the diameter can be changed every 40 km, because of the expectation that the optimum diameter may change as the pressure decreases, as shown in Fig. 18-11.

(a) Compute the diameter that results in a pressure drop from 3000 kPa to 2200 kPa in a 40 km length, and thus fill in the gap in Table 18.5. The basis for the calculation of the pressure drop in compressible, isothermal flow in a long gas pipeline comes from the combination of the following equations:

  momentum
  relationship of drag to the pressure drop equation
  continuity
  perfect gas (constant temperature of 20°C)

FIGURE 18-11
Pipeline in Prob. 18.5 where diameter can be changed every 40 km.

TABLE 18.5
Diameters resulting in the pressure drop indicated.

Press. Ent. Sect.	Pressure leaving the section, kPa										
	2000	2200	2400	2600	2800	3000	3200	3400	3600	3800	4000
2200	.733										
2400	.632	.720									
2600	.578	.621	.708								
2800	.541	.568	.612	.697							
3000	.513		.560	.603	.687						
3200	.491	.505	.525	.552	.594	.678					
3400	.472	.484	.498	.517	.544	.587	.670				
3600	.457	.466	.477	.492	.511	.538	.580	.662			
3800	.443	.450	.460	.471	.485	.505	.531	.573	.654		
4000	.431	.437	.445	.454	.465	.480	.499	.526	.567	.648	
4200	.420	.425	.432	.439	.449	.460	.474	.493	.520	.561	.641

and is
$$p_1{}^2 - p_2^2 = \frac{V_1{}^2 p_1}{1000(v_1)}\left[\frac{fL}{D} + 2\ln\left(\frac{V_2}{V_1}\right)\right]$$

where p = pressure, kPa
V = velocity, m/s
v = specific volume, m³/kg
f = friction factor = 0.02
L = length = 40,000 m
D = diameter, m

(b) The diameters, $D_1, \ldots D_5$ are to be chosen such that the pipeline requires the minimum mass of metal. Any pipe diameter is available, because an order of 40 km of pipe is sufficient that the steel mill will manufacture any specified diameter. The thickness of the pipewall is constant regardless of the diameter, so the problem resolves to one of determining the diameters giving a minimum value of the summation:

$$\sum_{i=1}^{5} D_i$$

Use dynamic programming to determine $D_1, \ldots D_5$.
   Ans.: 2.902 m.

18.6. Ice builders are thermal storage devices that freeze ice during periods of the day when the cooling requirements are low and melt the ice to provide cooling during periods when a demand charge must be paid for electricity. Most ice builders consist of a water tank in which are immersed circuits of pipe. Refrigerant flowing through these pipes evaporates at a low temperature during the freezing portion of the cycle.

   In the operation of a certain ice builder, 50 mm thickness of ice is to be built up in 10 hours on tubes that are 30 mm in diameter. The total tube length is 800 m. The evaporating temperature is held constant during each of the five two-hour periods during the buildup. Dynamic programming is to be used to determine each of these evaporating temperatures such that minimum refrigeration compressor energy is required during the 10-hour buildup time.

A simple model that yields realistic results is that the ice freezes at the outside surface at 0°C and the latent heat of fusion must pass by conduction through the ice to the refrigerant, as illustrated in Fig. 18-12. The only resistance to heat transfer that need be considered is that of the ice, since the resistances of the tube and the boiling coefficient are relatively small. The conduction relations for a thick cylinder must be used.

Data

Ice properties

$$\text{density} = 900 \text{ kg/m}^3$$

$$\text{latent heat of fusion} = 340 \text{ kJ/kg}$$

$$\text{thermal conductivity} = 0.0023 \text{ kW/(m} \cdot \text{K)}.$$

Coefficient of performance (COP) of the refrigeration system is 60% that of the Carnot COP at the prevailing evaporating and condensing temperatures. The condensing temperature is constant at 35°C.

(*a*) Table 18.6 shows the energy in kJ required by the refrigeration compressor for two-hour periods in building the ice from one thickness to another. Verify one value on the table, for example, the building of ice from a thickness of 20 mm to 30 mm.

(*b*) Using dynamic programming with the energy costs shown in Table 18.6, determine the thicknesses at the end of each of the five two-hour periods that build the total thickness of 50 mm with minimum total energy required of the refrigeration plant.

Ans.: 823.1 MJ

18.7. The constrained optimization for the startup of the turbine in Problem 18.3 is to be solved by dynamic programming. The rotative speeds ω at 250, 500,

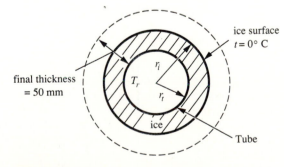

FIGURE 18-12
Freezing ice on a tube in an ice builder.

TABLE 18.6
Energy in MJ required by the compressor of the refrigeration system in Problem 18.6 for freezing ice in two-hour periods from initial to final ice thicknesses shown

Initial thickness, mm	Final thickness, mm																
	10	12.5	15	17.5	20	22.5	25	27.5	30	32.5	35	37.5	40	42.5	45	47.5	50
0	68.0	92.2	120.4	153.0	191.0	235.3											
10					111.0	150.9	197.5	252.0									
12.5						123.1	167.7	220.0	281.4								
15							135.8	185.5	244.0	312.9							
17.5							102.3	149.1	204.3	269.5	346.5						
20								111.5	163.0	224.1	296.6						
22.5									121.0	177.6	245.1	325.3					
25										130.9	192.9	267.1	355.7				
27.5											141.2	209.0	290.4	388.0			
30												151.9	225.7	314.9	422.0		
32.5												97.2	163.0	243.2	340.6	458.0	
35													103.5	174.5	261.6	367.6	496.0
37.5													48.7	110.0	186.4	280.7	396.0
40															116.8	198.8	300.6
42.5																123.7	211.7
45																56.8	130.8
47.5																	59.7

TABLE 18.7
Cost in a stage, Problem 18.7

Speed at start, rps	Speed at end of stage 1, rps					Speed at end of stage 2, rps					Speed at end of stage 3, rps					Stage 4
	19.0	19.5	20.0	20.5	21.0	36.5	37.0	37.5	38.0	38.5	49.5	50.0	50.5	51.0	51.5	60.0
0.0	175.2	184.0	193.0	202.4	212.0											
19.0						147.1	155.6	164.3	173.3	182.5						
19.5						139.0	147.2	155.6	164.3	173.3						
20.0						131.2	139.1	147.2	155.7	164.4						
20.5						123.7	131.3	139.1	147.4	155.8						
21.0						116.5	123.9	131.5	139.4	147.5						
36.5											81.6	87.9	94.4	101.2	108.2	
37.0											75.6	81.7	87.9	94.4	101.1	
37.5											69.9	75.7	81.7	88.0	94.5	
38.0											64.5	70.0	75.8	81.8	88.0	
38.5											59.3	64.5	70.1	75.8	81.8	
49.5																53.3
50.0																48.4
50.5																43.9
51.0																39.5
51.5																35.4

TABLE 18.8
Revolutions in a stage, Problem 18.7

Speed at start, rps	Revolutions in stage 1					Revolutions in stage 2					Revolutions in stage 3					Stage 4
	19.0	19.5	20.0	20.5	21.0	36.5	37.0	37.5	38.0	38.5	49.5	50.0	50.5	51.0	51.5	60.0
0.0	2521	2563	2604	2646	2688											
19.0						6969	7021	7073	7125	7177						
19.5						7052	7104	7156	7208	7260						
20.0						7135	7188	7240	7292	7343						
20.5						7219	7271	7323	7375	7427						
21.0						7302	7354	7406	7458	7510						
36.5											10823	10879	10934	10990	11045	
37.0											10896	10951	11007	11063	11118	
37.5											10969	11024	11080	11135	11191	
38.0											11042	11097	11153	11208	11264	
38.5											11115	11170	11226	11281	11337	
49.5																13750
50.0																13819
50.5																13889
51.0																13958
51.5																14028

and 750 s from the start are to be determined such that the summation

$$\text{Cost, } \$ = \sum_{i=1}^{4} \text{Cost}_i$$

is a minimum, where Cost_i is the integral of $120 \, (\omega')^2$ in a time stage. The constraint requires the total number of revolutions in the 1000 s startup to be 35,000,

$$\sum_{i=1}^{4} \text{Rev}_i = 35,000$$

where Rev_i is the number of revolutions in a time stage. The solution of Problem 18.3 by calculus of variations determined that $\lambda = 0.0144$ \$/revolution, and this value may be used in the dynamic programming solution to guide the tradeoff between cost and number of revolutions. Table 18.7 provides values of Cost_i and Table 18.8 the number of revolutions Rev_i in a 250 s stage between selected initial and final values of ω.

(a) Verify by integration of $120(\omega')^2 dt$ and $\omega \, dt$ one entry in each of Table 18.7 and 18.8, respectively (for example between the initial ω of 20 rev/s at 250 s and the final ω of 37.5 rev/s at 500 s). Use a quadratic function for ω, namely $\omega = At^2 + Bt$, where $t = $ time in s from the startup.

(b) Perform the dynamic programming optimization to determine values of ω at the end of each 250 s interval.

Ans.: total cost $= \$467.20$ with 34,938 revolutions.

CHAPTER
19

PROBABILISTIC
APPROACHES
TO DESIGN

19.1 PROBABILISTIC VS. DETERMINISTIC

Most designs and analyses are made under the assumption that numerical magnitudes on which the determinations are made are known precisely. Any projection into the future cannot be certain, however, so the best that can be predicted is the most likely situation. Once it is admitted that the prediction is not certain, the next question is the degree of uncertainty. It is valuable for the decision maker to know whether high confidence can be placed in data or whether the distribution curve is flat (see Fig. 1-4).

The application of probabilistic approaches to thermal systems is an effort still in its infancy even though probability theory and numerical applications of the theory are well established. Also, certain engineering disciplines, such as structural engineering and reliability analysis, have progressed in using these tools. The purpose of this chapter is to present a few of the principles and then show some situations where probabilistic approaches apply to thermal systems.

One of the major challenges is to alert thought processes so that situations that are probabilistic can be recognized. The next challenge is to translate the problem to a form that can be addressed by probability theory. An intended contribution of this chapter is to present some situations and their solutions in order to illustrate the distinction between deterministic and probabilistic approaches.

498

19.2 DISCRETE EVENTS VS. CONTINUOUS DISTRIBUTION

One of the classifications into which many of the applications divide is whether they are discrete or continuous. The distinction between the two categories may be crudely expressed as the number or density of events. As the density increases the representation shifts from discrete to continuous. The characterization of discrete events appears in probability texts in examples of rolling dice, selection of cards from a deck, and drawing balls of various colors from urns. These applications should not suggest that the usefulness of the principles of discrete probability is limited to games.

If a distinction between the analysis of discrete events and continuous representations is the density of events, it should be possible to progress from the discrete-event equations to continuous ones. Indeed, such a development will be explained starting in Sec. 19.12. First, however, the probability of several elementary situations involving random events will be explored.

19.3 PROBABILITY

Not all plans for the future involve probability. Predicting the day of the week on which July 4 will fall in the year 2004 projects into the future, but lies outside the realm of probability because we know with certainty (barring a change in the calendar) on what day of the week July 4 will fall. We are concerned about the situation where there is both the possibility of an event happening and not happening.

The probability is defined as the number of favorable ways an event can occur divided by the number of different ways the event occurs. What is the probability of drawing the ace of spaces from a 52-card deck? The probability p is 1/52:

$$p = \frac{1 = \text{number of favorable ways}}{52 = \text{number of different ways}}$$

Probabilities can vary from zero to unity, with zero probability representing the limiting case of no possibility of a favorable occurrence, and the probability of 1.0 indicating certainty of the favorable occurrence.

19.4 INDEPENDENT EVENTS

A set of events is independent if the occurrence of any one is not influenced by the occurrence of the others. The probability of a specific combination of two independent events is the product of the probabilities of the separate events.

Example 19.1. What is the probability of achieving the head-2 combination when a coin is flipped and a die rolled?

Solution. The probability p_1 of the head appearing is 1/2 and the probability p_2 of the 2 appearing on the die is 1/6, so the probability of the combination is $(1/2)(1/6) = 1/12$.

19.5   SUCCESSIVE EVENTS

An extension of the concept of independent events is that of successive events where the first event alters the probability of the second event.

Example 19.2. What is the probability of drawing two aces in succession from a 52-card deck?

Solution. The probability of the first event p_1 is 4/52 since there are originally four aces. The probability of the second event is 3/51 if the first event is favorable. Therefore, the probability of drawing two aces in succession is $(4/52)(3/51) = 1/221$.

19.6   DISTINGUISHABILITY OF *X* THINGS TAKEN *Y* AT A TIME

Suppose that eight objects are present and these objects are numbered 1 through 8—thus they are distinguishable. What is the probability of making 3 extractions and drawing the objects 2, 5, and 7? At first it may seem that the probability of drawing the 2-5-7 combination is $(1/8)(1/7)(1/6)$, since the probability of drawing the 2 is 1/8, then of drawing the 5 the probability is 1/7, and finally of drawing the 7 the probability is 1/6. This approach overlooks the possibility of drawing a 2 first, next a 7, and finally a 5. In fact there are 3! combinations of 2-5-7, each of which have the $(1/8)(1/7)(1/6)$ probability appearing. These combinations are 2-5-7, 2-7-5, 5-2-7, 5-7-2, 7-2-5, and 7-5-2. The probability of the result of three drawings including the 2-5-7 objects, then, is

$$p = (3!)\left(\frac{1}{8}\right)\left(\frac{1}{7}\right)\left(\frac{1}{6}\right) = \frac{3!5!}{8!}$$

Another way of thinking of the probability in the above situation is to say that the probability is the reciprocal of the number of ways that 8 distinguishable objects can be taken 3 at a time,

$$p = \frac{1}{{}_nC_r}$$

where $_nC_r$ is the number of ways that n objects can be taken r at a time, and where

$$_nC_r = \frac{n!}{(n-r)!r!} \tag{19.1}$$

Example 19.3. What is the probability of drawing 4 white balls from an urn containing 10 white, 4 black, and 3 red balls?

Solution. The number of ways that 4 white balls can be chosen from the 10 white balls is equal to the number of combinations of 10 things taken 4 at a time,

$$_{10}C_4 = \frac{10!}{4!\ 6!}$$

The total number of ways in which 4 balls can be chosen from the 17 available is

$$_{17}C_4 = \frac{17!}{4!\ 13!}$$

The probability of success is

$$p = \frac{_{10}C_4}{_{17}C_4} = \frac{10!\ 4!\ 13!}{4!\ 6!\ 17!} = \frac{3}{34}$$

19.7   NON-OCCURRENCE OF AN EVENT

What is the probability of the ace appearing at least once in n throws of a die? The probability of the ace appearing in one roll is 1/6, but the probability of rolling the ace at least once in two rolls is not (1/6)(1/6), because the chances should be improved, not reduced. Nor is the probability of rolling the ace in two rolls 1/6 + 1/6 as the absurd situation after 7 rolls indicates.

The key to the correct procedure is to work with "non-occurrence." The probability of the ace *not* occurring on one roll is 5/6. The probability of an ace not occurring in either of two rolls is (5/6)(5/6), or the probability of the ace appearing is 1 − (5/6)(5/6).

Example 19.4. The probability that Paul will solve a problem is 1/2 and the probability that John will solve it is 1/3. What is the probability that the problem will be solved if both work at it independently?

Solution

$$p_{succ} = 1 - p_{fail} = 1 - \left(\frac{1}{2}\right)\left(\frac{2}{3}\right) = 2/3$$

FIGURE 19-1
Probability of getting from A to B.

19.8 MUTUALLY EXCLUSIVE EVENTS

Events are called mutually exclusive if the occurrence of one prevents the occurrence of the others.

> **Example 19.5.** A hitchhiker whose destination is Point B in Fig. 19-1 has been sleeping and is let out of a car at Point A. When he regains his senses he is faced with a choice of going in one of four directions. Unknown to him is that if he takes route X he will face another choice of three possibilities, or if he chooses Y he will face an unmarked fork. The other two choices at A don't lead to B.
>
> What is the probability that he will select a path that will lead to B?
>
> **Solution.** If the hiker chooses path X his probability of making the correct choice at the subsequent intersection is 1/3, so the probability of the $\{x,1\}$ combination of choices is $(1/4)(1/3) = 1/12$. Correspondingly, the probability of the $\{y,2\}$ successful combination is $(1/4)(1/2) = 1/8$. The probability of success in getting from A to B is, then, the *sum* of the two probabilities:
>
> $$1/12 + 1/8 = 5/24$$
>
> The principle applicable to mutually exclusive events is that if p_1, $p_2, \ldots p_n$ are the probabilities of an event occurring if each operation were independent, then the probability of the event occurring, p is
>
> $$p = p_1 + p_2 + \cdots = p_n$$
>
> if making any one of the choices excluded all other choices.
>
> Example 19.5 considers alternate routes and carries with it some of the flavor of mutually exclusive events. Having once started on one route, the chance of trying any other routes is excluded.
>
> **Example 19.6.** An urn contains 2 white balls and 3 black balls. Two balls are drawn in succession. What is the probability that one is white and one black?
>
> **Solution.** The probability of drawing a white ball first and then a black ball, as shown by route X in Fig. 19-2, is $(2/5)(3/4) = 3/10$. Similarly

FIGURE 19-2
Probability of drawing one white and one black ball.

the probability of drawing a black ball first and then a white ball, which is route Y, is $(3/5)(2/4) = 3/10$ also. These are mutually exclusive events, so the probability of one white and one black ball as the result is the sum of the independent probabilities,

$$3/10 + 3/10 = 3/5$$

19.9 REPEATED AND INDEPENDENT TRIALS

This section deals with the question of the probability of an event occurring exactly once in n trials when the probability of the occurrence of that event in a single trial is known.

Example 19.7 What is the probability of rolling exactly one 1 in one throw, two throws, three throws, etc., of a single die?

Solution. The probability of rolling exactly one 1 in one roll is 1/6.

Two throws: (*a*) $p = 1/6$ of rolling a 1 on the first throw
                   $p = 5/6$ of no 1 on the second throw
                   $p_1 = (1/6)(5/6) = 5/36$
           (*b*) $p = 5/6$ of rolling no 1 on the first throw
                   $p = 1/6$ of rolling a 1 on the second throw
                   $p_2 = (5/6)(1/6) = 5/36$

Sequences (*a*) and (*b*) are mutually exclusive, so the probability of rolling exactly one 1 in two throws is $5/36 + 5/36 = 5/18$.

Three throws: The three mutually exclusive events each have a probability of $(1/6)(5/6)(5/6) = 25/216$, so the probability of rolling exactly one 1 in three throws is $3(25/216) = 25/72$

Four throws: $p = 4[(1/6)(5/6)(5/6)(5/6)] = 250/648$

Table 19.1 shows probabilities for one to ten throws.

TABLE 19.1
Probabilities of rolling
exactly one 1 in a
given number of throws

Number of throws	Probability
1	0.1667
2	0.2778
3	0.3472
4	0.3858
5	0.4019
6	0.4019
7	0.3907
8	0.3721
9	0.3489
10	0.3230

19.10   THE BINOMIAL LAW

The rule applicable to the situation in Example 19.7 and to certain more general probabilities as well is called the "binomial law" and is as follows: If the probability of an event occurring in a single trial is p, then the probability that it will occur exactly r times in n independent trials is

$$p_r = {}_nC_r \; p^r \; (1-p)^{(n-r)} \qquad (19.2)$$

Example 19.8. Use the binomial law to compute the probability of rolling exactly one 1 in 8 throws of a die.

Solution

$$p = 1/6, \quad n = 8, \quad r = 1$$

$$_nC_r = {}_8C_1 = \frac{8!}{7! \; 1!} = 8$$

so

$$p_r = (8)(1/6)^1 \; (1 - 1/6)^{(8-1)} = 0.3721$$

Example 19.9. What is the probability that a 1 will appear exactly four times in the course of 10 throws of a die?

Solution

$$p = 1/6, \quad n = 10, \quad r = 4$$

$$p = \frac{10!}{4! \; 6!} \; (1/6)^4 (5/6)^6 = 0.0543$$

19.11   THE PROBABILITY DISTRIBUTION

The binomial law is the calculative tool that leads to the development of the probability distribution curve. Consider an example of an electric utility that knows that 40% of the heat pumps in its area are of Brand A and 60% of Brand B, but does not know in which homes a given brand is installed. It wishes to make an inspection of the heat pumps and would like to predict in random sampling what its probability is of choosing one brand or the other. Figure 19-3 shows the predictions from the binomial law of the number of

FIGURE 19-3
Probability of selecting Brand A heat pumps if the probability of a single event is 0.40.

Brand A it will find if the total number of samples $n = 5$, 10, and 30. The probability of the single event of selecting Brand A is 0.4. Figure 19-3 shows the probability p_r of selecting a given number, r, of Brand A for the various totals n. Also shown is the maximum probability. As the number of samples increases the shape of the probability curve begins to approach that of the normal probability curve.

19.12 CONTINUOUS REPRESENTATION OF BINOMIAL LAW

Through a series of approximations the results of the binomial law, Eq. (19.2), can be represented by a continuous function

$$p_r = \frac{1}{\sqrt{2\pi n p (1 - p)}} \exp\left\{ \frac{-(r - np)^2}{[2np(1 - p)]} \right\} \tag{19.3}$$

The development of Eq. (19.3) from Eq. (19.2) is available in many books on probability, and the details will be bypassed here. Instead, Table 19.2 will show some comparative values of p_r calculated from the two equations. At least two conclusions can be drawn from an examination of Table 19.2:

TABLE 19.2
Values of p_r computed from the binomial law compared with values from the continuous representation. The form of the table is (p_r from binomial/p_r from continuous)

p	r	$n = 10$	$n = 20$	$n = 30$	$n = 40$
0.2	1	0.2684/0.2308	0.0577/0.0547	0.0093/0.0135	0.0013/0.0034
	2	0.3020/0.3154	0.1369/0.1194	0.0337/0.0344	0.0065/0.0095
	3	0.2013/0.2307	0.2054/0.1908	0.0785/0.0713	0.0205/0.0224
	4	0.0881/0.0904	0.2182/0.2230	0.1325/0.1200	0.0475/0.0452
	5	0.0264/0.0189	0.1746/0.1908	0.1723/0.1641	0.0854/0.0781
	6	0.0055/0.0021	0.1091/0.1194	0.1795/0.1821	0.1246/0.1154
	7	0.0008/0.0001	0.0546/0.0547	0.1538/0.1641	0.1513/0.1458
	8	0.0000/0.0000	0.0222/0.0183	0.1106/0.1200	0.1560/0.1577
	9	0.0000/0.0000	0.0074/0.0045	0.0676/0.0713	0.1387/0.1458
	10	0.0000/0.0000	0.0020/0.0008	0.0355/0.0344	0.1075/0.1154
0.4	5	0.2007/0.2091	0.0747/0.0713	0.0042/0.0050	0.0001/0.0002
	6	0.1115/0.1119	0.1244/0.1200	0.0115/0.0122	0.0005/0.0007
	7	0.0425/0.0395	0.1659/0.1641	0.0263/0.0262	0.0015/0.0019
	8	0.0106/0.0092	0.1797/0.1821	0.0505/0.0489	0.0040/0.0046
	9	0.0016/0.0014	0.1597/0.1641	0.0823/0.0796	0.0095/0.0100
	10	0.0001/0.0001	0.1171/0.1200	0.1152/0.1126	0.0197/0.0198
	11	0.0000/0.0000	0.0710/0.0713	0.1396/0.1387	0.0357/0.0350
	12	0.0000/0.0000	0.0355/0.0344	0.1474/0.1487	0.0576/0.0560
	13	0.0000/0.0000	0.0146/0.0135	0.1360/0.1387	0.0827/0.0806
	14	0.0000/0.0000	0.0049/0.0043	0.1101/0.1126	0.1063/0.1045
	15	0.0000/0.0000	0.0013/0.0011	0.0783/0.0796	0.1228/0.1222

(1) the agreement between the two equations becomes progressively better with large values of n and r, and (2) the peak value of p_r in each of the eight blocks of Table 19.2 occurs when $r = np$. Thus, for $p = 0.2$, the maximum value of p_r occurs at $r = 2, 4, 6,$ and 8 for $n = 10, 20, 30,$ and 40, respectively.

The symbols used in the normal probability distribution curve are usually different than those shown in Eq. (19.3), and while there is no complete standardization, a set of symbols often seen is

$$p_r = \frac{1}{\sigma\sqrt{2\pi}} e^{-(x-x_0)^2/2\sigma^2} \qquad (19.4)$$

When Eq. (19.4) is graphed, as in Fig. 19-4, the curve takes on the characteristic bell shape with the peak value of $p_r = 1/\sigma\sqrt{2\pi}$ when $x = x_0$. The curve is broad for large values of σ and is narrow with a high peak when σ is small.

It should be pointed out that the ordinate of Fig. 19-4 is not probability, but probability density. In other words, the ordinate alone does not generally provide practical information. The area under the p_r curve indicates the probability; for example, the area between x_1 and x_2 in Fig. 19-4 represents the probability of the event occurring between these x values. Even for the discrete probabilities shown in Fig. 19-3, areas were implied because

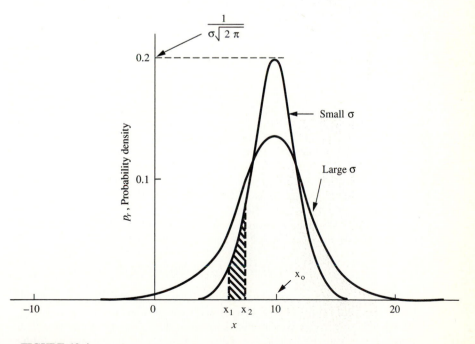

FIGURE 19-4
Normal probability distribution curve.

the numbers on the abscissa represented a band between the vertical lines. The total area under the normal probability curve is the integral

$$\frac{1}{\sigma\sqrt{2\pi}} \int_{-\infty}^{\infty} [e^{-(x-x_0)^2/2\sigma^2}]dx = 1 \tag{19.5}$$

which can be shown through the application of the definite integral

$$\int_{-\infty}^{\infty} e^{-a^2x^2}\, dx = \frac{\sqrt{\pi}}{2a}$$

The interpretation of Eq. (19.5) is that the event is certain to occur if the full range of x is spanned.

The normal probability curve and the area under it are sometimes useful tools in making judgments with respect to physical situations of a probabilistic nature. In such cases there is a need to translate between the physical data and the mathematical representation. Consider first the case where the normal distribution curve or the equation is available and is to be translated into probabilities of the occurrence of an event. As pointed out previously, the probability of occurrence in a certain range of x is represented by the integral of the equation between those limits. While it was possible to evaluate the definite integral, as in Eq. (19.5), an explicit solution is not available when the limits are not infinity. The integration must be performed numerically; the values from such integrations are presented in Table 19.3. The table lists values of the integral of the center portion of the curve from $x_0 - z\sigma$ to $x_0 + z\sigma$.

Example 19.10. At 3:00 PM on a June day the probability density of the outdoor temperature, $t°C$, in a certain locality is

$$p_r = \frac{0.15}{\sqrt{\pi}} e^{-0.0225(t-26)^2}$$

(a) What is the probability of the temperature being 22°C?
(b) What is the probability of the temperature being above 30°C?

Solution. (a) This is an improper question, because it asks the probability of the temperature being precisely 22°C. Since the area under the curve of the infinitely thin line is zero, there is no likelihood of the temperature being precisely 22°C. A better question would be this: what is the probability of the temperature being between 21.5 and 22.5°C? This probability is

$$\int_{21.5}^{22.5} p_r \, dt \approx (1)(p_r \text{ at } 22) = 0.059$$

(b) Table 19.3 will be used to evaluate this integral. By matching the equation for the temperature probability density with the standard form shown in Eq. (19.4), the value of σ is determined to be 4.71°C. The area sought is that from $x_0 + z\sigma$ to $+\infty$ where $z = (30 - 26)/4.71$, as shown in Fig. 19-5. From Table 19.3 with $z = 4/4.71 = 0.8493$, $P = 0.6043$. The area on either

TABLE 19.3
Probability P (area under normal probability curve)
between $x_o - z\sigma$ and $x_o + z\sigma$

z	P	z	P	z	P
0.00	0.0000	1.00	0.6827	2.00	0.9545
0.05	0.0399	1.05	0.7063	2.05	0.9596
0.10	0.0797	1.10	0.7287	2.10	0.9643
0.15	0.1192	1.15	0.7499	2.15	0.9684
0.20	0.1585	1.20	0.7699	2.20	0.9722
0.25	0.1974	1.25	0.7887	2.25	0.9756
0.30	0.2358	1.30	0.8064	2.30	0.9786
0.35	0.2737	1.35	0.8230	2.35	0.9812
0.40	0.3108	1.40	0.8385	2.40	0.9836
0.45	0.3473	1.45	0.8529	2.45	0.9857
0.50	0.3829	1.50	0.8664	2.50	0.9876
0.55	0.4177	1.55	0.8789	2.55	0.9892
0.60	0.4515	1.60	0.8904	2.60	0.9907
0.65	0.4843	1.65	0.9011	2.65	0.9920
0.70	0.5161	1.70	0.9109	2.70	0.9931
0.75	0.5467	1.75	0.9199	2.75	0.9940
0.80	0.5763	1.80	0.9281	2.80	0.9949
0.85	0.6047	1.85	0.9357	2.90	0.9963
0.90	0.6319	1.90	0.9426	3.00	0.9973
0.95	0.6579	1.95	0.9488	4.00	0.99994
1.00	0.6827	2.00	0.9545	5.00	0.9999994

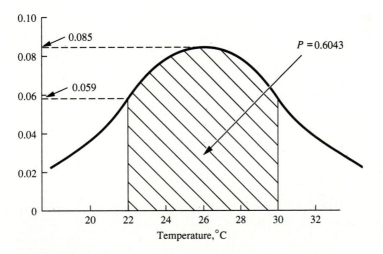

FIGURE 19-5
Area showing P of temperature exceeding $30°C$ in Example 19.10.

510

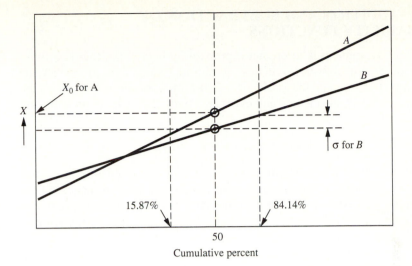

FIGURE 19-7
Two different normal distributions shown on the arithmetic probability chart.

side of this central area is $1 - 0.6043 = 0.3957$, and one-half of this, or 0.1979, represents the probability of the temperature exceeding 30°C.

The σ term appears so frequently in probability work that it is given a special name. It is called the "standard deviation." The area under the normal probability curve extending one standard deviation below to one standard deviation above the most probable value of x is 0.6827, thus embracing 68.27% of the possibilities.

The process of moving in the opposite direction, namely from physical data from previous experience to a normal probability curve or equation, is based on the relationship between z and P as presented in Table 19.3. A graphic tool for the translation is arithmetic probability paper, a sample of which is shown in Fig. 19-6.

The cumulative percentages are computed and plotted as shown on the skeleton diagram in Fig. 19-7 for two different distributions. If the line is straight, then the distribution is a normal probability. The point at which the line crosses the 50% cumulative percentage gives the x_0 value. The standard deviation is the difference in x values from 50% to 15.87% or 84.14%. The line with the greater slope has the larger standard deviation.

If arithmetic probability paper is not available, it can be constructed by scaling $100(P/2)$ from Table 19.3 in both directions from the 50% cumulative percentage.

FIGURE 19-6
Arithmetic probability paper.

19.13 ADDITION AND SUBTRACTION OF PROBABILITY FUNCTIONS

A frequent assignment is to predict a probability expression that is a function of other probabilistic variables. A simple example is to predict the probability function for the profit of an enterprise where the profit is the difference between the income and expenses, both of which are probabilistic expressions,

$$\text{profit} = \text{income} - \text{expenses}$$

A more difficult situation is where several imposed variables on a system simulation are probabilistic in nature and have a complex influence on the simulation. We first consider the most fundamental and simplest case where two probabilistic functions are combined through addition or subtraction. Even though this case falls short of answering the challenge of a system simulation, the results indicate some characteristics of combinations of probabilities. The objective will be to determine z in the equation

$$z = y - w \qquad . \tag{19.6}$$

where y and w are probabilistic functions, as shown in Fig. 19-8. More specifically, we seek the probability density p_z as a function of two normal probability density distributions p_y and p_w.

$$p_w = \frac{1}{\sigma_w \sqrt{2\pi}} e^{-(w-w_0)^2/2\sigma_w^2}$$

and

$$p_y = \frac{1}{\sigma_y \sqrt{2\pi}} e^{-(y-y_0)^2/2\sigma_y^2}$$

For a difference between y and w of z_1, as shown in Fig. 19-8,

$$p_{z_1} = \int_{-\infty}^{\infty} p_w \, p_{y=w+z_1} \, dw$$

or in general

$$p_z = \int_{-\infty}^{\infty} p_w \, p_{y=w+z} \, dw$$

so

$$p_z = \frac{1}{\sigma_w \sqrt{2\pi}} \int_{-\infty}^{\infty} e^{-(w-w_0)^2/2\sigma_w^2} \frac{1}{\sigma_y \sqrt{2\pi}} e^{-(w+y-y_0)^2/2\sigma_y^2} \, dw \tag{19.7}$$

w, y, z (Same units)

FIGURE 19-8
Difference of two probability functions.

For convenience, let $W = w - w_0$, $dw = dW$, and $\Delta = y_0 - w_0$.

$$p_z = \frac{1}{\sigma_w \sigma_y \sqrt{2\pi}} \int_{-\infty}^{\infty} e^{-W^2/2\sigma_w^2} \; e^{-(W+z-\Delta)^2/2\sigma_y^2} \, dW \qquad (19.8)$$

Next perform the following steps: (1) combine the exponents of e, (2) complete the square for W, and (3) move the nonvariable terms outside the integral sign to obtain

$$p_z = \frac{1}{\sigma_w \sigma_y \sqrt{2\pi}} \exp\left[-\frac{(z-\Delta)^2}{2(\sigma_y^2 + \sigma_w^2)}\right] \int_{-\infty}^{\infty} \exp\left[-\left(\frac{\sigma_y^2 + \sigma_w^2}{2\sigma_y^2 \sigma_w^2}\right)(W+B)^2\right] dW$$

$$(19.9)$$

Let

$$B = \frac{(z-\Delta)\sigma_w^2}{\sigma_y^2 + \sigma_w^2}$$

For the terms within the integral, let $W* = W + B$, so $dW = dW*$ with no change of limits. This integral is of the form

$$\int_{-\infty}^{\infty} e^{-a^2 x^2} \, dx = \frac{\sqrt{\pi}}{a}$$

which when integrated yields the desired expression,

$$p_z = \frac{1}{\sqrt{2\pi}\sqrt{\sigma_y{}^2 + \sigma_w{}^2}} \exp\left\{ -\frac{[z-(y_0-w_0)]^2}{2(\sigma_y{}^2 + \sigma_w{}^2)} \right\} \qquad (19.10)$$

Of special importance are three conclusions that can be drawn from an examination of Eq. (19.10):

1. The difference between (or the sum of) two probability density functions is also a normal probability distribution curve.
2. The peak value of z, $z_0 = y_0 - w_0$.
3. The standard deviation of the sum or difference σ_z is

$$\sigma_z = \sqrt{\sigma_y{}^2 + \sigma_w{}^2} \qquad (19.11)$$

The consequences of Eq. (19.11) are that the peak probability density of z is either equal or less than the peaks of either y or w. If y is a definite value, for example, then $\sigma_y = 0$ and $\sigma_z = \sigma_y$. When both y and w are probabilistic functions, the peak probability density of z is less than either that of y or w.

A decision maker would prefer a low σ, but Eq. (19.10) shows that the confidence in a final result is degraded when the inputs to the deciding variable are probabilistic.

Equation (19.6) may be deceiving, because it could suggest that if $z = y - w$, that y could be restored by adding w to the z thus found. But an equation of the form of Eq. (19.11) will act once again, so that σ_w of the reconstructed w will be much higher than the original one. The process is not a reversible one.

The details will not be shown, but Haugen[1] showed for another arithmetic operation, a multiplication of probabilistic functions,

$$z = yw$$

that the value of z where the peak probability occurs is

$$z_0 = y_0 w_0$$

and the standard deviation σ_z degrades according to the expression

$$\sigma_z = w_0{}^2 \sigma_y{}^2 + y_0{}^2 \sigma_w{}^2 + \sigma_y{}^2 \sigma_w{}^2$$

19.14 FUTURE OF PROBABILISTIC APPROACHES TO THERMAL DESIGN

The design process projects into the future, and the future is uncertain. Designers and managers instinctively know that they are making decisions with limited information. The tool of probability analysis is already in use in some fields[1,2,3,4] to assist in making decisions when available information is uncertain or at best probabilistic. In structural systems, civil engineers know that the probability of a failure always exists, but the designer must assure that this probability is prohibitively low. The primary concern of a designer of thermal systems is to provide adequate capacity in an economic manner and a failure of this goal is usually not catastrophic. The designers of a thermal system for a building could be well served by the ability to predict the probability of the load exceeding the capability of the system. If several choices of system sizes were available having differing capacities and costs, the owner would be in a position to balance additional cost against frequency that air temperature in the space exceeded a comfortable level.

The rate at which applications of probabilistic approaches to design of thermal systems advances will be influenced by the pace at which designers recognize that they are dealing with probabilistic situations, and by the pace at which calculative tools develop.

PROBLEMS

19.1. Two sources of natural gas, shown as A and B in Fig. 19-9, each have adequate capacity to supply the city. Sources A and B are each outfitted with a compressor, 1 and 2, respectively. In addition, compressor 3 is located at the junction, and this compressor must operate along with either or both compressors 1 and 2. The probabilities of the compressors failing are: compressor 1, 0.008; compressor 2, 0.0012; and compressor 3, 0.00002. What is the probability of the city being without gas?

19.2. A chain measuring 30 lengths is composed of 10 single links and some links that are not as strong, so they will be doubled in each length, as in Fig. 19-10. During an unusually heavy load the probability of any of the single links failing is 0.0002, and the probability of a weaker link failing if it were used as a single element is 0.012. The double link doesn't divide the load, but one link either sustains the load or breaks and transfers the load to its partner. What is the probability of the chain breaking during the heavy load?

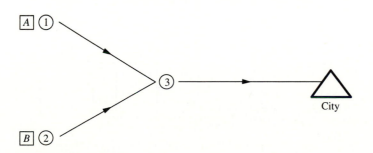

FIGURE 19-9
Gas-supply system in Prob. 19.1.

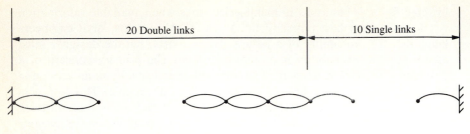

FIGURE 19-10
Chain in Prob. 19.2.

19.3. A condensate line, Fig. 19-11*a*, serves 10 steam coils which at design conditions have a steam flow rate of 0.03 kg/s each. The condensate traps on each of these coils can dump at the rate of 0.15 kg/s, so that the flow rate from one trap has the form shown in Fig. 19-11*b*. What are the probabilities of the simultaneous dumping of 0 trap, 1 trap, 2 traps, . . . 10 traps? (Further reference: "Sizing a Condensate Return System by Monte Carlo Simulation," by W. C. Huang and Y. K. Lee, *Heating, Piping and Air Conditioning*, vol. 45, no. 7, pp. 44–46, July 1973.

19.4. In laboratory buildings equipped with fume hoods, the air conditioning system must be designed to treat the makeup air entering the building when the fume hoods operate. In a certain laboratory building there are 30 hoods, and experience shows that any hood is in operation approximately 1/3 of the time and in a random pattern. What is the probability that 15 or more hoods will be operating at a given time?

  Ans.: 0.0435

19.5. A refrigerated warehouse has 12 large doors through which product is transported. It is important when designing the refrigeration plant to know how many doors are open at a time in order to provide adequate refrigeration capacity. The operating pattern at this warehouse is that any given door is opened on the average every 10 minutes. When opened, there is a 0.4 probability of it staying open 15 s and a 0.6 probability of it being opened for 20 s. What is the probability of 2 or more doors being open at a time?

FIGURE 19-11
(*a*) Steam traps (*b*) flow profile for condensate line in Prob. 19.3.

TABLE 19.4
Distribution of monthly electric bills in Problem 19.6

Monthly bill, dollars	Number of customers	Monthly bill, dollars	Number of customers
Less than 16	1	80-84	7
32-36	0	84-88	6
36-40	1	88-92	5
40-44	1	92-96	5
44-48	1	96-100	4
48-52	2	100-104	4
52-56	3	104-108	3
56-60	3	108-112	2
60-64	5	112-116	1
64-68	5	116-120	1
68-72	5	120-124	1
72-76	6	124-128	1
76-80	7	Over 128	0

19.6. The number of customers in each of the ranges of monthly electric bills is shown in Table 19.4. Plot the cumulative distribution on arithmetic normal paper and (a) determine whether the distribution is normal, and (b) if normal, determine x_0 and the standard deviation.

19.7. At a certain time of the day there is a 30% probability that the electric kitchen range in any of the apartments in an 80-unit building will be turned on. (a) Derive the continuous distribution equation of the probability density as a function of the total number of ranges turned on at one time. (b) What is the probability of more than 35 ranges operating at any one time?
      Ans.: (b) 0.00367.

19.8. A manufacturer in one month anticipates the most probable number of units that it can manufacture is 2200 with a standard deviation of 300. The most probable selling price is $380 with a standard deviation of $40 and the most probable manufacturing plus sales cost is $275 with a standard deviation of $25. What is the most probable profit for the month on this product and the standard deviation when expressing the function as (a) (number of units)(price − cost), and (b) (number of units)(price) − (number of units)(cost). (c) Propose different operations that represent the two different expressions of parts (a) and (b).

REFERENCES

1. E. B. Haugen, *Probabilistic Approaches to Design*, John Wiley, New York, 1968.
2. D. F. Rudd and C. C. Watson, *Strategy of Process Engineering*, John Wiley, New York, 1968.
3. A. H-S. Ang and W. H. Tang, *Probability Concepts in Engineering Planning and Design*, John Wiley, New York, 1975.
4. W. H. King, *Probability for Management Decisions*, John Wiley, New York, 1968.

APPENDIX
I

COMPREHENSIVE
PROBLEMS

This section of the appendix presents some sample projects which apply principles studied in the text, e.g., economics, equation fitting, simulation, optimization, or a combination of them. These problems may be used as projects accompanying the study of the text material and running as a part-time effort all term. Many instructors devise their own similar problems based on their own engineering experiences. Some of the problems may carry over from one term to the next, with one team of students picking up the work where the preceding group left off.

Engineering students become proficient in solving short problems such as homework problems which require 45 min, but most professional engineering problems are long-term, requiring weeks or months for completion. It is therefore appropriate for senior-level or graduate-student engineers to gain some experience with comprehensive problems which require discipline to maintain progress over a longer period of time. Also, at the beginning of any long-term project there is the period of deliberation on how to start the problem—how to find the handle. Inexperienced engineers spend considerable time spinning their wheels and making false starts before focusing on a valid solution. Experience with comprehensive projects is the best means of developing proficiency in thought and work habits. Written or oral reports make good targets for completion and have their own benefit as well.

The projects that follow consist of a statement of the problem, which contains some or all of the required data. In certain cases property data will have to be extracted from handbooks. With some of the problem statements will be included some brief comments on the experiences of groups who have worked on the problem.

A-1. Optimum temperature distribution in a multistage flash-evaporation desalination plant

A-2. Heat recovery from exhaust air with an ethylene glycol runaround system

A-3. Design of a fire-water grid

A-4. Optimum thickness of insulation in a refrigerated warehouse

A-5. Simulation of a liquefied-natural-gas facility

A-6. Optimization of a natural-convection air-cooled condenser

A-7. Enhancing the heat-transfer coefficient of boiling in tubes

A-8. Recovery of heat from exhaust air using a heat pump

A-9. Simulation of a dehumidifier for industrial drying

A-10. Optimum gas pipeline when recovering power at terminus of the pipeline

A-11. Conserving energy by using a refrigerant mixture

A-12. Optimum air ejector for a steam-power-plant condenser

A-13. Optimizing a hot-oil loop in a petrochemical plant

A-14. Optimum heat pump for pasteurizing milk

A.1 OPTIMUM TEMPERATURE DISTRIBUTION IN A MULTISTAGE FLASH-EVAPORATION DESALINATION PLANT

Flash Desalination

One of the methods for water desalination is a distillation process using multistage flash evaporators, as shown schematically in Fig. A-1. Seawater flows first through heat exchangers, on which vapor condenses to form fresh water. After this preliminary heating of the seawater in the condensers, a steam heater elevates the temperature of the seawater to the maximum permitted by corrosion limitations. In passing through the first throttling valve some of the hot brine vaporizes, and this water vapor condenses on the tubes of the condenser and drops to the freshwater collection pan. From the first stage the brine and the fresh water flow through pressure-reducing valves into the next stage.

Figure A-1 shows a two-stage plant, but commercial plants have many more stages. A Foster-Wheeler plant in San Diego, for example, has nine stages.[1]

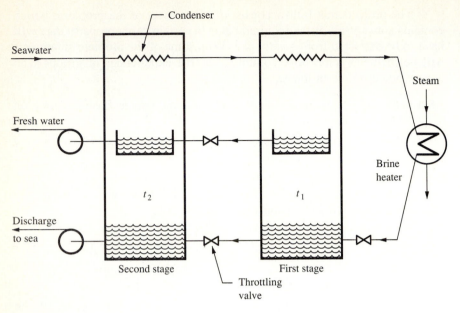

FIGURE A-1
Multiflash desalination facility.

The temperature in a stage is defined as the condensing temperature of the vapor, which in turn is dictated by the amount of heat-transfer area in each stage of the condenser. The assignment is to specify the optimum number of stages and the heat-transfer area in each of the condensers for a minimum total cost over the 20-year life of the desalination plant, for a plant with a freshwater capacity of 50 kg/s.

Design data

Life, 20 years.

Interest on investment, 8 percent.

Heat-exchanger cost, $95 per square meter for both condensers and the brine heater.

U value for both condensers and brine heater, 2800 W/(m² · K).

Seawater temperature, 15°C.

Steam temperature in brine heater, 150°C.

Maximum permitted temperature of brine leaving the brine heater, 120°C.

Steam for the brine heater costs 90 cents per gigajoule; each additional flashing stage costs $12,000.

Assume that the thermal properties of the brine are the same as those of pure water.

Discussion

One approach is to start with a single-stage plant, which provides experience in setting up the thermodynamic and heat-transfer equations. An interesting observation is that the minimum stage temperature is the average of the seawater temperature and the maximum brine temperature, thus (15 + 120)/2 = 67.5°C. This minimum temperature occurs with a condenser of infinite area. The optimum condenser area results in a temperature that is a fraction of a degree higher than 67.5°C.

For a two-stage plant, infinite area in the two condensers results in stage temperatures that again divide the 15 to 120°C range equally, but this time into thirds. Thus the minimum temperature of the first stage is 85°C and in the second stage 50°C.

This problem can be extended into a two-level optimization. The first level is to determine the optimum temperatures in, say, a two-stage plant. The next level is to optimize the number of stages and for each different number of stages perform a new temperature optimization.

A.2 HEAT RECOVERY FROM EXHAUST AIR WITH AN ETHYLENE GLYCOL RUNAROUND SYSTEM

Introduction

Most large buildings have ventilating requirements which bring in outdoor air and exhaust an equal amount. When the outdoor temperature is low, the cost of heating the outdoor air before introducing it into the building is appreciable. This heating cost can be reduced by recovering some heat from the exhaust air. One method of recovering heat is to place a finned-coil heat exchanger in the exhaust airstream, another in the outdoor airstream and to pump a fluid between the coils, as shown in Fig. A-2. Water would be the first choice for a heat-transfer fluid, but to guard against freezeup at low outdoor temperature, an antifreeze such as ethylene glycol must be added.

If the system serves any purpose at all, it must save more money in heating costs than it requires for its own amortization and operation. The first costs that are to be amortized include the costs of the coils, pump, piping, wiring, additional ductwork or revisions thereof, and additional cost of larger fans or motors, if needed. The operating costs include the power for the pump and additional fan power.

The potential savings should be evaluated for an optimum heat-recovery system. This optimum system consists of the most favorable combination of the following variables:

1. Length of the tubes in the coil
2. Height (or number of tubes high) of the coil
3. Number of rows of tubes deep (parallel to the path of the airflow)

FIGURE A-2
Heat-recovery system.

4. Fin spacing

5. Glycol flow rate

The qualitative effects of these five variables are as follows. Increasing the length and number of tubes high increases the heat-transfer area but also the cost. Furthermore, increasing the length and number of tubes high increases the cross-sectional area for airflow, which reduces the velocity and decreases the air-side heat-transfer coefficient but also decreases the power required of the fan to force the air through the coil. Increasing the number of rows of tubes deep increases the heat-transfer area but increases the first cost of the coil as well as both the air and the glycol pumping cost. Spacing the fins closer together increases the cost of the coil and the air-pressure drop but also increases the heat-transfer area. Finally, a high flow rate of the ethylene glycol increases the glycol-side heat-transfer coefficient but also increases the pumping cost.

Further Data and Assumptions

The size of the optimum system will pertain to a given airflow rate, and 3.0 m^3/s has been chosen for both the outdoor-air and exhaust-air flow rate. The coil circuiting chosen is that of vertical headers feeding horizontal tube circuits in parallel, as shown in Fig. A-2. The flow of the glycol through the U bends is counter to the flow of air. Assume pure counterflow in the coils; thus the equations that represent this runaround loop are those used in Prob. 5.14.

Further specifications

The copper tubes have an OD of 16 mm and a wall thickness of 1 mm.

The tube spacing is 41 mm vertically and horizontally.

The average outdoor temperature is 5°C for 250 days of 24-hr operation.

The life is 10 years, and the interest rate is 8 percent.

The electricity cost is 4 cents per kilowatthour.

The fan and pump motor efficiencies are 75 percent.

The Optimization

The net saving to be maximized is the difference between the reduction in heating cost and the annual cost (amortized first cost plus operating cost) of the recovery system. The building is heated electrically, so as far as the ventilation air is concerned, electric-resistance heat must warm the incoming outdoor air to the building temperature of 24°C. The first cost to be amortized includes that of the coils, the pump and motor (assumed constant at $400), and the interconnecting piping (assumed constant at $150). No allowance for capital cost for the fan is provided because fans already exist in the system, and it is assumed that they would not have to be enlarged to overcome the additional pressure drop of the heat-recovery coils.

Cost and Performance Equations

The equation for the first cost of each coil roughly reflects the proportionality of the cost to the mass of the coil,

$$\text{Cost} = 0.26[18 + (0.024)(\text{NF} + 500)W\,] \, [L(\text{NR} + 1)] \qquad \text{dollars}$$

where NF = fin spacing along the tube, fins/m
W = number of rows of tubes deep (in direction of airflow)
L = tube length, m
NR = number of layers of tubes high

The U value of the coil based on outside (air-side) area is

$$\frac{1}{U_o} = \frac{0.0226}{V_a^{0.58}} + 0.0032 + \frac{0.0088\text{NF} + 0.185}{430V_{eg}^{0.8}}$$

where U_o = U value, W/(m$^2 \cdot$ K)
V_a = face velocity of air through coil, m/s
V_{eg} = velocity of ethylene glycol in tubes, m/s

The thermal resistance of the tubes and fins is 0.0032 (m$^2 \cdot$ K)/W.

Air pressure drop (DPA) in pascals is

$$\text{DPA} = (4.1W\,)(0.25 + 0.0016\text{NF})(V_a^{1.75})$$

and the pressure drop of ethylene glycol (DPEG) in kilopascals is

$$DPEG = 5.2[0.15WL + 0.0875(W - 1) + 0.3](V_{eg}^{1.75})$$

In addition to the pressure drop of the ethylene glycol through the coil, some allowance should be made for the interconnecting piping, 15 kPa, for example.

Variables

The five variables for which optimal values are sought and the ranges of their values are

NF, fin spacing	300, 400, 500, 600 fins/m
L, tube length	0.4, 0.6, . . . , 3.4, 3.6 m
W, number of rows of tubes deep	2, 4, 6, 8, 10
NR, number of rows of tubes high	4, 5, 6, . . . to $L/0.06$
V_{eg}, velocity of ethylene glycol	0.4, 0.8, 1.2, 1.6, 2.0 m/s

Assignment

Determine the combination of the above variables that results in the optimum economic solution.

A.3   DESIGN OF A FIRE-WATER GRID

Fire Control

Refineries and other chemical plants that process flammable substances must provide elaborate measures to prevent and fight fires.[2] Almost all such plants must be equipped with a fire-water distribution system that is generally constructed underground to supply hydrants throughout the plant. Typically, the plant is subdivided into various areas, remote from each other so that if a fire breaks out in one area it can be contained in that area. To serve the entire plant, the fire-water grid should be capable of providing a specified rate of water flow to any one area at a time.

Two challenges face the designer. For a given pressure at the outlet of the pump

1. Select the pipe sizes so that the flow through the grid will supply the hydrants that surround a plant area with the required rate of water flow
2. Select the minimum first-cost combination of pipe sizes that meets the above requirement

Most designers of industrial fire-water grids are satisfied to achieve task 1, and even this assignment is a challenging one. Some designers use

FIGURE A-3
Fire-water grid.

specially constructed electrical analogs in combination with a cut-and-try
method of enlarging pipe sizes until each area individually can be blanketed
with specified water flow. The analog must be a special one, because the
fluid-flow conductor does not follow Ohm's law.

Objective

The goal of this project is to select pipe sizes for a fire-water grid to achieve
minimum first cost when the following conditions are specified:

1. Geometric layout of areas
2. Location of hydrants
3. Water pressure at the pump outlet
4. Minimum flow rate required by each operating area

Although in actual plants there may be a dozen operating areas with hydrants
distributed along the pipes, a suggested grid for this type of problem is
shown in Fig. A-3.

Operating and Cost Data

The supply pressure of the pump is 800 kPa. The grid must supply a minimum of 0.7 m³/s to area A and 0.8 m³/s to area B, but not simultaneously.
Area A is served by hydrants 1, 2, and 3, while area B is served by hydrants
2, 3, 4, and 5. The flow characteristics of the hydrants are

$$\text{Flow rate, m}^3/\text{s} = 0.01 \sqrt{\text{pressure, kPa}}$$

The installed cost of the pipe is 30 cents per meter of length for each
millimeter of diameter. The trenching cost is $80 per meter, but since it is
independent of the pipe size, it is a constant and does not affect the optimum
pipe-size selection. Neglect operating cost, because one hopes the number
of hours of pump operation will be negligible.

Discussion

Finding the optimum grid design reduces to two separate optimizations, optimizing one area at a time. The adequate system that serves area A alone will probably be inadequate for area B alone, so in this case the optimum design that serves area B should be checked to see whether it is adequate for area A.

The problem reduces to one involving equality constraints and thus is soluble by the method of Lagrange multipliers, although estimating the trial values of the λ's in the set of nonlinear equations is difficult.

The optimization is likely to shrink the size of some legs of the grid to zero, which may be undesirable from the security standpoint. For this reason another set of constraints is warranted which specifies the minimum diameter of all pipe sections to be 150 mm, for example.

Another refinement that adds more constraints is that there should be a minimum pressure at the hydrant, perhaps 400 kPa. The reason for this minimum is that adequate pressure must be available to send a stream of water at least into the center of the area, and perhaps farther, in order to provide some overlap of water coverage.

A.4   OPTIMUM THICKNESS OF INSULATION IN REFRIGERATED WAREHOUSES

Optimum Insulation Thickness

Deciding upon the optimum insulation thickness is a classic engineering problem. When heat transfer into or out of a controlled-temperature space results in cooling or heating cost, application of insulation will reduce these costs. Each additional unit thickness of insulation, however, is progressively less effective, and the point is reached where additional thickness of insulation is not compensated for by the reduction in operating cost.

A large food company suggested that some of the standard thicknesses of insulation used in refrigerated warehouses should be restudied. During recent years the cost of energy has increased at a more rapid rate than the cost of insulation and the first cost of the refrigeration plant. Thicker insulation than former standards may now be justified.

Cost and Engineering Data

The optimization will be performed for the insulation thickness in the roof, for which there is considerable latitude. (The walls are now usually constructed of prefabricated panels, where there is limited choice of insulation thickness.) The warehouse temperature is $-18°C$, and the average outdoor temperature during the year will be between 5 and 10°C, depending upon the climate.

The insulation board is available in 50-, 75-, 100-, and 125-mm thicknesses. To develop thicknesses greater than 125 mm, one or two additional layers of insulating board are required.

Additional data

Cost of insulation, 5 cents per square meter of area for each millimeter of thickness

Installation cost, first layer, $2.50 per square meter; second and third layers $1.50 per square meter

First cost of refrigeration equipment, $600 per kilowatt of refrigeration capacity

Power required by refrigeration plant, 0.6 kW of electric power per kilowatt of refrigeration

Cost of electric energy (depends upon local rates), usually between 3 and 6 cents per kilowatthour

Conductivity of insulation k, 0.04 W/(m · K)

The expected life of the plant is 15 years; the rate of interest is 9 percent; and for tax purposes the facility may be written off in 10 years (suggest straightline depreciation). Federal income tax is 50 percent.

Determine the optimum thickness.

Discussion

To determine the influence of the power cost and the average outdoor temperature on the optimum thickness, it is suggested that the optimization be performed for several values in the complete range of outdoor temperatures and power costs. The cost associated with the application of an additional layer of insulation is likely to inhibit the optimal thickness from exceeding 250 mm, or two full-size layers of insulation.

A.5 SIMULATION OF A LIQUEFIED NATURAL GAS FACILITY

Need for Liquefying Natural Gas

The principal reasons for liquefying natural gas are for shipping and for storage. When natural gas is shipped across an ocean, it is economical to convert it into liquid form since 600 times more gas can be contained in a given volume than in the gaseous form at standard atmospheric pressure. To hold liquefied natural gas (LNG) at atmospheric pressure, the liquid temperature must be approximately −155°C.

The need for storage of natural gas often arises in the following manner. Gas utility companies are distribution companies, and they purchase natural gas from transmission companies. The transmission companies assess demand charges, so it is often to the advantage of the distribution company to buy gas from the transmission line during periods of low demand and store the gas as LNG for use during the peak winter periods.

The LNG Plant

Figure A-4 shows a schematic diagram of a simplified LNG plant. In the cycle of Fig. A-4, gas from the transmission line enters the liquefaction plant at 1000 kPa and 290 K. The stream at point 1 joins with recycled gas from compressor 2 and is pumped up to the high pressure by compressor 1. The gas is then desuperheated in the propane evaporator and heat exchanger 1. Liquefaction takes place in the ethylene evaporator but at a high pressure. The saturated liquid at point 6 is subcooled in heat exchanger 2 by transferring heat to low-pressure fluid boiling in the other stream of the heat exchanger. The subcooled liquid at point 8 passes through a throttling valve and expands to atmospheric pressure. Some of this liquid flashes into vapor, and this vapor is recompressed by the boil-off compressor back to the transmission-line pressure.

Some pressures and temperatures are noted in Fig. A-4, and additional property equations for the gas (which is assumed to be methane) and performance characteristics of the components are provided below.

Properties

Enthalpy of saturated liquid

$$h_f, \text{ kJ/kg} = -219.6 + 7.025T - 0.034965T^2 + 0.00011321T^3$$

Enthalpy of saturated vapor

$$h_g, \text{ kJ/kg} = 790.13 - 4.0137T + 0.058268T^2 - 0.00019465T^3$$

Specific heat of superheated vapor

$$c_p, \text{ kJ/(kg · K)} = 2.114 + 0.000188p - (1.76 \times 10^{-8})p^2$$

Saturation pressure

$$\ln p = -\frac{1026}{T} + 13.8$$

where p = pressure, kPa
$\quad\quad T$ = temperature, K

Component performance

$$UA = \begin{cases} 2.64 \text{ kW/K} & \text{propane evaporator} \\ 4.22 \text{ kW/K} & \text{heat exchanger 1} \\ 5.275 \text{ kW/K} & \text{heat exchanger 2} \\ 42.2 \text{ kW/K} & \text{ethylene evaporator} \end{cases}$$

FIGURE A-4
LNG plant.

In the ethylene evaporator assume that the temperature of methane is at the saturation temperature throughout when determining the mean-temperature difference.

Compressor 1:
$$\frac{p_{disch}}{p_{inlet}} = 7.5 - 150.5 \frac{(w, \text{kg/s}) \sqrt{T_{inlet}, \text{K}}}{p_{inlet}, \text{kPa}}$$

Compressor 2:
$$\frac{p_{disch}}{p_{inlet}} = 5.6 - 142.8 \frac{(w, \text{kg/s}) \sqrt{T_{inlet}, \text{K}}}{p_{inlet}, \text{kPa}}$$

Discussion

Using the system-simulation procedures of Chap. 6 with the aid of a computer program, a team found several key results.

Liquefaction rate, 0.64 kg/s
High pressure, 3610 kPa
Pressure entering compressor, 2400 kPa

A.6   OPTIMIZATION OF A NATURAL-CONVECTION AIR-COOLED CONDENSER

Air-cooled condensers that use propeller fans to blow air over finned heat exchangers are commonly used in process industries, such as refineries. The electric motors, fans, bearings, and gears or belts, if they are used, require maintenance and sometimes fail if not properly maintained. Furthermore, the fan motors require energy, and natural-convection air-cooled condensers are now often used.[3] Figure A-5 shows a sketch of such a condenser, and Fig. A-6 is a sketch of the finned coil.

Objective

Design for minimum first cost an air-cooled condenser that rejects 140 kW of heat. Specifically select the height h, the dimensions of the square coil and stack b, and the number of rows of tubes high in the heat-exchanger section.

Performance data

Temperature of ambient air, 35°C.
Temperature of condensing fluid, 90°C.
Choices of number of rows of tubes high, 1 to 6.
Choices of b dimension, 1.4 to 3.0 m in 0.2-m increments.

FIGURE A-5
Natural-convection air-cooled condenser.

FIGURE A-6
Condenser coil.

Heat-transfer area on condensing-fluid side of heat exchanger per square meter of face area per row of tubes high, 1.3 m².

Effective heat-transfer area on the air side of the heat exchanger per square meter of face area per row of tubes high, 12.5 m².†

Neglect the resistance to heat transfer through the tube.

Heat-transfer coefficient of condensing fluid, 2.0 kW/(m² · K).

Heat-transfer coefficient on air side, $0.042V^{0.4}$ kW/(m² · K), where V is the face velocity of air in meters per second.

Pressure drop through the coil, Δp, Pa $= 2.2V^{1.8}(1.5 + 0.7n)$, where n is the number of rows of tubes.

Pressure drop due to friction of the air flowing through the stack must be considered; suggest a friction factor of 0.02.

Costs

Stack, including cost of supports, $45 per square meter of area
Coil, $(190)(b^{3.2})(n^{0.6})$ dollars

Discussion

An increase in the height h improves the stack effect and thus increases the airflow rate but adds to the cost. An increase in the coil and stack dimension b reduces the pressure drop through the coil and stack but increases the cost. An increase in the number of rows of tubes increases the heat-transfer area and outlet temperature of air t_o but increases the pressure drop. An increase in air velocity over the coil increases the heat-transfer coefficient on the air side but reduces t_o.

Not all combinations of b and n may be workable.

The stack effect due to the difference in densities of the ambient air and the air in the stack is responsible for airflow and compensates for the air-pressure drop through the coil and in the stack itself.

A.7 ENHANCING THE HEAT-TRANSFER COEFFICIENT OF BOILING IN TUBES

The heat-transfer coefficient of a boiling fluid inside a tube varies throughout the length of the tube as a function of the quality (fraction of vapor). A typical profile is shown in Fig. A-7. In order to improve the overall performance of the heat exchanger a modified evaporator, as shown in Fig.

---

† The actual surface area is greater than indicated, but the effective area accounts for the fin efficiency.

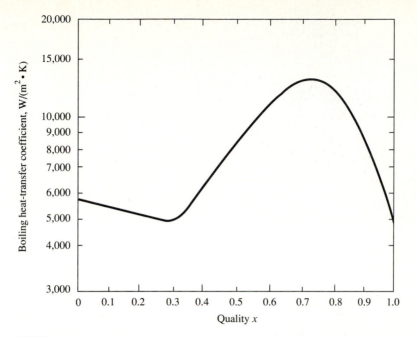

FIGURE A-7
Boiling heat-transfer coefficient inside a tube. (From J.M. Chawla, "A Refrigeration System with Auxilliary Liquid and Vapour Circuits," International Institute of Refrigeration, Meeting Comm. II and III, London, 1970; used by permission.)

A-8, extracts vapor at two positions and injects it elsewhere along the tube in order to maintain the quality near 0.7 or 0.8 and thus take advantage of the high heat-transfer coefficient that results.

Objective

Determine the positions of vapor extraction and the flow rates of vapor that result in the optimum mean heat-transfer coefficient if the local coefficients are as shown in Fig. A-8. Specifically, determine w_1/w, w_2/w, y/L, and z/L for the maximum mean coefficient. Assume that the vapor and liquid move at the same velocity.

Discussion

The analysis is complicated by the fact that the heat flux changes along the tube because the coefficient is not constant. Furthermore, the heat flux is influenced by the local temperature of the fluid being cooled. To simplify, assume that the rate of evaporation is uniform along the length; thus the quality varies linearly with distance from 0 to 1.0.

FIGURE A-8
Evaporator with enhanced heat-transfer coefficients.

A.8 RECOVERY OF HEAT FROM EXHAUST AIR USING A HEAT PUMP

In the interest of energy conservation, heat exchanges (rotary-wheel and run-around systems, as in Prob. A.2) are sometimes installed in buildings to recover some of the heat of exhaust air in winter and precool outdoor ventilation air in the summer. These heat exchangers function because the exhaust air is at a higher temperature than the incoming outdoor ventilation air in winter and at a lower temperature than the incoming air in summer.

To intensify the heat-transfer process, the installation of a heat pump is proposed as shown in Fig. A-9. The building is electrically heated, so

FIGURE A-9
Heat pump for recovering heat from exhaust air and preheating incoming ventilation air.

that if the heat pump does not bring the temperature of the incoming air up to 35°C (the supply air temperature), electric-resistance heat is used to supplement the heat pump.

Objective

Determine the optimum heat pump, namely, the size of compressor, condenser, and evaporator that provides a minimum total present worth of costs (first cost plus present worth of power costs).

Data

U values of condenser and evaporator coils, 25 W/(m² · K) based on air-side area

Cost of coils, $50 per square meter of air-side area

Compressor cost, including motor, $120 per motor kilowatt

Power cost, 3 cents per kilowatthour

Interest rate, 10 percent

Economic life, 10 years

Number of hours of operation per year, 4000

Average outdoor temperature during the 4000 h, 0°C

Air flow rate through both coils, 5 kg/s

The performance characteristics of the compressor can be expressed as a coefficient of performance (COP), where

$$COP = \frac{\text{refrigeration rate, kW}}{\text{electric power to compressor motor, kW}}$$

The COP is a function of the evaporating and condensing temperatures, $t_e°C$, and $t_c°C$, and can be represented by

$$COP = 7.24 + 0.352t_e - 0.096t_c - 0.0055t_et_c$$

Assignment

Determine the optimal combination of evaporator area, condenser area, and compressor motor input power that results in the minimum total present worth of costs. Also determine t_e, t_c, t_i, and t_o at this optimum.

Discussion

If the heat pump can be justified for heating operation alone, it has the possibility of providing additional savings by operating during the summer.

The temperature of the supply air entering the building is specified as 35°C, but during some periods of the heating season it may be desirable for this temperature to be above or below 35°C. This condition would revise the assignment somewhat.

The optimization is conducted on the basis of a constant outdoor air temperature of 0°C. During actual operation the COP of the system will be higher than average when the outdoor temperature is above 0°C and less heat recovery is needed. Conversely, when the outdoor temperature is below 0°C, the COP drops and more heat-recovery capacity can be used.

A.9 SIMULATION OF A DEHUMIDIFIER USED FOR INDUSTRIAL DRYING

A traditional method of drying many materials (food products, grain, paper, etc.) has been to heat ambient air and pass it over the material to be dried. Some industries have been forced to shift from heat obtained from natural gas and oil to the use of electricity. Instead of using the electricity in a resistance heater, some applications have employed a heat pump, as shown in Fig. A-10. The evaporator of the heat pump cools and dehumidifies the air and then reheats the air with the refrigerant condenser. Next the air passes over the material to be dried. Energy is added to the air by the heat pump and also by virtue of the water entering the airstream as vapor and leaving as liquid. A water-cooled heat exchanger removes this energy.

FIGURE A-10
Heat-pump dehumidifier for industrial drying.

Data

> Airflow rate, 10 kg/s
>
> Rate of addition of water vapor, 0.02 kg/s
>
> Temperature of water entering water-cooled heat exchanger, 20°C
>
> Flow rate of water entering heat exchanger, 4.0 kg/s
>
> UA of water-cooled heat exchanger, 10.5 kW/K
>
> UA of condenser, 15.3 kW/K

Compressor performance. The pumping capacity of the compressor expressed in terms of heat-transfer rate at the evaporator is

$$q_e = 126.9 + 7.683t_e - 1.134t_c - 0.05516t_e t_c \text{ kW}$$

and the power requirement of the compressor is

$$P = 15.05 - 1.266t_e + 0.283t_c + 0.0299t_e t_c \text{ kW}$$

where t_e = evaporating temperature, °C
$\quad\quad\ t_c$ = condensing temperature, °C

Evaporator. The evaporator is a heat exchanger where both sensible heat and mass are transferred from the air to the surface of the heat exchanger, which is wet because of the dehumidification. The heat then flows through the metal and thence to the refrigerant. Data on the evaporator are

> Refrigerant-side area A_r, 15 m²
>
> Refrigerant-side heat-transfer coefficient h_r, 2.2 kW/(m² · K)
>
> Neglect the heat-transfer resistance of the metal
>
> Air-side area A_a, 120 m²
>
> Air-side convection heat-transfer coefficient, h_a = 0.09 kW/(m² · K)

Heat and mass transfer in the evaporator. The processes occurring in the evaporator coil can be visualized as shown in Fig. A-11. The driving force for the transfer of sensible heat is the difference in dry-bulb temperatures of the air and the wetted surface. If the arithmetic-mean difference is assumed to be sufficiently accurate,

$$q_s = h_a A_a \left(\frac{t_3 + t_4}{2} - \frac{t_{ws3} + t_{ws4}}{2} \right)$$

FIGURE A-11
Heat and mass transfer at the evaporator coil.

where q_s is the rate of sensible-heat transfer in kilowatts. The rate of latent heat transfer[4] is

$$q_L = \frac{h_a}{c_p} A_a \left(\frac{W_3 + W_4}{2} - \frac{W_{ws3} + W_{ws4}}{2} \right) h_{fg} \text{ kW}$$

where W_3, W_4 = humidity ratio of air, (kg water)/(kg dry air)
   W_{ws3}, W_{ws4} = humidity ratio of saturated air at wetted-surface temperature, (kg water)/(kg of dry air)
   c_p = specific heat of air, kJ/(kg · K)
   h_{fg} = latent heat of water (suggest using a constant value of 2450 kJ/kg)

On the refrigerant side this combination of q_L and q_s equals the rate of heat absorbed by the refrigerant q_e, and

$$q_e = h_r A_r \left(\frac{t_{ws3} + t_{ws4}}{2} - t_e \right)$$

The temperature of the wetted surface, t_{ws3} for example, must adjust itself so that the combined rate flow of sensible plus latent heat equals the rate flow of heat to the refrigerant at that position.

Assignment

Simulate this system in the sense of computing all pertinent operating variables, including the conditions of the air throughout the cycle; heat-transfer rates at the evaporator, condenser, and water-cooled heat exchanger; and the power required by the compressor.

Discussion

The specification of the rate of moisture removal (0.02 kg/s) permits a simplification in this simulation in that the humidity ratios of the air throughout the cycle can be left free to float until the remainder of the variables are simulated. An extension of this problem is to express the rate of moisture removal as a function of the mass transfer of water from the product to the air. The humidity ratios of the air must then adjust to provide a water balance.

A.10   OPTIMUM GAS PIPELINE AND PUMPING FACILITY WHERE WORK CAN BE RECOVERED AT DESTINATION

In a natural-gas (methane) pipeline 120 km long (Fig. A-12) there is a need for power at the receiving point, so the system is being designed with a gas turbine at the receiving end to expand the gas from the pipeline back down

FIGURE A-12
Natural-gas pipeline where work is recovered at terminal position.

to 100 kPa.[5] Compression to a high pressure permits a high-pressure drop through the pipeline and also results in dense gas; both reduce the pipe size. On the other hand, there is additional energy required for compression, and not all this energy is recovered at the turbine due to inefficiencies of the electric motor, compressor, and turbine, as well as pipeline losses.

Data

> Methane enters the compressor at 100 kPa.
>
> Methane leaves the turbine at 100 kPa.
>
> All the power generated by the turbine-generator can be used.
>
> Gas enters the compressor at 30°C, is cooled after the compression, and remains at 30°C throughout the pipeline.
>
> Methane flow rate is 50 kg/s.
>
> The efficiencies of the electric motor and generator are 95 percent.
>
> The efficiency (with respect to the isentropic process) of the compressor is 80 percent and of the turbine is 75 percent.

Costs

> Electric motor and generator first cost, $50 per kilowatt output
>
> Compressor first cost, $125 per kilowatt input (input designated W_c)
>
> Value of electricity at compressor end of pipeline, 3 cents per kilowatthour
>
> Value of electricity at turbine end of pipeline, 4 cents per kilowatthour
>
> Turbine first cost, $150 per kilowatthour output (output designated W_t)
>
> Pipe cost in dollars per meter length, $300D^{1.6}$, where D is the pipe diameter in meters

Life of facility, 15 years

Interest rate, 10 percent

Operation, continuous

Pipeline pressure drop. Because the pressure changes as the gas flows through the pipeline, the density and velocity also change. For an element of length, dL, the pressure drop dp is expressed by

$$dp = \frac{f \, dL}{D} \frac{V^2}{2} \rho$$

where p = pressure, Pa

  f = friction factor, dimensionless (suggest $f = 0.02$)

  L = length, m

  D = diameter, m

  V = velocity, m/s

  ρ = density, kg/m^3

Assignment

Determine the optimum combination of sizes of compressor, turbine, and pipe to achieve the minimum sum of the first costs and present worth of operating costs.

A.11   CONSERVING ENERGY BY USING A REFRIGERANT MIXTURE

When a single refrigerant condenses and evaporates at a constant pressure between saturated liquid and saturated vapor, the temperature remains constant. A binary solution, on the other hand, experiences a change in temperature, as pointed out in Chap. 5. When a refrigeration system cools a fluid stream through a large temperature range or the system rejects heat to a fluid that increases in temperature as it passes through the condenser, it may be possible to conserve compression power by using a binary solution.[6] Some applications of this concept have been made in the cryogenic and petrochemical industries.[7]

The refrigeration cycle adaptable to refrigerant mixtures employs the usual components (compressor, condenser, evaporator, and expansion device) and a heat exchanger as well, as shown in Fig. A-13. A mixture of refrigerants R-12 and R-114 will be explored.

Data

$$UA = \begin{cases} 5 \text{ kW/K} & \text{evaporator} \\ 5 \text{ kW/K} & \text{condenser} \\ 0.3 \text{ kW/K} & \text{heat exchanger} \end{cases}$$

FIGURE A-13
Refrigeration system using a mixture of refrigerants.

The condenser is water-cooled; water enters at 25°C with a flow rate of 0.8 kg/s.

The evaporator fluid is cooled from −15 to −25°C. Its flow rate is 0.57 kg/s, and its specific heat is 3.5 kJ/(kg · K).

The refrigerant is saturated liquid at point 3 and saturated vapor at point 1.

The compressor has adjustable capacity which is regulated to provide the specified refrigeration rate in the evaporator.

For saturated pressure

R-12: $\ln p = 14.861 - 2498.3/T$

R-114: $\ln p = 15.407 - 2993.2/T$

where p = pressure, kPa
T = temperature, K

For enthalpy of saturated liquid

$$h_f = \begin{cases} 200 + 0.925t + 0.00081t^2 & \text{R-12} \\ 200 + 0.9545t + 0.00116t^2 & \text{R-114} \end{cases}$$

where h_f = enthalpy of liquid, kJ/kg
t = temperature, °C

For enthalpy of saturated vapor

$$h_g = \begin{cases} 351.5 + 0.4283t - 0.00071t^2 & \text{R-12} \\ 337.4 + 0.6234t - 0.000086t^2 & \text{R-114} \end{cases}$$

where h_g = enthalpy of saturated vapor, kJ/kg
t = temperature, °C

Work of compression

$$\Delta h = \begin{cases} 188 \left(1 - \dfrac{14.861 - \ln p_2}{14.861 - \ln p_1}\right) & \text{R-12} \\ 158 \left(1 - \dfrac{15.407 - \ln p_2}{15.407 - \ln p_1}\right) & \text{R-114} \end{cases}$$

where Δh = work of compression, kJ/kg
p_2 = discharge pressure (total pressure), kPa
p_1 = suction pressure (total pressure), kPa

When compressing a mixture, the work of compression is found by proportioning according to the mass fraction of the constituents in the mixture. The mixture is assumed to be ideal, and Dalton's and Raoult's laws apply. The molecular weights of R-12 and R-114 are 120.93 and 170.94, respectively.

Assignment

For this specified refrigeration duty (approximately 20 kW), determine the composition of the mixture that results in minimum power requirements at the compressor.

Discussion

The optimum mixture is in the neighborhood of 50 percent of each. The heat exchanger causes an appreciable improvement in the performance of a mixed-refrigerant cycle but is of little benefit in a single-refrigerant system. After determining the optimum mixture, a further optimization would be possible in which the distribution of heat-transfer areas between the condenser, evaporator, and heat exchanger could be performed.

A.12 OPTIMUM AIR EJECTOR FOR A STEAM-POWER-PLANT CONDENSER

The pressure of steam at the exhaust of the turbine and in the condenser of a steam power plant is below atmospheric, so air seeps into the system through any unavoidable leaks. The presence of air and other noncondensables results in a higher exhaust pressure from the turbine than would otherwise prevail, reducing both the power delivered by the turbine and the cycle efficiency. Provisions are made to remove air continuously by the use of such devices as an air-ejector system, shown in Fig. A-14. The function of the ejection system is to extract a sample of air and water vapor out of the main condenser and ultimately to reject the air in the mixture to the atmosphere. The removal of air is to be accomplished with a low loss of steam.

In the system shown in Fig. A-14, 0.12 kg/s of air-vapor mixture that contains 5 percent air by mass is extracted from the main condenser, and the mixture that is vented to atmosphere should contain 50 percent air, by mass. The main condenser operates with a total pressure of 6.6 kPa, and the aftercondenser operates at 101 kPa. Cooling water with a flow rate of 7 kg/s enters the intercondenser at a temperature of 35°C and passes in series through the intercondenser and then the aftercondenser.

Performance of Condensers

The performance of the intercondenser and the aftercondenser can be approximated by assuming that they are counterflow heat exchangers, where the temperature on the condensing side is that of the saturation temperature of water at the partial pressure of the water vapor. Thus, if the pressure at one point in the intercondenser is 20 kPa and the partial pressure of the air is

FIGURE A-14
Two-stage air ejector.

2 kPa, the condensing pressure will be the saturation pressure at 18 kPa or 57.8°C. Use the log-mean temperature difference in the condenser with an applicable U value of 1.4 kW/(m² · K).

Assignment

Select the heat-transfer areas of the intercondenser and aftercondenser, the flow rates of motive steam w_1 and w_2, and the intermediate pressure p_i so that the total present worth of all costs is a minimum.

Further Data

Condenser cost, $150 per square meter of heat-transfer area

Cost of motive steam, 60 cents per megagram

Interest rate, 10 percent.

Life of facility for economic evaluation, 6 years.

Hours of operation per year, 4000.

The ratios of motive steam flow to pumped flow for the ejectors are given by

Low stage: Ratio (kg/s)/(kg/s) $= 0.19\left(\dfrac{p_i}{6.6}\right)^{2.2}$

High stage: Ratio (kg/s)/(kg/s) $= 0.14\left(\dfrac{101}{p_i}\right)^{2.2}$

The combined first cost of the ejectors is approximately constant in the expected range of sizes to be examined.

The results of Prob. 4.18 may be useful.

Discussion

The optimal design is likely to be one where more heat-transfer area is placed in the intercondenser than in the aftercondenser, since whatever steam (both from the main condenser and from the low-stage ejector) can be condensed at the intercondenser will not have to be pumped by the high-stage ejector. Often in multistage compressions the intermediate pressure is the geometric mean of the suction and discharge pressures. The economics of this system will shift the optimum intermediate pressure to a value lower than the geometric mean.

A.13 OPTIMIZING A HOT-OIL LOOP IN A PETROCHEMICAL PLANT

One method of distributing heat in refineries and petrochemical plants is to heat the oil in a central furnace and pump it to various heat exchangers where heating is required. Figure A-15 shows a hot-oil loop that serves three reboilers. The furnace, reboilers, and pump are to be selected so that the total present worth of costs for the economic life of the facility is a minimum.

Oil data

> Oil chemically stable up to 370°C
> Specific heat, 2.1 kJ/(kg · K)
> Density, 850 kg/m³

Reboiler data

	Duty, kW	Temperature of fluid being heated, °C	U value, W/(m² · K)
Reboiler 1	1800	200	450
2	1600	150	450
3	2400	175	450

Temperature of fluid being boiled in a given reboiler is essentially constant.

Cost of the reboilers is $90 per square meter of heat-exchanger area.

FIGURE A-15
Hot-oil loop in a petrochemical plant.

Furnace data

Treat the furnace as a combination of a combustion chamber and a counterflow heat exchanger.

Combustion air is heated to temperature t_h in the combustion chamber and then flows over heat-exchanger surface.

U value of furnace heat exchanger, 30 W/(m$^2 \cdot$ K).

First cost of furnace, $30 per square meter of heat-exchanger area.

Cost of heating by natural gas, $1.20 per gigajoule.

Flow rate of combustion gas, 6.0 kg/s.

Specific heat of combustion gas, 1.1 kJ/(kg $\cdot$ K).

Pump, motor, and electric data

Efficiency of pump, 85 percent

Efficiency of electric motor, 90 percent

First cost of pump, $160 for each kilogram per second of oil flow

First cost of electric motor, $50 per kilowatt shaft-power output

Cost of electricity, 3.2 cents per kilowatthour.

The pressure buildup required of the pump is a function of the specific design of the furnace, reboilers, and piping system, but as an approximate measure of the pressure drop throughout the loop, use

$$\Delta p, \text{ kPa} = 8.2 w^{1.8}$$

where w is the system flow rate in kilograms per second.

Economic data

Plan on full-time operation for the 5-year economic life of the facility.

Interest rate, 10 percent

Assignment

Optimize the system so that the minimum total present worth of costs results for the economic life of the facility. Specifically, determine the optimum values of the total oil flow rate and the flow rate in each reboiler; the leaving temperature of oil from each reboiler; t_a, t_b, t_h, t_g; the system oil-pressure drop; and the size of the electric motor.

Discussion

Since the dominant cost is that of the gas fuel, the optimization will drive toward reduction of this cost. In particular the temperatures of the oil leaving the reboilers will drop very close to the reboiler temperatures in an effort to reduce t_a, which in turn permits a low t_g and thus a reduced loss of energy in the leaving flue gases. A further extension that designers of these systems would consider is a regenerative heat exchanger to recover more heat from the flue gases and thus reduce the fuel cost.

A.14 OPTIMUM HEAT PUMP FOR PASTEURIZING MILK

The essential requirement in the pasteurization process of milk is to bring the temperature up to 73°C and hold it for approximately 20 s. The milk arrives at the dairy from the tank truck at a temperature of 7°C and is to be delivered from the pasteurizing plant to the packaging operation at a temperature of 4°C.

The traditional pasteurizing cycle uses a steam or hot-water heater to bring the temperature of the milk to 73°C; then the milk flows through a water-cooled heat exchanger and a refrigerant evaporator. The heat from the refrigeration plant is rejected to the atmosphere by a cooling tower or an air-cooled condenser.

In the interest of conserving energy, the possibility of using a heat pump is now sometimes considered.[8] One possible cycle is shown in Fig. A-16. The incoming milk flows first through a regenerative heat exchanger and then to the heater, which is also the forecondenser of the heat pump. This forecondenser elevates the temperature of the milk to 73°C. Thereafter the milk flows through the other side of the regenerative heat exchanger and finally through the evaporator of the heat pump.

FIGURE A-16
Milk pasteurizing system using a heat pump.

Data

Flow rate of milk, 4 kg/s for 4 h/day

Specific heat of milk, 3.75 kJ/(kg · K)

Cost of all heat-exchanger surfaces (evaporator, condensers, and regenerative heat exchanger), $95 per square meter

U value of regenerative heat exchanger, 500 W/(m^2 · K)

U values of refrigerant evaporator, forecondenser, and aftercondenser, 600 W/(m^2 · K)

Coefficient of performance of heat pump (ratio of the evaporator capacity to the compressor power), 75 percent of the Carnot COP for prevailing evaporating and condensing temperatures

First cost of the compressor and motor, $120 per kilowatt of power

Cost of electricity, 3.5 cents per kilowatthour

Interest rate, 9 percent

Economic life of the plant, 6 years

Cooling water enters the aftercondenser at 30°C and leaves at 35°C; flow rate of water is set to remove desired rate of heat flow.

Assignment

Optimize the system to provide the minimum total present worth of costs for the economic life of the plant. Specifically, determine the areas of all heat exchangers, the power required by the compressor, and temperatures t_1 and t_2 that result in the optimum.

Discussion

The optimum should occur when most of the heat exchange is accomplished in the regenerator.

The overall energy balance of the cycle indicates that because energy flows into the system through the use of compression work and because the milk leaves at a lower temperature than it enters, there must be a corresponding rejection of energy elsewhere. The heat rejection takes place at the aftercondenser. An alternate cycle to explore is one where the heat is rejected by a water-cooled heat exchanger in the milk stream rather than in the refrigerant steam.

REFERENCES

1. W. F. Gardner, "Desalination Test Module," *Heat Eng.*, Foster-Wheeler Corp., March–April 1967 and September–October 1968.
2. A. M. Woodard, "How to Design a Plant Firewater System," *Hydrocarbon Process*, October 1973, pp. 103–106.

3. P. T. Doyle and G. J. Benkly, "Use Fanless Air Coolers," *Hydrocarbon Process.*, July 1973, pp. 81–86.

4. *ASHRE Handbook Fundamentals Volume*, chap. 3, American Society of Heating, Refrigerating, and Air-Conditioning Engineers, Atlanta, 1985.

5. G. E. Zarnicki, L. A. Repin, and V. A. Elema, "Refrigeration by Utilizing the Pressure of Natural Gas Pumped through Pipelines," *Cholod. Tech.*, no. 6, 1974, pp. 27–29.

6. G. G. Haselden and L. Klimek, "An Experimental Study of the Use of Mixed Refrigerants for Non-Isothermal Refrigeration," *J. Refrig.*, vol. 1, no. 4, pp. 87–89, May–June 1958.

7. V. Kaiser, C. Becdelievre, and D. Gilbourne, "Mixed Refrigerant for Ethylene," *Hydrocarbon Process.*, vol. 55, no. 10, pp. 129–131, October 1976.

8. D. R. Lascelles and R. S. Jebson, "Some Process Applications of Heat Pumps," *Int. Inst. Refrig. Comm. Meet.*, Melbourne, 1976.

APPENDIX
II

GENERALIZED SYSTEM SIMULATION PROGRAM

Structure

Main Program that provides input data and calls SIMUL

Subroutine EQNS that provides the simultaneous equations (many of which are probably non-linear). The required form is specified on the next page

Subroutine SIMUL that performs the Newton-Raphson solution of the equations in Subroutine EQNS with appropriate application of Subroutines PARDIF and GAUSSY

Subroutine PARDIF that extracts partial derivatives

Subroutine GAUSSY that solves a set of linear simultaneous equations

Main Program

```
PROGRAM EXAMP (INPUT, OUTPUT, TAPE6=OUTPUT)
DIMENSION V( ), DES( ), R( ), VCORR( ), VD( ), RD( ), PD( , )
NVAR = _____   Number of Equations
TLRNCE = _____   Suggest 0.001
ITMAX = _____   Suggest 10
DATA V/•• . ••, ••• . ••, ••• . ••, . . ./ Trial Values of the Variables
DATA DES/"TOUT", "TIN", ...,"W14"/ 4-Character or Less Alphanumeric
CALL SIMUL (NVAR, TLRNCE, ITMAX, V, DES, R, PD, VCORR, VD, RD)
STOP
END
```

Equations Subroutine

```
SUBROUTINE EQNS (NVAR,V,R)
DIMENSION V(NVAR),R(NVAR)
R(1)=V(10) - V(12) ** V(5)
   •            •
   •            •
   •            •
RETURN
END
```

```
      SUBROUTINE SIMUL (NVAR, TLRNCE, ITMAX, V, DES, R, PD,VCORR,VD,RD)
C
C     THIS IS THE NEWTON-RAPHSON SUBROUTINE WHICH IS COMBINED WITH
C     THE PARDIF AND GAUSSY SUBROUTINES
C
C     USER MUST PROVIDE THE MAIN PROGRAM AND THE EQUATION SUBROUTINE
C
C     GLOSSARY OF TERMS USED
C
C     ITER = NUMBER OF ITERATIONS
C     ITMAX = MAXIMUM NUMBER OF ITERATIONS TO BE PERMITTED
C     NVAR  = NUMBER OF UNKNOWNS = NUMBER OF EQUATIONS
C     PD(I,J) = PARTIAL DERIVATIVE OF FUNCTION I WITH RESP TO VARIABLE J
C     R( ) = RESIDUAL OF EQUATION
C     TLRNCE = MAXIMUM FRACTION OF VALUE OF VARIABLES PERMITTED BEFORE
C       ITERATION COMPLETE.  THUS TLRNCE = 0.01 REQUIRES CHANGE OF
C       ALL VARIABLES TO BE LESS THAN 1 PERCENT FOR CONVERGENCE
C     V( ) = VALUE OF THE VARIABLE
C     VCORR( ) = CORRECTION IN THE VARIABLE DURING THIS ITERATION
C     DES( ) = DESIGNATION OF VARIABLE IN A4 FIELDS
C
      DIMENSION VD(NVAR), RD(NVAR)
      DIMENSION V(NVAR),R(NVAR),PD(NVAR,NVAR),VCORR(NVAR), DES(NVAR)
C
C     WRITING OUT THE INPUT DATA
C
      WRITE(6,20) NVAR
 20   FORMAT(1H1,///," NUMBER OF VARIABLES = ", I4)
      WRITE(6,21) TLRNCE
 21   FORMAT(" MAXIMUM FRACTION CHANGE FOR CONVERGENCE =", F10.4, //)
      WRITE(6,22)
 22   FORMAT("0VARIABLE NUMBER AND ITS TRIAL VALUE")
      WRITE (6,23)  (J, DES(J), V(J), J = 1, NVAR)
 23   FORMAT (" V(", I2, ") = ", A4, " = ", F15.5)
C
C     INITIALIZING THE ITERATION COUNTER
      ITER = 1
C     CALLING SUBROUTINES TO CALCULATE VALUES OF RESIDUALS, PARTIAL
C   DERIVATIVES AND CHANGES IN VALUES OF VARIABLES
C
 30   CALL EQNS(NVAR, V, R)
      WRITE (6,33)
 33   FORMAT("0EQUATION NUMBER     RESIDUAL")
      WRITE (6,35)  (I , R(I),  I = 1,NVAR)
 35   FORMAT (I10, F20.5)
      CALL PARDIF(NVAR, V, R, PD, VD, RD)
C
C     PRINTING OUT NON-ZERO VALUES OF PARTIAL DERIVATIVES
```

```
C
        DO 54  I = 1, NVAR
        DO 53  J = 1, NVAR
        Z = ABS(PD(I,J))
C       REMOVE THE C FROM THE NEXT THREE CARDS IF PRINTOUT DESIRED
C       IF(Z - 0.00000001) 53, 53, 51
C 51    WRITE(6,52) I, J, PD(I,J)
C 52    FORMAT(" PD(",I2,",",I2,") =", F20.10)
   53   CONTINUE
   54   CONTINUE
        CALL GAUSSY(PD, R, VCORR, NVAR)
C       CORRECTING THE VALUES OF THE VARIABLES
        DO 44  L = 1,NVAR
   44   V(L) = V(L) - VCORR(L)
C
C       WRITING OUT RESULTS OF THIS ITERATION
C
        WRITE (6,32) ITER
   32   FORMAT("0RESULTS AFTER", I4, "  ITERATIONS")
        WRITE(6,34)
   34   FORMAT("0",4X,"VARIABLE",16X,"VALUE",9X,"CHANGE FROM PREVIOUS")
        WRITE (6,36)  (I, DES(I), V(I), VCORR(I), I = 1,NVAR)
   36   FORMAT (" V(", I2, ") = ", A4," = ", 2F21.5)
C
C       TERMINATING IF MAXIMUM NUMBER OF ITERATIONS REACHED OR OTHERWISE
C    INCREMENTING THE ITERATION COUNTER
        IF(ITER - ITMAX) 38, 99, 99
   38   ITER = ITER + 1
C
C       CHECK TO SEE IF CHANGE OF VARIABLE IS LESS THAN SPECIFIED TOLERANCE
C
        K = 1
   40   VAL = ABS(VCORR(K)) - ABS(TLRNCE*V(K))
        IF(VAL) 41, 30, 30
   41   IF(K - NVAR) 42, 99, 99
   42   K = K + 1
        GO TO 40
   99   RETURN
        END
```

```
      SUBROUTINE PARDIF (NVAR, V, R, PD, VD, RD)
      DIMENSION VD(NVAR), RD(NVAR)
      DIMENSION V(NVAR), R(NVAR), PD(NVAR,NVAR)
C
C     GLOSSARY FOR SUBROUTINE PARDIF
C     DV = FRACTION OF VARIABLE CHANGE USED IN TAKING PARTIAL
C     VD =  V + DELTA V  =  V + V*DV
C     RD = R EVALUATED AT VD
C     THE PARTIAL DERIVATIVE IS  (RD - R)/(V*DV)
C
      DV = 0.001
C     SETTING ALL VD = V
      DO 550  K = 1,NVAR
  550 VD(K) = V(K)
      DO 560  J = 1, NVAR
C     ADDING DELTA TO VD(J)
C     CIRCUMVENT CASE OF V(J) = 0
      IF(ABS(V(J)) - 10.**(-30)) 551, 551, 552
  551 VD(J) = V(J) + 0.001
      GO TO 553
  552 VD(J) =(1. + DV)*V(J)
  553 CALL EQNS(NVAR, VD, RD)
      DO 558 I = 1, NVAR
      IF(ABS(V(J)) - 10.**(-30)) 555, 555, 556
  555 PD(I,J) = (RD(I) - R(I))/0.001
      GO TO 558
  556 PD(I,J) = (RD(I) - R(I))/(V(J)*DV)
  558 CONTINUE
C     RETURNING VD(J) TO V(J) VALUE
      VD(J) = V(J)
  560 CONTINUE
      RETURN
      END


      SUBROUTINE GAUSSY (A, B, X, N)
C
C     SOLUTION OF SIMULTANEOUS EQUATIONS BY GAUSS ELIMINATION
C
      DIMENSION A(N,N), X(N), B(N)
C     BEGINNING OF ELIMINATION PROCESS
      DO 28  K = 1,N
C     MOVING LARGEST COEFFICIENT INTO DIAGONAL POSITION
      AMAX = 0.
      DO 4  I = K,N
      IF(ABS(A(I,K)) - ABS(AMAX)) 4, 4, 2
    2 AMAX = A(I,K)
      IMAX = I
    4 CONTINUE
```

```
C         TESTING FOR INDEPENDENCE OF EQUATIONS
          IF(ABS(AMAX) - 0.1E-15) 10, 10, 14
   10     WRITE (6,12)
   12     FORMAT ("0 EQUATIONS ARE NOT INDEPENDENT")
          RETURN
C         EXCHANGING ROW IMAX AND ROW K
   14     BTEMP = B(K)
          B(K) = B(IMAX)
          B(IMAX) = BTEMP
          DO 18  J = K,N
          ATEMP = A(K,J)
          A(K,J) = A(IMAX, J)
   18     A(IMAX,J) = ATEMP
C         SUBTRACTING A(I,K)/A(K,K) TIMES TERM IN FIRST EQ FROM OTHERS
          KPLUS = K + 1
          IF(K - N) 22, 28, 28
   22     DO 24  I = KPLUS,N
          B(I) = B(I) - B(K)*A(I,K)/A(K,K)
          ACON = A(I,K)
          DO 24  J = K,N
   24     A(I,J) = A(I,J) - A(K,J)*ACON/A(K,K)
   28     CONTINUE
C         BACK SUBSTITUTION
          L = N
   32     SUM = 0.0
          IF(L - N) 34, 38, 38
   34     LPLUS = L + 1
          DO 36  J = LPLUS, N
   36     SUM = SUM + A(L,J)*X(J)
   38     CONTINUE
          X(L) = (B(L) - SUM)/A(L,L)
          IF(L - 1) 42, 42, 40
   40     L = L - 1
          GO TO 32
   42     RETURN
          END
```

APPENDIX
III

```
SUBROUTINE NZERO(I,J,R,JA)
C
C    THIS SUBROUTINE STORES THE NONZERO ELEMENTS IN
C    THE PROPER ORDER TO BE USED WITH SUBROUTINE XGAUSS
C
C    GLOSSARY FOR SUBROUTINE NZERO
C    I      = ROW NUMBER OF NEW NONZERO ELEMENT
C    J      = COLUMN NUMBER OF NEW NONZERO ELEMENT
C    R      = COEFFICIENT OF NEW NONZERO ELEMENT
C    JA     = TEST VARIABLE, SET EQUAL TO ZERO AT
C               START OF CALLING SUBROUTINE
C
      COMMON /AREA31/ MAX,MT,NVAR
      COMMON /AREA41/ IROW(3)
      COMMON /AREA42/ JCOL(2,7)
      COMMON /AREA43/ A(7)
C
C    FIRST TIME ZERO IS CALLED MATRICES ARE INITIALIZED
C
      IF(JA .NE. 0) GOTO 15
      JA=JA+1
      MT=1
      DO 5 L=1,NVAR
    5 IROW(L)=0
      DO 10 L=1,MAX
      JCOL(1,L)=0
      JCOL(2,L)=L+1
   10 A(L)=0
   15 IF(IROW(I) .NE. 0) GOTO 20
```

```
C
C    NEW ELEMENT IS THE FIRST NONZERO ELEMENT FOUND
C    IN ROW I
C
     IROW(I)=MT
     JCOL(1,MT)=J
     A(MT)=R
     MT=JCOL(2,MT)
     JCOL(2,IROW(I))=0
     RETURN
C
C    SEARCH TO FIND PROPER LOCATION OF NEW ELEMENT IN
C    ROW I
C
  20 LCT=IROW(I)
     LCTOLD=0
  25 IF(J .LT. JCOL(1,LCT)) GOTO 35
     IF(JCOL(2,LCT) .EQ. 0) GOTO 30
     LCTOLD=LCT
     LCT=JCOL(2,LCT)
     GOTO 25
  30 JCOL(2,LCT)=MT
     JCOL(1,MT)=J
     A(MT)=R
     MT=JCOL(2,MT)
     JCOL(2,JCOL(2,LCT))=0
     RETURN
  35 IF(LCTOLD .EQ. 0) GOTO 40
     JCOL(1,MT)=J
     A(MT)=R
     JCOL(2,LCTOLD)=MT
     MT=JCOL(2,MT)
     JCOL(2,JCOL(2,LCTOLD))=LCT
     RETURN
  40 JCOL(1,MT)=J
     A(MT)=R
     IROW(I)=MT
     MT=JCOL(2,MT)
     JCOL(2,IROW(I))=LCT
     RETURN
     END
```

```
           SUBROUTINE XGAUSS (IROW,JCOL,A,B,X,N,MAX,MT)
C
C    SOLUTION OF SIMULTANEOUS LINEAR EQUATIONS BY
C    GAUSS ELIMINATION
C    THIS SUBROUTINE STORES ONLY NONZERO ELEMENTS
C    USING LINKED STORAGE
C    TO BE USED ON SPARSE MATRICES IN CONJUNCTION
C    WITH SUBROUTINE NZERO
C
C   GLOSSARY FOR SUBROUTINE XGAUSS
C    N          = NUMBER OF VARIABLES
C    MAX        = MAX NUMBER OF NONZERO ELEMENTS AT
C                 ANY TIME
C    MT         = NUMBER OF FIRST EMPTY LOCATION
C    X()        = CALCULATED VALUE OF VARIABLES
C    IROW()     = LOCATION OF FIRST NONZERO ELEMENT OF
C                 EACH ROW
C    JCOL(1, )= COL NO OF NONZERO ELEMENT OR 0 IF
C                 LOCATION EMPTY
C    JCOL(2, )= LOCATION OF NEXT NONZERO ELEMENT OR
C                 NO OF NEXT EMPTY LOCATION
C    A()        = VALUE OF COEFFICIENT
C
      DIMENSION IROW(N),JCOL(2,MAX),A(MAX),B(N),X(N)
      MTMAX=0
      DO 50 K=1,N
C
C    MOVING LARGEST COEFFICIENT INTO DIAGONAL POSITION
C
      AMAX=0
      DO 5 I=K,N
      IF (JCOL(1,IROW(I)) .NE. K) GOTO 5
      IF(ABS(AMAX) .GE. ABS(A(IROW(I)))) GOTO 5
      AMAX=A(IROW(I))
      IMAX=I
    5 CONTINUE
C
C    TESTING FOR INDEPENDENCE OF EQUATIONS
C
      IF (ABS(AMAX) .GT. 1.0E-16) GOTO 10
      WRITE(6,100)
  100 FORMAT('0 EQUATIONS ARE NOT INDEPENDENT')
      RETURN
```

```
C
C    EXCHANGING ROW IMAX AND ROW K
C
   10 BTEMP=B(K)
      B(K)=B(IMAX)
      B(IMAX)=BTEMP
      ITEMP=IROW(K)
      IROW(K)=IROW(IMAX)
      IROW(IMAX)=ITEMP
C
C    SUBTRACTING A(I,K)/A(K,K) TIMES TERM IN FIRST
C    EQ FROM OTHERS
C
      KPLUS=K+1
      IF (K .EQ. N) GOTO 50
      DO 45 I=KPLUS,N
      IF(JCOL(1,IROW(I)) .NE. K) GOTO 45
      LI=JCOL(2,IROW(I))
      LK=JCOL(2,IROW(K))
      LIOLD=IROW(I)
      B(I)=B(I) - (A(IROW(I))/A(IROW(K)))*B(K)
   15 IF(LK .EQ. 0) GOTO 40
   20 IF(LI .EQ. 0) GOTO 25
      IF(JCOL (1,LI) .EQ. JCOL(1,LK)) GOTO 35
      IF(JCOL (1,LI) .GT. JCOL(1,LK)) GOTO 25
      LIOLD=LI
      LI=JCOL(2,LI)
      GOTO 20
C
C    NO CORRESPONDING ELEMENT IN ROW I AS IN ROW K
C    CREATE NEW NONZERO ELEMENT
C
   25 LCT=MT
      MT=JCOL(2,MT)
      IF (MTMAX .LT. MT) MTMAX=MT
      IF (MT .LT. MAX) GOTO 30
      WRITE (6,200) MAX
  200 FORMAT(1H, 'ALLOCATED STORAGE EXCEEDED,
     1 MAX=',I5/1H, 'INCREASE MAX, THE ARRAYS
     2 JCOL(2,MAX) AND A(MAX) AND RERUN')
      RETURN
```

```
   30 JCOL(1, LCT)=JCOL(1,LK)
      A(LCT)=-(A(IROW(I))/A(IROW(K)))*A(LK)
      JCOL(2,LIOLD)=LCT
      JCOL(2,LCT)=LI
      LIOLD=LCT
      LK=JCOL(2,LK)
      GOTO 15
C
C   CORRESPONDING ELEMENT IN ROW I AS IN ROW K
C
   35 A(LI)=A(LI)-(A(IROW(I))/A(IROW(K)))*A(LK)
      LIOLD=LI
      LK=JCOL(2,LK)
      LI=JCOL(2,LI)
      GOTO 15
C
C   ELEMENT I,K IS NOW ZERO, ADD LOCATION TO LIST
C   OF EMPTY SPACES
C
   40 LCT=IROW(I)
      JCOL(1,LCT)=0
      A(LCT)=0
      IROW(I)=JCOL(2,LCT)
      JCOL(2,LCT)=MT
      MT=LCT
   45 CONTINUE
   50 CONTINUE
C
C   BACK SUBSTITUTION
C
      DO 60 I=1,N
      PART=B(N+1-I)/A(IROW(N+1-I))
      LCT=JCOL(2, IROW(N+1-I))
   55 IF (LCT .EQ. 0) GOTO 60
      PART=PART-A(LCT)*X(JCOL(1,LCT))/A(IROW(N+1-I))
      LCT=JCOL(2,LCT)
      GOTO 55
   60 X(N+1-I)=PART
      RETURN
      END
```

INDEX